KU-736-332

PROMISCUOUS PLASMIDS OF GRAM-NEGATIVE BACTERIA

PROMISCUOUS PLASMIDS OF GRAM-NEGATIVE BACTERIA

edited by

Christopher M. Thomas

School of Biological Sciences
University of Birmingham
P.O. Box 363
Birmingham
B15 2TT U.K.

POLYTECHNIC LIBRARY
WOLVERHAMPTON

ACS 491741

CONTROL

DATE -1 APR. 1991

CLASS L

579.3

THO

D22

RS

ACADEMIC PRESS

Harcourt Brace Jovanovich, Publishers
London San Diego New York Berkeley Boston
Sydney Tokyo Toronto

ACADEMIC PRESS LIMITED
24/28 Oval Road
LONDON NW1 7DX

United States Edition published by
ACADEMIC PRESS INC.
San Diego, CA 92101

Copyright © 1989 by
ACADEMIC PRESS LIMITED

All Rights Reserved
No part of this book may be reproduced in any form by photostat, microfilm or
by any other means, without written permission
from the publishers

British Library Cataloguing in Publication Data
Thomas, C.M.
  Promiscuous Plasmids of Gram-Negative
  Bacteria.
  1. Gram-negative bacteria. Genetics
  I. Title
  589.9'015

  ISBN 0-12-688480-3

Typeset by Electronic Village Ltd, Richmond, Surrey
Printed in Great Britain by Alden Press, Oxford

CONTRIBUTORS

Michael Bagdasarian *Department of Microbiology, Michigan State University and Michigan Biotechnology Institute, PO Box 27609, Lansing, MI 48909, USA.*

F. Christopher H. Franklin *School of Biological Sciences, University of Birmingham, PO Box 363, Birmingham B15 2TT, UK.*

Joachim Frey *Institute of Veterinary Bacteriology, University of Berne, Länggasstraße 122, CH-3012 Berne, Switzerland.*

Donald G. Guiney *Department of Medicine H811F, University of California, San Diego, Medical Center, 225 Dickinson Street, San Diego, CA 92103, USA.*

Volker Haring *Abt. Schuster, Max-Planck-Institut für Molekulare Genetik, Ihnestraße 73, 1 Berlin 33 (Dahlem), Federal Republic of Germany.*

Dieter Hass *Mikrobiologisches Institut, Eidgenossicshe Technische Hochschule, CH-8092 Zurich, Switzerland.*

Donald R. Helinski *Department of Biology, B-022, University of California, San Diego, La Jolla, CA 92093, USA.*

V.N. Iyer *Department of Biology and Institute of Biochemistry, Carleton University, Ottawa, Ontario, Canada K1S 5B6.*

Clarence I. Kado *Department of Plant Pathology, University of California Davis, CA 95616, USA.*

Erich Lanka *Abt. Schuster, Max-Planck-Institut für Molekulare Genetik, Ihnestraße 73, 1 Berlin 33 (Dahlem), Federal Republic of Germany.*

Eberhard Scherzinger *Abt. Schuster, Max-Planck-Institut für Molekulare Genetik, Ihnestraße 73, 1 Berlin 33 (Dahlem), Federal Republic of Germany.*

Reinhard Simon *Fakultat Biologie, Universitat Bielefeld, Postfach 8640, D-8400 Bielefeld 1, Federal Republic of Germany.*

Robert Spooner *Department of Biological Sciences, University of Warwick, Coventry CV4 7AL, UK.*

Christopher A. Smith *Department of Anatomy, University of Birmingham, PO Box 363, Birmingham B15 2TT, UK.*

Christopher M. Thomas *School of Biological Sciences, University of Birmingham, PO Box 363, Birmingham B15 2TT, UK.*

Carrie R.I. Valentine *Department of Microbiology and Immunology, Oral Roberts University, Tulsa, Oklahoma 74137-1297, USA.*

Abbreviations

Antibiotics

Ap	Ampicillin
Cb	Carbenicillin
Cm	Chloramphenicol
Gm	Gentamycin
Hg	Mercuric chloride
Km	Kanamycin
Nm	Neomycin
Sm	Streptomycin
Sp	Spectinomycin
Su	Sulfanilamide
Tc	Tetracycline
Te	Tellurite
Tb	Tobramycin
Tp	Trimethoprim

Other Terms

APH	Aminoglycoside phosphotransferase
bp	Nucleotide base pairs
bla	β-lactamase
CAT	Chloramphenicol acetyl transferase
Da	Dalton
IAA	Indoleacetic acid
kb	10^3 base pairs
kDa	Kilodaltons
LPS	Lipopolysaccharide
Md	Megadaltons
MTX	Methotrexate
ORF	Open Reading Frame
ori	Origin
p	Promoter
SD	Shine-Dalgarno sequence
Ti	Tumour-inducing
tra	Conjugal transfer

INTRODUCTION

Plasmids are the major vehicles for the transfer of genes between bacteria. The natural occurrence of this transfer plays a vital role in the evolution of bacterial populations. In theory, any gene can be transferred in this way and transposable elements play a crucial but not necessarily exclusive role in this process. The genes which are transferred may be only transiently plasmid associated or may become a relatively permanent component of the plasmid genome. The genes, which have come to reside on plasmids, confer a variety of phenotypes on their bacterial hosts, ranging from bacterial resistance and toxin production to symbiotic and metabolic determinants. They are thus very important for bacterial adaptation to a variety of ecological environments and many of these genes create significant medical and veterinary problems. However, plasmids have also proved immensely valuable for both *in vivo* and *in vitro* genetic manipulation. *In vivo*, they can be forced to insert into and excise from the bacterial chromosome and, in the process, they can pick up and transfer segments of the bacterial chromosome. This allows genetic exchange and recombination between bacteria as well as the creation of partially diploid strains for the analysis of dominance/recessivity properties of mutant phenotypes. For *in vitro* gene manipulation, their relatively small size and maintenance at multiple copies per chromosome has facilitated their exploitation as key vectors in gene cloning. However, the classical plasmid tools developed in *Escherichia coli*, namely sex factor F and cloning vectors such as those related to or derived from pBR322 are restricted in host range and can only be used for *E. coli* and closely related species. Thus many bacterial species of current interest are unable to maintain these plasmids. One solution to overcome this host range barrier is to construct so-called 'shuttle vectors' which are hybrids consisting of two plasmid replication systems, each specialized for a particular species. This allows the initial gene cloning to be carried out in *E. coli* with subsequent transfer to and maintenance in the second species. However, an alternative solution is provided by the existence of certain plasmids which have a more extended host range. These broad-host-range or promiscuous plasmids provide a simple system for maintaining cloned genes in a wide variety of species. For this reason, considerable effort has been directed towards characterizing these plasmids and developing ways to promote their promiscuity.

The purpose of this book is to provide an account of the basic biology of the best studied promiscuous plasmids of Gram-negative bacteria. Plasmids of Gram-positive bacteria are not described, although in recent years it has become clear that many of the plasmids of this major group of bacteria are also promiscuous. Indeed, a number of such plasmids are capable not only of replication in Gram-positive bacteria but also in certain Gram-negative species. Conversely, while plasmids of the IncP group are not able to replicate in Gram-positive bacteria, their conjugation systems are capable of transferring DNA from Gram-negatives to Gram-positives. These observations thus provide a picture of considerable potential for the spread of genes throughout the bacterial kingdom.

The specific plasmids described in this book belong to the incompatibility groups IncN, P, Q and W and the various chapters of this book describe not only their basic biology but also the ways that these plasmids can be exploited for genetic manipulation. Detailed study of these plasmids has revealed them to be very interesting elements which can justify study independently of their value as genetic tools. One intriguing question that remains unanswered is what is the basis for the broad-host-range of these plasmids? There certainly does not appear to be a single answer. In the case of IncN and W plasmids, studies to date have not revealed any features of the plasmid maintenance systems which particularly distinguish them from plasmids of other groups, including narrow-host-range plasmids. On the other hand, studies on the plasmids of the IncP and Q groups have revealed complexity unique among the plasmids studied to date. This may be related to their promiscuity. The complexity is of different sorts in the two cases. In the IncQ plasmids, complexity of the plasmid-encoded replication functions appears to provide a considerable degree of independence from host replication functions. In the case of the IncP plasmids, the complexity is twofold. First, there appears to be a degree of flexibility in the replication system provided, in part, by the alternative forms of the plasmid-encoded replication initiation protein. Second, the plasmid possesses a complex array of functions which are coregulated with the replication system and which appear to be involved in stable inheritance of the plasmid. Studies on the basic biology of these plasmids has had the bonus of providing an understanding of some of the properties of vectors derived from them and has aided in the rational design of new vectors.

To review recent progress in our understanding of these plasmids and to define the large number of questions still to be answered were the main reasons for the organization by Drs. Chris Thomas and Chris Franklin of an EMBO Workshop on *Promiscuous Plasmids of Gram-negative Bacteria* at Birmingham University in July 1987. While this book is in no way to be regarded as a book of the proceedings of that workshop, most of the chapters in this book are written by contributors to that meeting. I hope that these chapters, by reviewing current and in some cases unpublished work, will be useful both for plasmid biologists studying fundamental plasmid properties and for workers seeking to exploit these plasmids as tools for genetic manipulation.

Christopher Thomas

CONTENTS

CHAPTER 1

VEGETATIVE REPLICATION AND STABLE INHERITANCE OF IncP PLASMIDS

Christopher M. Thomas and Donald R. Helinski

I. INTRODUCTION

Plasmids classified as possessing IncP incompatibility can be divided into three main subgroups (see Chapter 3). Subgroups IncPα and IncPβ appear to possess a single replication system which displays the promiscuity normally associated with IncP plasmids while the third subgroup contains plasmids which fit into neither IncPα or β subgroups. Some of these other plasmids have been shown to be very restricted in host range, to exhibit asymetric incompatibility and to have at least one additional replication system. Originally it was thought that the comparison of these narrow host range IncP plasmids with the promiscuous IncP plasmids might provide a clue to the basis of promiscuity. However, it has recently been shown that in the case of at least one of these plasmids, pHH502.1, only part of the IncP replicon is present, perhaps explaining why the plasmid lacks a host range characteristic of "normal" IncP plasmids (82). Because of their relatedness in host range properties and structural features, this chapter will deal with replication and maintenance only of IncPα and β plasmids.

Plasmids belonging to both IncPα and β subgroups are capable of transfer between and stable maintenance in almost all Gram-negative bacterial species. Species and genera known to maintain IncP plasmids are: *Acetobacter xylinum*

Promiscuous Plasmids of Gram-Negative Bacteria
ISBN 0-12-688480-3
© 1989 Academic Press Limited
All rights of reproduction in any form reserved

(101), *Achromobacter parvulus* (87), *Acinetobacter* spp. (58). *Aeromonas* spp.(87), *Agrobacterium* spp. (18), *Alcaligenes* spp. (58), *Anabaena* spp. (106), *Azospirrillum brazilense* (66), *Azotobacter* spp. (58), *Bordetella* spp. (40), *Caulobacter* spp. (19), Enterobacteriaceae (13,22,39,41), *Haemophilus influenzae* (Dreich, personal communication), *Hypomycrobium* X (21), *Legionella pneumophila* (24), *Methylophilus methyltrophus* (105), *Methylosinus trichosporium* (102), *Myxococcus xanthus* (10,61) *Neisseria* spp. (58), *Paracoccus denitrificans* (60), *Pseudomonas* spp. (18,19,58), *Rhizobium* spp. (18,19,8), *Rhodopseudomonas* spp. (58), *Rhodospirillum* spp. (58), *Thiobacillus* spp. (20,51), *Vibrio cholerae* (58), *Xanthomonas* spp. (52), *Zymomonas mobilis* (11), *Yersinia enterocolitica* (41). The only Gram-negative genera known to be nonpermissive for IncP plasmid maintenance are *Bacteroides* (36,76) and *Myxococcus* (10). However, IncP plasmids can transfer to both *Bacteroides* and *Myxococcus* indicating that it is the IncP replicon which limits host range. This is also the case with narrow host range plasmids of IncFI and IncIα groups whose transfer systems are less restricted than their replicons (35,9). In addition, while all transconjugants to *M. xanthus* appear to have only an integrated copy of RP4, the plasmid will transfer out again intact, occasionally bringing with it a piece of chromosomal DNA (10). Thus, while gene expression may be a barrier to maintenance in *Bacteroides* it does not seem to be in *Myxococcus xanthus* since the conjugal transfer system and antibiotic resistance genes are known to function in this species. It appears that there is a reversible integration system which allows RP4 to be maintained as part of the *Myxococcus* chromosome (J. F. Guespin-Michel, personal communication). It is quite possible that IncP plasmids are maintained in this state in other species which have not been studied in detail.

There are also reports of bacterial strains from genera which are capable of IncP plasmid maintenance, being nonpermissive for IncP plasmids (53). While the basis for this effect is not generally known, a specific prophage B3 in *Pseudomonas aeruginosa* can render this species nonpermissive for IncP plasmid maintenance (86).

IncPα and β plasmids show a similar genetic organization when segments, rich in restriction sites and coding for a variety of phenotypic markers (which in some cases have been shown to be transposable elements) are ignored (see Chapter 3). This IncP backbone consists of a contiguous block of conjugal transfer functions joined to what appears to be a contiguous block of replication and maintenance functions, Figure 1. These latter functions consist of *oriV* (vegetative replication origin); *trfA* (transacting replication function necessary to activate *oriV*); *inc* and *cop* functions (involved in incompatibility and copy number control); *kil* and *kor* genes (determinants discovered by Figurski and coworkers (28) who found that certain regions containing a *kil* gene could not be cloned unless other IncP determinants designated *kor* or *kil*-override were present in the same cell); and the recently discovered *kcr* and *kfr* genes (regulated by Kor C and Kor F respectively, see below). This chapter will describe the organization and function of these regions as well as describing other loci which have been implicated in plasmid stability.

II. THE MINIMAL REPLICATION SYSTEM

For replication in a wide range of species RK2, which is representative of the Inc Pα plasmids (see Chapter 3), requires only *ori V*, from which replication has been shown in *Escherichia coli* to proceed unidirectionally (55) (Figure 1), and the *trfA* gene whose product(s) are essential to activate *ori V* (26,74,90).

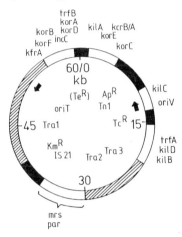

Figure 1. Map of RK2 showing the postions of loci (which are defined in the text) which are known and proposed to be involved in stable inheritance (open segments). Regions encoding conjugal transfer functions are shown as hatched segments while resistance determinants and transposable elements are shown as solid segments. A more detailed physical/genetic map of RK2 is presented in Chapter 3.

Deletion analysis, however, originally implicated a third region in replication and this was designated *trfB* (96) because it also appeared to code for a *trans*-acting product. Subsequent analysis has shown that this requirement is due to the fact that the KilD$^+$ (or KilB1$^+$) phenotype is associated with the *trfA* operon (65,65,78) so that a wild-type *trfA* region cannot be maintained without *korD*, which is enoded by the *trfB* region and is identical to *trfB* and *korA* (28). By attempting to clone *trfA* without *trfB/korA/korD*, mutant *trfA* regions have been isolated which are phenotypically KilD$^-$ but are TrfA$^+$ (64,90). The one such whose sequence has been reported to date differs from the wild-type by a T » C transition at the downstream end of the *trfA* promoter, decreasing its homology to the *E. coli* Eσ^{70} promoter consensus at one of the highly conserved positions and reducing promoter strength approximately ten-fold (75,83) (Figure 4). Since Kor D (the product of *korD*) represses transcription from the *trfA* promoter ten-fold, this point mutation reduces the unregulated level of transcription to about the same level as from the wild-type *trfA* promoter in the presence of *korD*. This result and other reported observations (64,65,72) suggested that expression of the Kil D $^+$ phenotype is not essential for replication. Studies involving plasmid constructions that do not exhibit the Kil D $^+$ phenotype provide direct evidence for the non-essential nature of this function for RK2 replication (25,70,71). The replication and maintenance properties of various derivatives of RK2 were examined in nine Gram-negative bacterial species and while the origin of replication and the *trfA* gene are sufficient for replication in all nine species tested, stable maintenance (less than 0.3% loss per generation without selection) in most of these species requires additional regions of RK2 (71). The broad host range maintenance requirements of RK2 are, therefore, encoded by multiple functions and the requirement for one or more of these varies among bacterial species. A description of the various RK2 regions affecting copy number, regulation of *trfA* expression and stable maintenance of the plasmid is presented below.

A. The *oriV* region

While the minimal DNA segment essential in *cis* to provide *oriV* activity is approximately 400 bp (17,85,98), sequences upstream and downstream of this segment seem to modulate *oriV* function so that the *oriV* region occupies approximately 1000 bp. The main features of this region as determined by DNA sequence analysis are summarized in Figure 2. Their probable functional importance is suggested by their being conserved between Inc Pα and β plasmids despite an overall sequence divergence of approximately 38% (81). A major structural feature is a series of repeats of a 17 bp sequence (85) to which Trf A protein binds (see below). The minimal *oriV* contains five tandemly repeated copies of this sequence arranged with 22–23 bp between the starts of adjacent repeats. Upstream are a single copy (103) and an unevenly spaced set of three repeats. All of these nine repeats are highly conserved and in a single orientation, while downstream there are at least two less well conserved copies in both direct and inverted orientation (81; CMT and JP Ibbotson, unpublished). While the effect of these downstream sequences on replication has not been analysed, the upstream repeats are known to be involved in incompatibility expression and copy number control. When cloned at high copy number, repeats 1–9 of RK2 cause rapid elimination of Inc P plasmids (98). A similar effect is observed for repeats 1–4 but repeats 5–9 show a much weaker effect which is only significant towards plasmids which produce limiting amounts of Trf A protein (16,98). Since deletion analysis suggests that the incompatibility exhibited by this *oriV* segment is due to the direct repeats, the group of five repeats in the minimal *ori V* must behave in both a qualitatively and a quantitatively different way from the upstream repeats. The presence of direct repeats in the origin of plasmid RK2 and the requirement for the *trfA* gene for activity of this origin are characteristic features of the majority of plasmids (narrow-host-range or broad-host-range) in Gram-negative bacteria. These various plasmids including R6K, R485, RK2, pSC101, F, P1, Rts1, lambda dv, pSa and RSF1010 contain direct repeats varying from 17 to 24 bp in length within and/or near the replication origin.

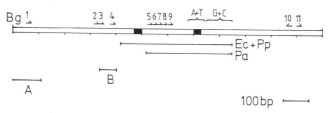

Figure 2. Organization of the *ori V* region of RK2 showing the direct repeats numbered 1 to 11. The DNA, divided into 100 bp sections, runs counter-clockwise from the *Bgl*II site (Bg) at 13.2 kb on RK2. Regions showing homology to Dna A consensus sequences are shown as solid blocks. A and B represent deletions defining *incA/copA* and *incB/cop*B respectively. E. c, P. p and P. a represent the minimal origins in *E. coli*, *P. putida* and *P. aeruginosa* respectively.

Additional features of the *ori V* region of RK2 are the A+T-rich and G+C-rich segments of the minimal *ori V* which are conserved between RK2 (IncPα) and R751 (IncPβ) despite considerable base substitution and deletions/insertions (81). The G+C-rich region has much potential secondary structure. One inverted repeat is particularly highly conserved between RK2 and R751. It is possible that this could represent a primase recognition signal involved in origin function. However, use of a mutant M13 phage to probe for such signals has failed to detect a primase signal acted upon by either host or IncP encoded primase in this region (DG Guiney, personal communication).

It has also been proposed that the minimal origin may contain transcriptional initiation signals but attempts to detect these using a promoter probe inserted at a series of positions throughout the origin have failed to date (16). This may be consistent with the apparent insensitivity of IncP plasmid DNA replication *in vitro* to rifampicin (63). Comparison of the *ori V* nucleotide sequences for RK2 and R751 have not only shown the direct repeats and the A+T-rich and G+C-rich regions to be conserved features but suggest that a putative inverted pair of DnaA binding sites upstream of repeats 5–9 may be important in replication. Mini plasmids derived from RK2 are unable to replicate in *dnaA* nul strains (32,63) suggesting that IncP plasmid replication depends on DnaA, and this has been confirmed through analysis of plasmid replication *in vitro* (63) and *in vitro* binding of DnaA to *ori V* fragments (32).

A second Dna A binding site has been proposed to exist within the A+T rich region and disruption of this site inactivates replication (32). However in this case *ori V* activity can be restored by sequences which do not restore efficient binding of Dna A (32) suggesting that binding of Dna A at this secondary site may not be essential for *ori V* activity, which is consistent with it not being highly conserved in the Inc Pβ plasmid R751 (81). The dependence of RK2 on Dna A places it in the same category as F and certain other plasmids since despite this dependence on Dna A RK2 is able to suppress the temperature-sensitivity of *dna A*ts mutations by integration into the chromosome (54). Possible explanations are either that the *dna A* mutation suppressed were leaky or that Inc P plasmids carry a function that can substitute for Dna A.

A 393 bp *Hpa*II fragment in the *ori V* region is functional as an origin in *E. coli* (98). Insertions of a 35 bp DNA segment by transposon mutagenesis within this region resulting in loss of *ori V* function (17). While the same region is also required in *Pseudomonas putida*, insertions upstream from repeat 5 do not inactivate *ori V* in *P. aeruginosa* (15,17,49,57) but do affect stability of maintenance suggesting that the detailed functioning of *ori V* may show some variations from host to host depending possibly on host proteins and/or other components of the cellular environment. The smaller *Hpa*II *ori V* region (Figure 2) also has been found to function less effectively that the larger *Hae*II origin region in *P. putida* (70).

B. The *trf A* gene

The *trf A* gene is the second cistron in an operon, designated the *trf A* operon (Figure 3). The *trf A* gene specifies two products, of 382 (43 kDa) and 285 (32 kDa) amino acids, respectively, resulting from alternative translational starts within the same ORF (74,79). The predicted amino acid sequences of both polypeptides indicate that they are basic proteins, consistent with the finding that the purified Trf A polypeptides show considerable nonspecific DNA binding activity (63). A search of the amino acid sequence reveals a region which may adopt the α helix-turn-α helix structure found in a number of well characterized DNA-binding proteins (79) although this region does not score highly on the systematic method for searching for such regions recently reported (23). DNA fragment retardation experiments in agarose gels utilizing a 80–90 % pure mixture of the two Trf A polypeptides have shown that Trf A can bind DNA nonspecifically; forming complexes which will not enter the gel. Footprinting experiments performed at a Trf A concentration below that at which such complexes are formed demonstrate the specific binding to the direct repeat sequence of the RK2 *ori V* region (63).

The full significance of the production of two Trf A polypeptides differing by 97 amino acids at their N-terminal end is not known. In *E. coli*, *Rhizobium meliloti*, *Agrobacterium tumefaciens*, *Azotobacter vinelandii* and *P. putida*

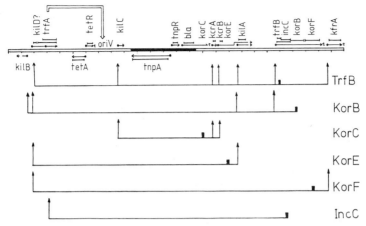

Figure 3. Organization of the *kil*, *kor*, *trf* and related operons. Solid circles represent promoters; horizontal arrows represent transcripts; t represents transcriptional terminators; bars represent functional ORFs. Solid regulatory circuits are negative; open block circuits are positive.

the smaller product, P_{285}, is sufficient for replication from *oriV*, while mini IncP plasmids carrying *trfA* but not producing P_{382} cannot be maintained in *P. aeruginosa* (25,73). This may suggest that initiation of replication at *oriV* is adapted to specific environments. It is possible that in certain bacterial hosts the 32 kDa protein alone is incapable of efficient interaction with host proteins. In these hosts the 43 kDa protein may serve to assure efficient interaction with host proteins. It is interesting to note that the minimal *oriV* requirements also differed *P. aeruginosa*.

The likelihood that different bacterial hosts present different cellular environments for the functioning of the TrfA protein(s) may explain the observation that several IncP ts mutants (38,68,100,104) which by complementation analysis in *E. coli* have all been shown to be *trfA*ts mutants (90; CMT, unpublished), exhibit species-dependent temperature-sensitivity (42,99). Thus varying ionic environments or protein-protein associations in different species may result in different consequences of amino acid substitution in the TrfA protein. Indeed the temperature sensitivity of one such *trfA*ts mutant, pME301, in *Pseudomonas aeruginosa* but not in *E. coli* appears to depend on the presence of a functional Tc^R gene (67). This may be of particular interest since one consequence of Tc^R is a reduced concentration of divalent cations in the cell membranes.

III. *IN VITRO* REPLICATION OF IncP PLASMIDS

To fully understand the process of initiation of DNA replication at *oriV* and the role of TrfA as well as host proteins in this process it is necessary to be able to analyse replication events *in vitro*. The first RK2 *in vitro* replication system was developed using DNA-membrane complexes isolated from plasmid-containing minicells, by the M-band technique in which the DNA-membrane complexes are attached to Sarkozyl crystals and can be separated by sucrose density gradient centrifugation (31). Incorporation of ^{3}H-TTP into such complexes has many of the characteristics expected of DNA replication including a semi-conservation replication mechanism and dependence on the *trfA* gene product. However, one or more protein components may not be present in sufficient quantity in this system since both transcription and translation (protein synthesis) are necessary for optimum incorporation rates (47). The membrane nature, low activity and requirement for protein synthesis in this *in vitro* system have restricted its utility. A problem with this membrane-associated system is that electron microscopic analysis of 'replicative intermediates' revealed D-loops originating not only in the *oriV* region but also in the Km^R and Tc^R determinants of the mini plasmid used as template (30) raising a question over whether or not the DNA synthesis observed really represents an IncP-specific process.

Recently, a soluble system which should lend itself better to analysis has been developed using extracts of plasmid-free *E. coli* bacteria prepared by standard protocols and supplementing them with purified plasmid DNA and purified TrfA protein(s), and standard substrates for DNA replication (63). A key point in the successful development of this system appears to have been to use as template, DNA of a high copy number mini RK2 derivative, pCT461, which lacks repeats 1–3 in the origin region (91). The *in vitro* activity shows dependence on Mg^{2+} (21% of complete activity remains when this is component is omitted), energy in the form of ATP replenished by creatine phosphate and creatine phosphokinase (0.7%), NAD (44%), polyethylene glycol (9–44% depending on what sort of extract is used), dNTPs (18%). A strong requirement for NTPs has not been demonstrated possibly because of the difficulty of producing extracts completely free of these compounds.

When a streptomycin sulphate precipitation step is included to remove endogenous plasmid DNA activity is abolished unless a DnaA-enriched extract is also added (M. Pinkney and R. Diaz, unpublished). Presumably endogenous DnaA protein is precipitated along with the DNA at this stage. This apparent requirement for DnaA is consistent with *in vivo* results (see above) as well as the presence of a pair of inverted DnaA binding sites in oriV which is conserved between both IncPα (RK2) and IncPβ (R751) plasmids (81).

Specific antisera have been used to demonstrate that also *E. coli* DnaB (helicase) and DnaG (primase) are essential for IncP replication. RP4 primase stimulates the *in vitro* system and can substitute for DnaG suggesting the possibiltiy of a specific role of the IncP primase in vegetative replication (63). DNA gyrase is also essential since replication in the *in vitro* system is inhibited by both nalidixic acid and novobiocin while the inhibition observed by araCTP suggests that DNA Pol III is required. However, rifampicin does not inhibit suggesting that $E\sigma^{70}$ RNA polymerase is not essential for initiation, consistent with the inability to demonstrate significant transcription in the *ori V* region (16). Unlike the membrane derived system described above, chloramphenicol does not inhibit replication (63).

Inhibition of *in vitro* replication with progressively higher levels of ddATP to block replication during the elongation cycle demonstrated that initiation must occur in or around *ori V* as defined genetically, providing further evidence that the incorporation of radioactivity observed *in vitro* does represent genuine replication (63).

IV. EXPRESSION AND REGULATION OF THE trfA OPERON

The *trfA* gene is the second cistron in an operon transcribed from a promoter which has been identified by RNA polymerase footprinting and reverse transcriptase mapping of the 5' end of the mRNA (62,83). The 5' end of the mRNA is identical in *E. coli*, *P. putida* and *P. aeruginosa* suggesting that the same *trfA* promoter is active in all three species. This is further supported by the observation that point mutations in the *trfA* promoter region have similar effects on promoter strength in all three species (62). The importance of *trfA* being expressed from a broad-host-range promoter is possibly illustrated by the properties of a Tn7 insertion mutant of R18, selected on the basis of its maintenance in *P. aeruginosa* but not in *E. coli* (15). The insertion separates *trfA* from its promoter (50). Expression of *trfA* in this mutant is apparently from a promoter internal to Tn7 which is strong enough in *P. aeruginosa* but not in *E. coli* to give a sufficient level of *trfA* expression for replication (50).

Although the *trfA* promoter is a strong promoter, which is consistent with its homology to the *E. coli* RNA polymerase $E\sigma^{70}$ consensus sequence (Figure 4), it is not normally expressed constitutively, transcription initiation being repressed by a number of gene products. This has been most clearly demonstrated for the protein products, TrfB and KorB, of two genes, *trfB* (= *korA* and *korD*) and *korB*, which are in a second operon, designated the *trfB* operon (28,72,75). It appears that the sites of action for these repressors are inverted repeats, designated O_A(TrfB/KorA/KorD operator) and O_B(KorB operator) (83). These assignments are based both on the correlation between the presence of these sequences in and transcriptional repression of *trfA*, *trfB*, *kilA*, *kilB*, *kcrA* and *kfrA* promoters (83,96a,109) as well as the nucleotide sequence of operator-constitutive mutants which exhibit changes in these putative operator sequences (88; CMT and A. Brown, unpublished), (Figure 4). An additional transcriptional repressor has recently been discovered to be encoded by a third gene, designated *korF*, within the *trfB* operon. Its action has been characterized with respect to the *kfrA* operon but it also appears to act on the *trfA* operon at an as yet undefined operator sites (V Shingler and CMT unpublished). Similarly, the product of yet another recently described regulatory gene *korE* (108) appears to act on *trfA* expression in an as yet undefined way (D Figurski, personal communication).

In addition, recent studies have suggested that the level of the protein products of the *trfA* operon may be regulated by IncC, the polypeptide product of a fourth gene in the *trfB* operon (56,93).

V. THE *trfB* OPERON

The *trfB* operon is located on the opposite side of RK2 from the *trfA* operon. The promoter for this operon, which shows substantial homology to the *trfA* promoter (Figure 4), has been defined by deletion analysis and RNA polymerase footprinting (12,80; M Pinkney, unpublished). Multiple RNA transcripts of this operon have been observed by Northern blot analysis and appear to be approximately 1500, 2500–2800 and 4100 bases in size (6,7). All except the smallest of these appear to hybridize to both *korA* and *korB* probes. This is consistent with the previous observations that *korB* is transcribed from the *trfB* promoter (80). Superimposing the sizes of these transcripts on the current map of this region shows that the 1500 base transcript would terminate just past the end of *incC*, the 2500–2800 base transcript would terminate just past *korB* and the 4100 base transcript would contain enough room for an additional polypeptide of approximately 400 amino acids. This fits well with the recent discovery of a fourth regulatory gene in this operon which has been designated *korF* (CMT unpublished). From the DNA sequence a possible rho-dependent transcriptional termination signal or RNA degradation pause site is found just after the end of *korB* (89) while a strong putative rho-independent terminator is present at about coordinate 55.3 kb (CMT unpublished) consistent with the end of the operon predicted from the longest of these transcripts.

Transcription of the *trfB* operon is doubly autogenously regulated, being repressed by the protein products of both *trfB/korA* and *korB* genes (88,109). The genes known to be encoded by the *trfB* operon are organized as shown in Figure 3. The first gene consists of an ORF with two translational starts which produce polypeptides of 364 and 259 amino acids (97). It is not certain whether both of these polypeptides can mediate the incompatibility associated with the *incC* locus (56) although a point mutation in the first 105 amino acids of the large product does not abolish activity, while truncation at the C-terminal end does (80). The TrfB/KorA/KorD$^+$ phenotype is associated with a 101 amino acids polypeptide coded by an ORF (6) which starts after the first *incC* start codon and continues in a different frame to a stop codon, TGA, which overlaps the second translational start of *incC*. The overlap appears to result in translational coupling between *trfB* and the second start in *incC* (97).

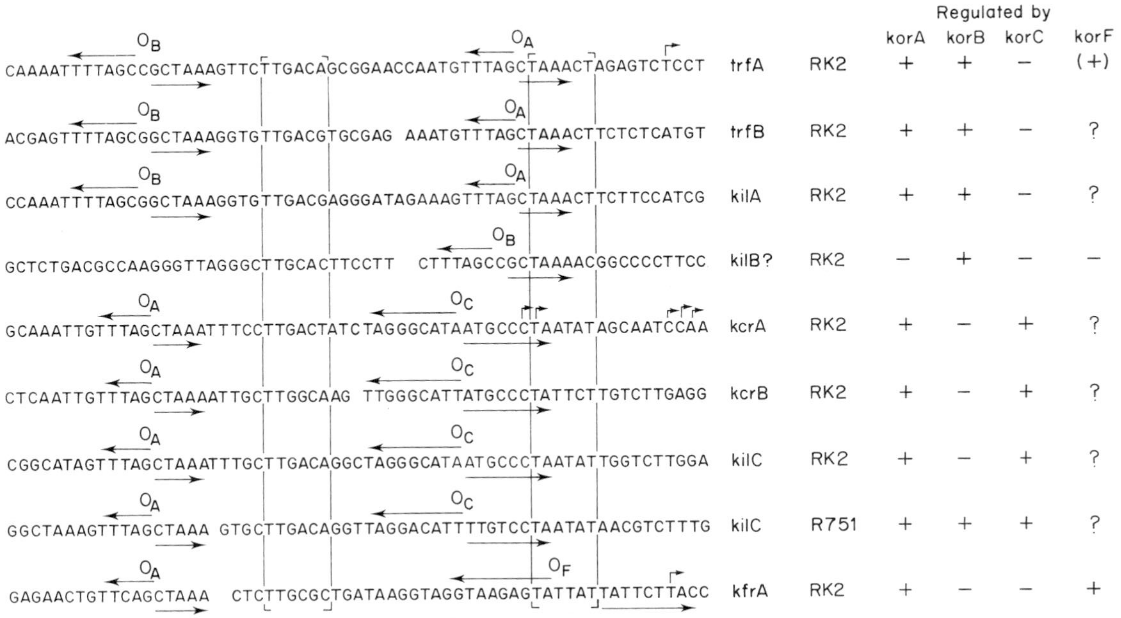

Figure 4. Comparison of promoters of the *kil*, *kor*, *trf* and related operons and the loci known to regulate transcription from these promoters.

Examination of the polypeptide sequence of TrfB reveals a 21 amino acid segment which shows considerable similarity to the α helix-turn-α helix structure of a number of well characterized DNA binding proteins (59) (Figure 5). That this region of TrfB is important is indicated by a mutation which inactivates transcriptional repression by TrfB and which results in a change val » met (Figure 5) in a conserved position in this putative DNA binding domain. This primary mutation can be suppressed by a val » ala second site mutation (CMT, unpublished). Further studies of this sort may provide additional evidence of the intramolecular structure of TrfB.

The *korB* product KorB (7,80,89) is much larger than TrfB, being predicted from DNA sequence analysis to contain a monomeric polypeptide of 324 amino acids (39.015) kDa) (48,89). Given the similarity in the putative operator sites for TrfB and KorB it is interesting that no apparent homology exists between these polypeptides (88) ruling out the possibility of recent evolution, via duplication, from a common ancestral regulatory protein as has been proposed for certain other regulatory proteins whose operator sites are similar but show different spacing between the two half sites. However KorB does also contain regions which could adopt an α helix-turn-α helix structure characteristic of other DNA binding proteins (Figure 5) and while the overall net charge on KorB is−21 this local region has a net positive charge of +1 consistent with a role in binding DNA. There is still considerable scope for further work to characterize TrfB and KorB DNA binding and to determine whether there is any cooperativity in activity on those promoters which are regulated by both proteins.

As described below, TrfB and KorB are involved in regulating a series of operons, at least some of whose products are deleterious to host bacterial growth. These genes may play an important role in stable inheritance of IncP plasmids (28). Thus, by virtue of their roles in the regulation of these different operons and the regulation of plasmid replication, *trfB* and *korB* probably play an important central role in coordinating the behaviour of IncP plasmids in a range of host bacteria. Since the genes that TrfB and KorB regulate may be expressed at different levels or their protein products may have different effects in different species, it is perhaps not surprising that transposon insertions in the *trfB–korB* operon can affect host range of IncP plasmids. Thus in the case of the plasmid pRP761 Tn76 is inserted into *incC* and while this plasmid can be maintained, albeit unstably, in *E. coli*, it cannot be maintained in *P. aeruginosa* (4,5). The Tn76 insertion apparently inactivates completely only *incC* (48) but since *incC* may influence the level of *trfA* operon gene products the effect on *incC* combined with reduced *korB* expression due to polarity may be sufficient to explain the poor growth and instability of pRP761 in *E. coli*. A similar effect with derivatives of a mini IncP plasmid, pRK2501. A temperature-sensitive mutant pRK2501ts3, defective in *trfB* can be maintained unstably in *E. coli* and *P. aeruginosa* while an *incC* deletion derivative

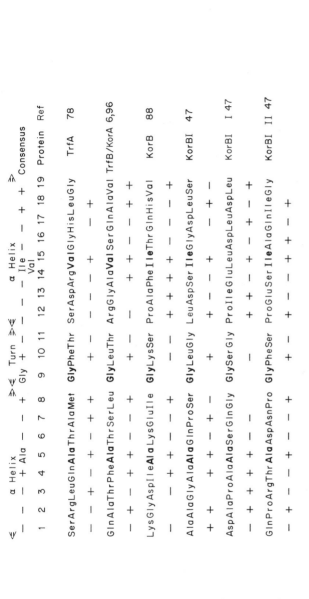

Figure 5. Compilation of possible DNA binding domains present in TrfA, TrfB/KorA and KorB. The consensus is shown with hydrophilic and hydrophobic residues being shown respectively as − and +.

of pRK2501ts3 is quite stable and no longer temperature-sensitive in *E. coli* but cannot be introduced at all into *P. aeruginosa* (CMT, unpublished). This again suggests a species-dependent variation in the importance of *incC*.

VI. CONTROL OF COPY NUMBER

The copy number of RK2 is estimated to be 4–7 copies per chromosome equivalent in *E. coli*, based on CsCl density gradient separation of RK2 DNA (average G+C content 60%) from chromosomal DNA (average G+C content 50%) (27). These estimates do not depend on the plasmid DNA remaining in CCC DNA form which may explain why the above estimate is higher than those carried out by CsCl/ ethidium bromide gradients (34) since RK2 is found partly in the form of a relaxation complex (37) which can be induced to the open circular form of RK2 DNA by the lysis procedures. Estimates for RP4 copy number in *P. aeruginosa* and *P. fluorescens* based on gene dosage levels of chromosomal genes cloned on RP4 suggest a copy number of about 3 per chromosome equivalent in these species (3,44).

Mini derivatives of RK2 like pRK229 which have lost *oriT* and the relaxation complex, but retain *korB* have a similar copy number to RK2 while smaller derivatives which have lost *korB* exhibit a higher copy number—approximately 10–11 copies per chromosome equivalent (95). Since KorB represses *trfA* expression by a factor of ten-fold over the repression by TrfB alone it seems likely that the *korB* effect on copy number is due to its effect on the level of TrfA protein. This conclusion is strengthened by the observation that the inhibitory effect of *korB* in *trans* on IncP replication can be overcome by supplying TrfA from a *trfA* gene expressed from a *korB*-independent promoter (72,95). Since the effect of KorB on *trfA* expression is only small in the absence of TrfB, the reduction of TrfA supply to a low level must depend on both *trfB* and *korB* which are therefore both major copy number control elements.

However, when TrfA concentration is not limiting, IncP replication can be controlled by elements in the *oriV* region. Thus deletions outside of the minimal origin, removing DNA segments containing either direct repeat 1 or direct repeat 2 and 3 (Figure 2; defining respectively *copA* and *copB*), result in a copy number of 17–20, while a combination of both deletions gives a copy number of 35–40 copies per chromosome equivalent (94). While wild-type *oriV* cloned in *trans* inhibits IncP replication (incompatibility) the deletions outside the minimal origin reduce this incompatibility, defining the loci *incA* and *incB*. It seems likely that it is the deletion of repeats, which is significant in causing an increase in copy number or reducing incompatibility. However, it is not yet clear how the repeats modulate copy number. If their effect is due simply to titration of TrfA and hence a lower effective TrfA concentration then the effect of the repeats might be expected to be highest when TrfA is limiting. While this does appear to be the case for the *trans* incompatibility effect, it is not true for control of copy number in *cis*. Thus when both *trfB* and *korB* are present in mini replicons the *oriV* region deletions have very little effect on copy number (95) leading to the conclusion that limitation of Trf A is the overriding copy control and that the repeats may act in some other way, possibly by allowing an inhibitory complex between Trf A and *oriV* repeats to form which is favoured when Trf A is not limiting, i. e. under conditions of excess Trf A.

The use of these *copA* and *copB* deletions has also allowed the demonstration that *incC* is involved in reducing Trf A concentration to a level limiting for replication (CMT, unpublished). Thus, while $TrfB^+IncC^+KorB^+$ mini replicons show very little change in plasmid copy number with *copA* and *copB* deletions, the copy number of $TrfB^+IncC^-KorB^+$ mini replicons is increased by *copA* and *copB* deletions (CMT, unpublished).

While there is indirect evidence for the dependence of plasmid RK2 copy number on the concentration of the Trf A protein(s), at least over a certain range, mutant *trfA* genes have been isolated that exhibit a high copy number phenotype (20–25 fold higher copy number than wild-type) for the RK2 origin when present in *trans* or in *cis* (R. Durland, personal communication). This phenotype is suppressed by the presence of the wild-type *trfA* gene in *trans*. These effects have been observed with the *trfA* gene transcribed from a heterologous, constitutive promoter, suggesting that the Trf A protein(s) may have both positive and negative activity in the control of RK2 copy number. Positive and negative activities for the R6K π replication protein in the control of plasmid R6K copy number in *E. coli* has been demonstrated by the isolation and characterization of π protein mutants that are similar in properties to the copy up Trf A mutant proteins and the finding that excess levels of the π protein in *E. coli* decrease copy number specifically of R6K derivatives (29,43,84). It is also of interest that quantitation of R6K plasmid copy number in *E. coli* cells producing various levels of the R6K π protein indicated that the plasmid copy number is not directly proportional to π protein concentration (29). It is important to carry out the direct quantitation of levels of RK2 *oriV* derivatives in bacteria producing various levels of the Trf A protein in order to assess whether the RK2 copy number is simply a function of Trf A protein concentration or involves a more complex mechanism of control although preliminary results for plasmids in which *trfA* expression is driven by the *trpE* promoter of *E. coli* show a low copy number when *trp* p is repressed and a high copy number when it is induced (92).

The existence of plasmids containing both *oriV* and *trfA* gene transcribed constitutively demonstrate that copy control is possible in the absence of regulation of *trfA* transcription. The copy mutants described above suggest that the *trfA* gene itself plays a role in that control. It remains to be seen what relationship the various proposed copy control elements (*trfB*, *korB*, *incC*, *korE*, *korF*, *trfA*, *copA* and *copB*) have to one another when all are present and functioning. It is conceivable that one or more of the *kil/kor* genes play a direct role in the initiation of replication in certain bacteria in addition to affecting *trfA* gene expression.

VII. OPERONS COREGULATED WITH *trfA* AND *trfB*—THE *kil* AND *kor* GENES

In addition to the *trfA* and *trfB* operons there are additional operons which form part of a coregulated network. These include: the *kilA* operon whose promoter is repressed by both TrfB and KorB and has strong sequence homology with the *trfB* promoter (CMT, unpublished) (Figures 3,4); a putative *kilB* operon whose promoter is repressed by KorB but not TrfB (Figures 3,4); an operon designated *kfrA* because all that is known about it at present is that it is regulated by KorF in addition to TrfB (Figures 3,4); and a series of operons regulated by both TrfB and KorC (Figures 3,4). These latter include an operon (*kcrC*) which encodes the KilC phenotype as well as two operons (*kcrA* and *kcrB*) which lie between *korC* and *kilA* and are transcribed towards *kilA* (95a). The *kcrB* promoter appears to be identical to a secondary *kilA* promoter (*kilA*p$_2$) identified by J Kornacki and D Figurski (personal communication). This second promoter apparently transcribes the *korE* gene (see below) before *kilA* and may explain an apparent role of *korC* observed by these workers in regulating the expression of the KilA$^+$ phenotype. The finding that TrfB is a second transriptional repressor of KorC-repressed operons explains the previous observation (107) that *korA* (*trfB*/*korD*) is needed along with *korC* to fully repress the KilC$^+$ phenotype. The observation that this requirement for *korA* to help *korC* can be overcome in a *rho E. coli* mutant (107) is most probably explained by the recent finding that *korC* is normally transcribed from the *bla* promoter of Tn*1* (95a) and that this promoter was not included in the original *korC* clone, thus rendering its expression dependent on vector promoters, read-through from which could be boosted if transcriptional termination was defective.

At least part of the *kcrA* and *kcrB* operons appear to have arisen by duplication since the first ORF in each operon predicts polypeptides of 77 and 76 amino acids respectively, which exhibit considerable homology to each other (31/77 fully conserved amino acids when a single gap is introduced into the *kcrB* product) (95a). The function of these polypeptides is not yet known. The only function as yet mapped to this general region is *korE* whose product acts together with Kor A to repress transcription of *kilA* (108) and *trfA* (H Schreiner and D Figurski, personal communication). The mode of action of *korE* is not understood.

The role of the *kil* genes is not understood either. While mini-derivatives of RK2 become progressively less stable as they lose regions containing *kil* genes it has not been demonstrated that it is loss of the *kil* genes which is responsible for the decreased stability. Their function may to be host lethal under certain circumstances, like for example the *ccd* genes of F (45) and the *hok/sok* genes of R1 (33), although it is not clear as to how they could be activated in a plasmid-free segregant since the control circuits discovered to date, except that mediated by IncC, involve transcriptional repression. Equally possible, their lethality may simply reflect the effect of over-production of a gene product that is normally made in relatively small quantities and which the plasmid may encode as a maintenance function for certain hosts.

VIII. OTHER STABLE INHERITANCE LOCI

In addition to the stable inheritance functions which might be provided by the *kil* genes RP4 encodes further functions involved in stability. In derivatives lacking Tn*1* and hence the multimer resolution system encoded by *tnpR* of Tn*1*, it can be demonstrated that a locus or loci, *mrs* (for multimer resolution system), in the *Pst*I fragment from coordinates 30.8–37.2 kb can resolve multimers and increase plasmid stability (N Grinter and PT Barth, personal communication). The primase (*pri*) gene or its product also seems to influence multimer formation (N Grinter and PT Barth, personal communication). Interestingly, even in the presence of the Tn*1* *tnpR* gene, the region from coordinates 30.8–37.2 kb has been found to influence plasmid stability, suggesting the presence of a second stability function in this region which has been designated *par* (69) and has been shown to increase the stability of high copy number plasmids like pACYC177 as well as unstable mini IncP plasmids (in a variety of bacterial species). However, IncPβ plasmids like R751 lack sequences homologous to this region as indicated by Southern blotting (110) and so must have other genes which provide the equivalent functions. While R751 may of course have heterologous genes that provide equivalent functions, it is interesting to note that a significant proportion of R751 DNA is found in the form of multimers suggesting a deficiency in this function (PT Barth, personal communication). Nevertheless, R751 is inherited stably and, therefore, has some means to overcome this deficiency.

Whether or not such a function is provided by the *kil* and *kor* genes it appears likely that IncP plasmids do carry genes which render them lethal to plasmid-free segregant bacteria. Thus *trfA*ts RP1 and RP4 plasmids after shift to the nonpermissive temperature cause a plateau in colony forming units after the plasmid is predicted to be diluted to approximately one copy per cell (38). The population produced under such circumstances are similar to those produced by an Fts plasmid (R. D'Ari, personal communication). In addition R751, which is displaced by the incompatibility expressed by *ori*V_{RK2} cloned in pBR322 possesses a function which can inhibit growth of host in bacteria in which R751 replication is inhibited (C.A. Smith and C.M. Thomas, unpublished). The loci responsible for these observations have not yet been mapped, largely due the instability of the phenotype.

IX. CONCLUSIONS

Considerable progress has recently been made in our understanding of the stable inheritance mechanisms of Inc P plasmids although many questions still remain. They appear to carry all the stability functions expected of a low copy number plasmid. However, of plasmids studied to date, they are unique in carrying a complex array of operons coordinately regulated with the replication functions. Two major questions are particularly intriguing. First, what is the role of the *kil* genes which are coregulated with the replication genes? It remains to be demonstrated whether or not they play a role in stability and if they do then whether they play a direct role in replication or gene expression or an indirect role by supplying auxilliary inheritance apparatus. Second, why is the replication of Inc P plasmids subject to so many regulatory circuits? It is hoped that the current research described in this chapter will soon shed more light on these questions.

ACKNOWLEDGEMENTS

We would like to thank all those who have allowed us to cite unpublished observations and to numerous colleagues who have made helpful comments on the contents of the chapter. Research on this subject in the authors' laboratories are supprted by the British Medical Research Council and by the National Science Foundation of the USA.

REFERENCES

1. Allen LN, Hanson RS (1985) Construction of broad-host-range cosmid cloning vectors: identification of genes necessary for growth of *Methylobacterium organophilum*. *J Bacteriol* **161**, 955–63
2. Alexander JL, Jollick JD (1977) Transfer and expression of *Pseudomonas* plasmid RP1 in *Caulobacter*. *J Gen Microbiol* **99**, 325–31
3. Barry GF (1986) Permanent insertion of foreign genes in the chromosomes of soil bacteria. *Biotechnology* **4**, 446–9
4. Barth PT (1979) RP4 and R300B as wide host-range cloning vehicles. *In* Plasmids of medical, environmental and commercial importance, (Timmis KN, Pühler A, eds.) Amsterdam: Elsevier/North Holland. pp. 399–410
5. Barth PT, Ellis K, Bechhoffer DH, Figurski DH (1984) Involvement of *kil* and *kor* genes in the phenotype of a host-range mutant of RP4. *Mol Gen Genet* **197**, 236–43

6. Bechhofer DH, Figurski DH (1983) Map location and nucleotide sequence of *korA*, a key regulatory gene of promiscuous plasmid RK2. *Nucl Acids Res* **11**, 7453-69

7. Bechhofer DH, Kornacki JA, Firshein WB, Figurski DH (1986) Gene control in broad-host-range plasmid RK2: expression, polypeptide product and multiple regulatory functions of *korB*. *Proc Natn Acad Sci USA* **83**, 394-8

8. Beringer JE (1974) R-factor transfer in *Rhizobium leguminosarum*. *J Gen Microbiol* **84**, 188-98

9. Boulnois GJ, Varley JM, Sharpe GS, Franklin FCH (1985) Transposon donor plasmids, based on Col Ib-P9, for use in *Pseudomonas putida* and a variety of other Gram-negative bacteria. *Mol Gen Genet* **200**, 65-7

10. Breton AM, Jaoua S, Guespin-Michel J (1985) Transfer of plasmid RP4 to *Myxococcus xanthus* and evidence for its integration into the chromosome. *J Bacteriol* **161**, 523-8

11. Browne GM, Skotnicki ML, Goodman AE, Rogers PL (1984) Transformation of *Zymomonas mobilis* by a hybrid plasmid. *Plasmid* **12**, 211-4

12. Burkardt H-J, Wohlleben W (1981) RNA polymerase binding sites on the broad-host-range plasmid RP4. *J Gen Microbiol* **125**, 189-93

13. Cho JJ, Panopoulos NJ, Schroth MN (1975) Genetic transfer of *Pseudomonas aeruginosa* R-factors to plant pathogenic *Erwinia* species. *J Bacteriol* **122**, 1141-56

14. Coetzee JN (1978) Mobilization of the *Proteus mirabilis* chromosome by R-plasmid R772. *J Gen Microbiol* **108**, 103-9

15. Cowan P, Krishnapillai V (1982) Tn7 insertion mutations affecting the host range of the promiscuous IncP-1 plasmid R18. *Plasmid* **8**, 164-74

16. Cross MA (1985) Molecular genetic studies on the *oriV* region of the broad-host-range plasmid RK2. PhD Thesis, University of Birmingham, UK

17. Cross MA, Warne SR, Thomas CM (1986) Analysis of the vegetative replication origin of broad-host-range plasmid RK2 by transposon mutagenesis. *Plasmid* **15**, 132-46

18. Datta N, Hedges RW (1972) Host-range of R-factors. *J Gen Microbiol* **70**, 453-60

19. Datta N, Hedges RW, Shaw EJ, Sykes RB, Richmond MH (1971) Properties of an R-factor from *Pseudomonas aeruginosa*. *J Bacteriol* **108**, 1244-9

20. Davidson MS, Summers AO (1983) Wide-host-range plasmids function in the Genus *Thiobacillus*. *App Env Microbiol* **46**, 565-72

21. Dijkhuizen L, Harder W, De Boer L, Van Boven A, Clement W, Bron S, Venema G (1984) Genetic manipulation of the restricted facultative methylotroph *Hypomicrobium* X by the R-plasmid mediated introduction of the *Escherichia coli pdh* genes. *Arch Microbiol* **139**, 311-8

22. Dixon RA, Cannon FC, Konorosi A (1976) Construction of a P-plasmid carrying nitrogen fixation genes in *Escherichia coli*. *Nature* **260**, 268-71

23. Dodd IB, Egan JB (1987) Systematic method for the detection of potential lambda cro-like DNA-binding regions of proteins. *J Mol Biol* **194**, 557–64

24. Dreyfus IA, Iglewski BH (1985) Conjugation mediated genetic exchange in *Legionella pneumophilia*. *J Bacteriol* **161**, 80–4

25. Durland RH, Helinski DR (1987) The sequence of the 43 kilodalton *trfA* protein is required for efficient replication or maintenance of minimal RK2 replicons in *Pseudomonas aeruginosa*. *Plasmid* **18**, 164–9

26. Figurski D, Helinski DR (1979) Replication of an origin-containing derivative of plasmid RK2 dependent on a plasmid function provided in *trans*. *Proc Natn Acad Sci USA* **76**, 1648–52

27. Figurski D, Meyer R, Helinski DH (1979) Suppression of ColE1 replication properties by the IncP-1 plasmid RK2 in hybrid plasmids constructed *in vitro*. *J Mol Biol* **133**, 295–318

28. Figurski DH, Pohlman RF, Bechhofer DH, Prince AS, Kelton CA (1982) The broad-host-range plasmid RK2 encodes multiple *kil* genes potentially lethal to *Escherichia coli* host cells. *Proc Natn Acad Sci USA* **79**, 1935–9

29. Filutowicz M, McEachearn MJ, Helinski DR (1987) Positive and negative roles of an initiator protein at an origin of replication. *Proc Natn Acad Sci USA* **83**, 9645–9

30. Firshein W, Caro L (1984) Detection of displacement ('D') loops with the properties of a replicating intermediate synthesized by a DNA/membrane complex derived from the low copy number plasmid RK2. *Plasmid* **12**, 227–32

31. Firshein W, Strumph P, Benjamin P, Burnstein K, Kornacki J (1982) Replication of a low copy number plasmid by a plasmid DNA-membrane complex extracted from mini cells of *Escherichia coli*. *J Bacteriol* **150**, 1234–43

32. Gaylo PJ, Turjman N, Bastia D (1987) DnaA protein is required for replication of the minimal replicon of the broad-host-range plasmid RK2 in *Escherichia coli*. *J Bacteriol* **169**, 4703–9

33. Gerdes K, Rasmussen PB, Molin S (1986) Unique type of plasmid maintenance function: post-segregational killing of plasmid-free cells. *Proc Natn Acad Sci USA* **83**, 116–20

34. Grinter NJ (1984) Replication control of IncP plasmids. *Plasmid* **11**, 74–81

35. Guiney DG (1982) Host-range of conjugation and replication functions of the *Escherichia coli* sex plasmid F'lac. *J Mol Biol* **162**, 699–703

35. Guiney DG (1982) Host-range of conjugation and replication functions of the *Escherichia coli* sex plasmid F'lac. *J Mol Biol* **162**, 699–703

36. Guiney DG, Hasegawa P, Davis CE (1984) Plasmid transfer from *Escherichia coli* to *Bacteroides fragilis*: differential expression of antibiotic resistance phenotypes. *Proc Natl Acad Sci USA* **81**, 7203–6

37. Guiney DG, Helinski DR (1979) The DNA-protein relaxation complex of the plasmid RK2: location of the site-specific nick in the region of the proposed origin of transfer. *Mol Gen Genet* **176**, 183-9

38. Harayama S, Tsuda M, Iino T (1980) High frequency mobilization of the chromosome of *Escherichia coli* by a mutant of plasmid RP4 temperature sensitive for maintenance. *Mol Gen Genet* **180**, 47-56

39. Hedges RW (1975) R-factors from *Proteus mirabilis* and *P. vulgaris*. *J Gen Microbiol* **87**, 301-11

40. Hedges RW, Jacob A, Smith JT (1974) Properties of an R-factor from *Bordetella bronchiseptica*. *J Gen Microbiol* **84**, 199-204

41. Heeseman J, Laufs R (1983) Construction of a mobilizable *Yersinia enterocolitica* virulence plasmid. *J Bacteriol* **155**, 761-7

42. Hooykaas PJJ, Dulk-Ras HD, Schilperoort RA (1982) Phenotypic expression of mutations in a wide-host range R-plasmid in *Escherichia coli* and *Rhizobium meliloti*. *J Bacteriol* **150**, 395-7

43. Inuzuka M, Wada Y (1985) A single amino acid alteration in the initiation protein is responsible for the DNA overproduction phenotype of copy number mutants of plasmid R6K. *EMBO J* **4**, 2301-7

44. Itoh Y, Watson JM, Haas D, Leisinger T (1984) Genetic and molecular characterization of the *Pseudomonas* plasmid pVS1. *Plasmid* **11**, 206-20.

45. Jaffe A, Ogura T, Hiraga S (1985) Effects of the *ccd* function of the F plasmid on bacterial growth. *J Bacteriol* **163**, 841-9

46. Jeyaseelan K, Guest JR (1979) Transfer of antibiotic resistance to facultative methylotrophs with plasmid R68.45. *FEMS Microbiol Lett* **6**, 87-9

47. Kornacki JA, Firshein W (1986) Replication of plasmid RK2 *in vitro* by a DNA-membrane complex: evidence for initiation of replication and its coupling to transcription and translation. *J Bacteriol* **167**, 319-26

48. Kornacki JA, Balderes PJ, Figurski DH (1987) Nucleotide sequence of *korB*, a replication control gene of broad-host-range plasmid RK2. *J Mol Biol* **198**, 211-22

49. Krishnapillai V (1986) Genetic analysis of bacterial plasmid promiscuity. *J Genet* **65**, 103-20

50. Krishnapillai V, Wexler M, Nash J, Figurski DH (1987) Genetic basis of a Tn*7* insertion mutation in the *trfA* region of the promiscuous IncP-1 plasmid R18 which affect its host range. *Plasmid* **17**, 164-6

51. Kulpa CF, Roskey MT, Travis MT (1983) Transfer of plasmid RP1 into chemolithotrophic *Thiobacillus neapolitanus*. *J Bacteriol* **156**, 434-6

52. Lai M, Panopoulos NJ, Shaffer S (1977) Transmission of R-plasmids among *Xanthomonas* spp. and other plant pathogenic bacteria. *Phytopathology* **67**, 1044-50

53. Lam ST, Lam SB, Stroebel G (1985) A vehicle for the introduction of transposons into plant-associated pseudomonads. *Plasmid* **13**, 200-4

54. Martin RR, Thornton CL, Ungler, L (1981) Formation of *Escherichia coli* Hfr strains by integrative suppression with P-group plasmid RP1. *J Bacteriol* **145**, 713–21

55. Meyer R, Helinski DR (1977) Unidirectional replication of the P-group plasmid RK2. *Biochim Biophys Acta* **487**, 109–13

56. Meyer R, Hinds M (1982) Multiple mechanisms for expression of incompatibility by broad-host-range plasmid RK2. *J Bacteriol* **152**, 1078–90

57. Nash J, Krishnapillai V (1987) DNA sequence analysis of host range mutants of the promiscuous IncP-1 plasmids R18 and R68 with Tn7 insertions in *oriV*. *Plasmid* **18**, 35–45

58. Olsen RH, Shipley P (1973) Host range of the *Pseudomonas aeruginosa* R-factor R1822. *J Bacteriol* **113**, 772–80

59. Pabo CO, Sauer RT (1984) Protein-DNA recognition. *Ann Rev Biochem* **17**, 293–321

60. Paraskeva C (1979) Transfer of kanamycin resistance mediated by plasmid R68.45 in *Paracoccus denitrificans*. *J Bacteriol* **139**, 1062–4

61. Parish JH (1975) Transfer of drug resistance to *Myxococcus* from bacteria carrying drug resistance factors. *J Gen Microbiol* **87**, 198–210

62. Pinkney M, Theophilus BDM, Warne SR, Tacon WCA, Thomas CM (1987) Analysis of transcription from the *trfA* promoter of broad-host-range plasmid RK2 in *Escherichia coli*, *Pseudomonas putida* and *Pseudomonas aeruginosa*. *Plasmid* **17**, 222–32

63. Pinkney M, Diaz R, Lanka E, Thomas CM (1988) Replication of mini RK2 plasmid in extracts of *Escherichia coli* requires plasmid-encoded protein TrfA and host-encoded proteins DnaA, B, G, DNA gyrase and DNA polymerase III. *J Mol Biol* **204** (in press)

64. Pohlman RF, Figurski DH (1983) Conditional lethal mutants of the *kilB* determinant of broad-host-range plasmid RK2. *Plasmid* **10**, 82–95

65. Pohlman RF, Figurski DH (1983) Essential genes of plasmid RK2 in *Escherichia coli*: *trfB* region controls a *kil* gene near *trfA*. *J Bacteriol* **156**, 584–91

66. Polsinelli M, Baldanzi E, Bazzicalupo M, Gallori E (1980) Transfer of pRD1 from *Escherichia coli* to *Azospirillum brazilense*. *Mol Gen Genet* **178**, 709–11

67. Rella M, Watson JM, Thomas CM, Haas D (1987) Deletions in the tetracycline resistance determinant reduce the thermosensitivity of a *trfA*(Ts) derivative of plasmid RP1 in *Pseudomonas aeruginosa*. *Annals Inst Pasteur* **138**, 151–64

68. Robinson MK, Bennett PM, Falkow S, Dodd HM (1980) Isolation of a temperature-sensitive derivative of RP4. *Plasmid* **3**, 343–9

69. Saurugger PN, Hrabak O, Schwab H, Lafferty RM (1986) Mapping and cloning of the *par*-region of broad-host-range plasmid RP4. *J Biotech* **4**, 333–43

70. Schmidhauser TJ, Filutowicz M, Helinski DR (1983) Replication of derivatives of the broad-host-range plasmid RK2 in two distantly related bacteria. *Plasmid* **9**, 325–30

71. Schmidhauser TJ, Helinski DR (1985) Regions of broad-host-range plasmid RK2 involved in replication and stable maintenance in nine species of Gram-negative bacteria. *J Bacteriol* **164**, 446–55

72. Schreiner HC, Bechhofer DH, Pohlman RF, Young C, Borden PA, Figurski DH (1985) Replication control in promiscuous plasmid RK2: *kil* and *kor* functions affect expression of the essential replication gene *trfA*. *J Bacteriol* **163**, 228–37

73. Shingler V (1984) Molecular genetic analysis of the *trfA* region of the broad-host-range plasmid RK2. PhD Thesis, University of Birmingham.

74. Shingler V, Thomas CM (1984) Analysis of the *trfA* region of broad-host-range plasmid RK2 by transposon mutagenesis and identification of polypeptide products. *J Mol Biol* **175**, 229–50

75. Shingler V, Thomas CM (1984) Transcription in the *trfA* region of broad-host-range plasmid RK2 is regulated by *trfB* and *korB*. *Mol Gen Genet* **195**, 523–9

76. Shoemaker NB, Getty C, Gardner JF, Salyers AA (1986) Tn*4531* transposes in *Bacteroides* spp. and mediates the integration of plasmid R751 into the *Bacteroides* chromosome. *J Bacteriol* **165**, 929–36

77. Skotnicki ML, Tribe DE, Roger PL (1980) *Appl Environ Microbiol* **40**, 7–12

78. Smith CA, Thomas CM (1983) Deletion mapping of *kil* and *kor* functions in the *trfA* and *trfB* regions of broad-host-range plasmid RK2. *Mol Gen Genet* **190**, 245–54

79. Smith CA, Thomas CM (1984) Nucleotide sequence of the *trfA* gene of broad-host-range plasmid RK2. *J Mol Biol* **175**, 251–62

80. Smith CA, Thomas CM (1984) Molecular genetic analysis of the *trfB* and *korB* region of broad-host-range plasmid RK2. *J Gen Microbiol* **130**, 1651–62

81. Smith CA, Thomas CM (1985) Comparison of the nucleotide sequence of the vegetative replication origins of broad-host-range IncP plasmids R751 and RK2 reveals conserved features of probable functional importance. *Nucl Acids Res* **13**, 557–72

82. Smith CA, Thomas CM (1987) Narrow-host-range IncP plasmid pHH502–1 lacks a complete IncP replication system. *J Gen Microbiol* **133**, 2247–52

83. Smith CA, Shingler V, Thomas CM (1984) The *trfA* and *trfB* promoter regions of broad-host-range plasmid RK2 share common potential regulatory sequences. *Nucl Acids Res* **12**, 3619–30

84. Stalker D, Filutowicz M, Helinski DR (1983) Release of initiation control by a mutation in the plasmid encoded protein required for R6K DNA

replication. *Proc Natn Acad Sci USA* **80**, 5500–5

85. Stalker DM, Thomas CM, Helinski DR (1981) Nucleotide sequence of the region of the origin of replication of the broad-host-range plasmid RK2. *Mol Gen Genet* **181**, 8–12

86. Stanisich VA, Ortiz JM (1976) Similarities between plasmids of the P-incompatibility group derived from different bacterial genera. *J Gen Microbiol* **94**, 281–89

87. Tardiff G, Grant RB (1980) Characterization of the host range of the N plasmids. *In* Plasmids and Transposons (Stuttard C, Rozee KR, eds) New York, Academic Press, pp. 351–59

88. Theophilus BDM, Cross MA, Smith CA, Thomas CM (1985) Regulation of the *trfA* and *trfB* promoters of broad-host-range plasmid RK2. Identification of sequence essential for regulation by *trfB/korA/korD*. *Nucl Acids Res* **13**, 8129–42

89. Theophilus BDM, Thomas CM (1987) Nucleotide sequence of the transcriptional repressor gene *korB* which plays a key role in regulation of the copy number of broad-host-range plasmid RK2. *Nucl Acids Res* **15**, 7443–50

90. Thomas CM (1981) Complementation analysis of replication and maintenance functions of broad-host-range plasmids RK2 and RP1. *Plasmid* **5**, 2377–91

91. Thomas CM (1983) Instability of a high copy number mutant of a mini plasmid derived from broad-host-range IncP plasmid RK2. *Plasmid* **10**, 184–95

92. Thomas CM (1984) Genetic evidence for the direction of transcription of the *trfA* gene of broad-host-range plasmid RK2. *J Gen Microbiol* **130**, 1641–50

93. Thomas CM (1986) Evidence for the involvement of the *incC* locus of broad-host-range plasmid RK2 in plasmid maintenance. *Plasmid* **16**, 15–29.

94. Thomas CM, Cross MA, Hussain AAK, Smith CA (1984) Analysis of copy number control elements in the region of the vegetative replication origin of broad-host-range plasmid RK2. *EMBO J* **3**, 57–63

95. Thomas CM, Hussain AAK (1984) The *korB* gene of broad-host-range plasmid RK2 is a major copy number control element which may act together with *trfB* by limiting *trfA* expression. *EMBO J* **3**, 1513–9

95a. Thomas CM, Ibbotson JP, Wang N, Smith CA, Tipping R, Loader NM (1988) Gene regulation on broad-host-range plasmid RK2: identification of three novel operons whose transcription is repressed by both KorA and KorC. *Nucl Acids Res* **16**, 5345–59

96. Thomas CM, Meyer R, Helinski DR (1980) Regions of the broad-host-range plasmid RK2 which are essential for replication and maintenance. *J Bacteriol* **141**, 213–22

97. Thomas CM, Smith CA (1986) The *trfB* region of broad-host-range

plasmid RK2: the nucleotide sequence reveals *incC* and key regulatory gene *trfB/korA/korD* as overlapping genes. *Nucl Acid Res* **14**, 4453–69

98. Thomas CM, Stalker DM, Helinski DR (1981) Replication and incompatibility properties of the origin region of replication of broad-host-range plasmid RK2. *Mol Gen Genet* **181**, 1–7

99. Tsuda M, Harayama S, Iino T (1984) Tn*501* insertion mutagenesis in *Pseudomonas aeruginosa* PAO. *Mol Gen Genet* **196**, 494–500

100. Urlapova SV, Myakinin VB, Stepanov AI (1979) Temperature-sensitive mutant of the plasmid RP1. *Russ Genet* **15(3)**, 433–43

101. Valla S, Coucheron DH, Kjasbakken S (1986) Conjugative transfer of the naturally occurring plasmids of *Acetobacter xylinum* by IncP-plasmid mediated mobilization. *J Bacteriol* **165**, 336–39

102. Warner PJ, Higgins IJ, Drozd JW (1980) Conjugative transfer of antibiotic resistance to methylotrophic bacteria. *FEMS Microbiol Lett* **7**, 181–5

103. Waters SH, Grinsted J, Rogowsky P, Altenbuchner J, Schmitt R (1983) The tetracycline resistance determinants of RP1 and Tn*1721*: nucleotide sequence analysis. *Nucl Acids Res* **11**, 6089–105

104. Watson J (1980) Replication mutants of the IncP-1 plasmid RP1. *Experimentia* **36**, 1451

105. Windass JD, Worsey MJ, Pioli EM, Pioli D, Barth PT, Atherton KT, Dart EC, Byrom D, Powell K, Senior PJ (1980) Improved conversion of methanol to single-cell protein by *Methylophilus methylotrophus*. *Nature* **287**, 396–401

106. Wolk CP, Vonshak A, Kehoe P, Elhai J (1984) Construction of shuttle vectors capable of conjugative transfer from *Escherichia coli* to nitrogen-fixing cyanobacteria. *Proc Natl Acad Sci USA* **81**, 1561–5

107. Young C, Bechhofer DH, Figurski DH (1984) Gene regulation in plasmid RK2: Positive control by *korA* in the expression of *korC*. *J Bacteriol* **157**, 247–52

108. Young C, Burlage RS, Figurski DH (1987) Control of the *kilA* gene of the broad-host-range plasmid RK2: involvement of *korA*, *korB* and a new gene, *korE*. *J Bacteriol* **169**, 1315–20

109. Young C, Prince AS, Figurski DH (1985) *korA* function of promiscuous plasmid RK2: an autorepressor that inhibits expression of host-lethal gene *kilA* and replication gene *trfA*. *Proc Natl Acad Sci USA* **82**, 7374–8

110. Yusoff K, Stanisich VA (1984) Location of a function on RP1 that fertility inhibits IncW plasmids. *Plasmid* **11**, 178–81

CHAPTER 2

CONJUGATIVE TRANSFER OF Inc P PLASMIDS*

Donald G. Guiney and Erich Lanka

* Dedicated to Professor Heinz Schuster on the occasion of his 60th birthday

I. INTRODUCTION

Bacterial conjugation is the process of genetic exchange that requires intimate cell-to-cell contact. Conjugation systems are widely distributed in the bacterial world, having been found in numerous Gram-negative and Gram-positive species. In Gram-negative organisms, all well-characterized conjugation systems are encoded by plasmids, either in an autonomous state or integrated into the chromosome. In Gram-positives, plasmids also specify transfer systems, but additional units of transfer, termed 'conjugative transposons', have been described. Most detailed studies of the mechanism of conjugative DNA transfer have involved Gram-negative plasmids; to determine whether the general mechanisms of Gram-positive and Gram-negative gene transfer are similar have yet to be evaluated.

Conjugation provides a means for gene exchange between different bacterial species as well as within a given species. Conjugation-mediated recombination between strains must be a major factor in the evolution of bacterial species. In addition, the less frequent but highly significant exchange of genes between widely different bacteria has probably played a major role in bacterial evolution. Conjugation also appears to be a crucial mechanism for the dissemination and survival of plasmids, since nearly all naturally-occurring plasmids are either self-transmissible or possess mobilization systems. The role

Promiscuous Plasmids of Gram-Negative Bacteria
ISBN 0-12-688480-3
© 1989 Academic Press Limited
All rights of reproduction in any form reserved

of conjugation in bacterial evolution is dramatically illustrated by the world-wide dissemination of drug resistance plasmids following widespread use of antibiotics.

Most conjugation systems in Gram-negative bacteria have an extended host range, being capable of transfer not only within their native species, but to related organisms as well. Transfer systems of the IncP plasmids have an extremely broad-host-range, and appear capable of mediating DNA transfer into virtually any Gram-negative bacterium. This versatile property makes conjugation better adapted for genetic manipulations of diverse species than transduction or transformation, which generally require closely related organisms for efficient gene transfer. The IncP plasmid conjugation systems are widely used as tools for gene transfer and genetic studies in a great variety of Gram-negative bacteria (see Chapters 7 and 8). In addition, the IncP conjugative machinery serves as a model for studying the mechanism of broad host range plasmid transfer.

Earlier studies on the mechanism of plasmid DNA transfer during conjugation centered on the F-plasmid of *Escherichia coli* (reviewed in reference 28). Based primarily on extensive experiments with the F transfer system, a general model of conjugation has evolved (63). The biochemistry and genetics of conjugation systems are complex, requiring multiple plasmid-encoded gene products. Mating requires the coordinated interaction of a number of cell surface components (sex pili, outer and inner membrane proteins) to form a mating bridge between donor and recipient cells. The detailed mechanism of this process, termed mating pair formation (Mpf, reference 62), is poorly understood. However, many essential features of the second phase of transfer, termed DNA transfer and replication (Dtr), have been elucidated (reviewed in reference 63). A specific strand of the plasmid is transferred with the 5' terminus leading. A site-specific nick in the transferred strand is supposed to initiate transfer at the site designated *oriT* (origin of transfer). Associated with, but not required for DNA transfer, complementary DNA strands are synthesized in the donor and recipient to yield two complete plasmid molecules.

Conjugation involves a specialized system of DNA replication which is clearly different in its basic mechanism from vegetative plasmid replication. Detailed study of DNA metabolism in IncP plasmid transfer is providing insight into the fundamental mechanism of conjugative DNA replication and the adaptations for efficient broad-host-range transfer.

II. MAPPING OF TRANSFER FUNCTIONS ON IncP PLASMIDS

The regions of the RK2/RP1/RP4 plasmids required for conjugative transfer have been mapped by deletion analysis and transposon insertion mutagenesis. Early work by Figurski *et al.* (12) used Mu phage insertions to generate new restriction sites in RK2 for use in the construction of deletions. A region located counterclockwise of the *Hind*III site (toward *Eco*RI), deleted in pRK214.1 and pRK215.1, was shown to be required for transfer. The specific location of transfer regions was identified in a series of studies by Barth and co-workers (2–4), who used Tn*7* and Tn*76* insertions to map conjugation-deficient plasmid mutants. These experiments identified three basic conjugation regions of RK2/RP4, designated Tra1, 2 and 3 (Figure 1). Tra1 is located between the kanamycin resistance gene and the *Eco*RI site, in the region deleted to form

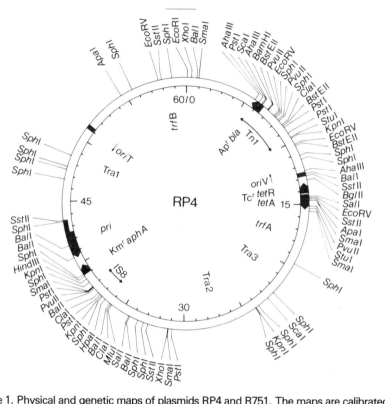

Figure 1. Physical and genetic maps of plasmids RP4 and R751. The maps are calibrated in kilobases and have been redrawn from (45,34). Abbreviations: *oriV* (origin of vegetative DNA replication), *trf* (*trans*acting replication function), *rep* (replication function), Tra (regions required for conjugal transfer), *oriT* (origin of transfer), *pri* (DNA primase), *dfr* (dihydrofolate reductase, Tp^r), *bla* (β-lactamase; Ap^r, ampicillin resistance), *tetA* (tetracycline resistance determinant, Tc^r), *tetR* (repressor gene), *aphA* (aminoglycoside 3'-phosphotransferase; Km^r, kanamycin resistance). The beginning of the coding regions for the two *pri* gene products (RP4) of 118 and 80 kDa are indicated at coordinates 43.5 and 42.4 by the staggered arrangement of the bold line. Delimitations for Tn*1* and IS*8* are marked inside the circle by arrows; for Tn*402* (R751) a line ending with dots indicates that delimitations are not published yet.

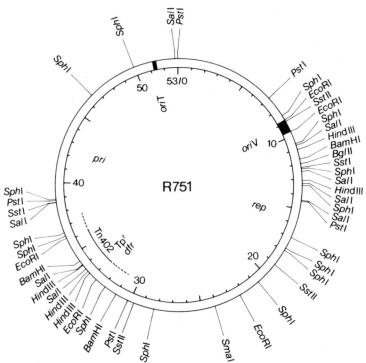

Figure 1 (continued)

pRK214.1 and pRK215.1 (12). Current work on Tra1, described in detail in Section III, precisely maps the region between coordinates 40 and 55 kb. Tra1 contains most of the transfer deficient Tn7 insertions isolated by Barth and Grinter (3), and could be divided into four complementation groups. Guiney and co-workers (23,25) subsequently mapped the origin of transfer (*oriT*) and the relaxation nick at a coincident site at the end of Tra1. Subcloning of *oriT* located the exact position on the map at 51 kb (Figure 1, references 25,15). The orientation of transfer of plasmid RP4 has been determined (1,17) to occur counterclockwise with regard to the standard RP4 map so that most of Tra1 enters the recipient cell last.

Tn7 insertions that affect the host range of R18 (similar or identical to RK2/RP4) have been reported to map very close to *oriT* (6,30,50). One set of inserts decreases the transfer frequency from *Pseudomonas aeruginosa* into *E. coli* (pM0510–512) while a different insert decreases transfer to *Pseudomonas stutzeri*. An 800 bp DNA fragment, which maps in the region of coordinate 53 kb, is able to complement the transfer deficiency of pMO511 (50). The mechanism of the recipient-specific transfer defects in these mutants is unknown. A number of Tn7 insertion mutations in the DNA primase gene of RP4 and R18, located at the other end of Tra1, have been isolated and shown to affect plasmid transfer into certain species (30,31). Based on sensitivity to IncP pili-specific phage it was established that Tra1 also contains genes involved in sex pilus synthesis (4).

Unlike Tra1, the Tra2 and Tra3 regions are not as well-characterized. Tra2 is defined by five Tn7 inserts and one Tn76 mutation, all mapping between the KpnI site at 24 kb and the PstI site at 31 kb (2,4). Three inserts located toward the PstI end were defective in P pili production, and three other inserts toward the KpnI site were moderately deficient in surface exclusion. Tra3 is defined by a single Tn7 insert at about 21 kb on the map. This mutant did not express any surface exclusion. Due to the lack of transposon insertion mutants between the regions Tra2 and Tra3, it is not possible to determine at present whether the regions are contiguous or genetically separate.

To date, detailed studies of the organization of the Tra regions have only been conducted on the RK2/RP4 group of plasmids (Figure 1) which, as discussed in Chapter 3, belong to the IncPα subgroup. Of the other major subgroup, the IncPβ plasmids, R751 (Figure 1) is the best studied. Although restriction sites differ markedly between RK2/RP4 and R751, the similar organization of the major genetic regions (17,54; see Chapter 3 for detailed discussion) strongly indicates that the IncPα and β groups evolved from a common ancestor.

Study of the functional relationship between the RK2 and R751 plasmids revealed that R751 could not mobilize plasmids containing the 760 bp oriT fragment of RK2, consistent with the considerable sequence divergence observe between the oriT regions of these two plasmids (65). However, provision of several kilobases of additional RK2 DNA from the Tra1 region allowed mobilization by R751 in trans showing that portions of the transfer systems of α and β plasmids are interchangeable.

III. GENETIC ORGANIZATION OF THE Tra1 REGION AND FUNCTIONAL CORRELATION

A. The Detailed Genetic Structure of the Tra1 Region

The coincident location of the relaxation nick site and the origin of transfer in the Tra1 region of RK2/RP4/RP1 suggests that it may contain the plasmid-encoded information involved in DNA metabolism during conjugal transfer (25,58,34). The resistance of some Tn7 insertions in Tra1 to pili-specific phage indicates that this region also contains genes involved in pili synthesis (4). Mapping of additional restriction sites and the construction of a restriction fragment bank of RP4 (34) were the basic requirements for systematically approaching the dissection of the Tra1 region providing a detailed picture of the genetic organization of Tra1. The general scheme of this approach is shown in Figure

2. Fragments from the Tra1 region were cloned in stable recombinant plasmids using primarily pBR-type vectors, although R300B (RSF1010) derivatives were employed with some constructions. A study of potential promoter sites in the Tra1 region provided important information regarding the subcloning of intact structural genes into expression vectors. These sites were mapped as RNA polymerase binding sites by electron microscopical analysis of protein-DNA complexes (35,13,46). The expression vectors contained the inducible *tac*-promoter (10), a pBR- or RSF1010-type replicon and the *lac*-repressor gene (*lac*I^Q). The latter ensures a very low level of transcription in the non-induced state (14). The molecular cloning of defined Tra1 fragments into the autoregulated expression vectors facilitated overproduction of Tra gene products (13,46,14,15). Proteins were isolated and purified by conventional column ion exchange chromatography, gel filtration and chromatography on heparin Sepharose. The latter has proven extremely useful in the purification of proteins involved in DNA metabolism (37,61). The purified proteins were then used to elicit specific antisera in rabbits. These antibodies demonstrated on immunoblots that the antigens correspond to RP4 gene products and do not represent cloning artifacts. Using the solid phase immunoblotting technique, crude extracts of cells containing deletion derivatives were tested for truncated gene products to precisely localize the corresponding transfer genes. Another important result obtained by expression vector cloning was to assign the direction of transcription relative to the standard RP4 map. Since specific Tra$^-$ mutants (nonsense, missense or nonpolar transposon insertion mutants, reference 58) are not available, the functional analysis of Tra1 still relies on *in vitro* characterization of the purified proteins (Section III, D./E.).

The gene organization of Tra1 of RP4 and portions of the R751 Tra1 region are shown in Figure 3. Tra1 of RP4 maps between coordinates 40 kb and 55 kb (Figure 1) and is bordered on one side by the kanamycin resistance determinant, *aphA* (46). The Tn7 inserts which affect the host range of conjugal transfer, described in Section II, map on the right side of the *Apa*I site in Tra1. In addition, a Tn7 insertion mutant pRP831 (4,34) conferring a Tra$^+$ phenotype maps just to the right of *Apa*I. This indicates that the right end of Tra1 is probably formed by the operon transcribed from promoters P$_{R2}$/P$_{R3}$(Figure 3). Barth *et al.* (4) determined four complementation groups for Tra1; some of these groups contain more than one phenotypic class of mutants and represent more than one cistron because of the possible polar and deletion effects of Tn7 insertions (4). The existence of four different promoter regions determined as RNA polymerase binding sites (35,13) suggest that Tra1 consists of at least four operons, three of which are polycistronic. It is not known if all four operons are expressed constitutively or if some are regulated. The RP4 Tra1 region encodes at least 16 proteins. Initial attempts to utilize antibodies against RP4 Tra1 proteins for the detection of R751 proteins by cross reaction reveal a striking similarity to the gene products of the

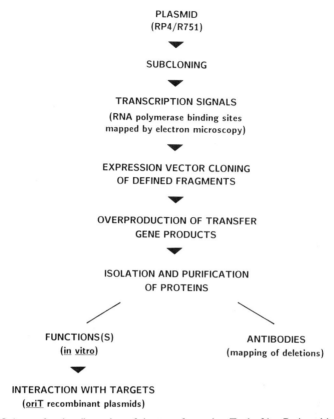

Figure 2. Scheme for the dissection of the transfer region Tra1 of IncP plasmids. Details are explained in the text.

Tra1 region of RP4 (Figure 3). However, the R751 proteins vary in the strength of immunological cross reactivity and differ in size from the corresponding RP4 gene products. The proteins tested so far include the primases (see Section III E.), the 85 kDa, the 11 kDa, and the 26 kDa protein. With exception of the antiserum against the 11 kDa protein, which has a reduced cross reacting activity to the corresponding 12.5 kDa product of R751, antisera against the other proteins reveal strong cross reaction. This result suggests that they share homologous regions. The low level of relationship between the two analogous DNA binding proteins—11 kDa (RP4) and 12.5 kDa (R751)—reflect sequence differences which apparently determine the *oriT* specificities of the Pα and Pβ transfer systems (see Section III C.). The identical order of genes in the transfer region Tra1 of RP4 and R751 demonstrates that the overall genetic organization of IncP Tra1 appears to be highly conserved (Figure 3).

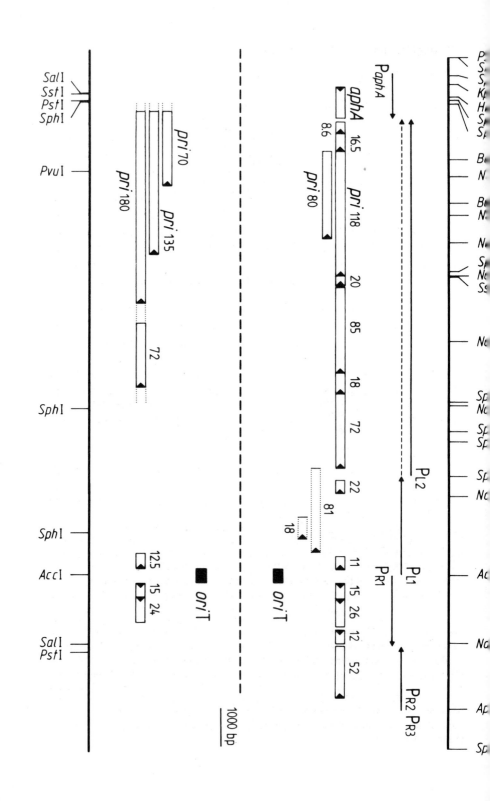

B. Definition of the origin of transfer

In the general model of conjugation outlined previously, the origin of transfer (*ori T*) represents the site on the transferred strand where the transfer process is initiated by the single-strand nicking event. The origin of transfer is also defined genetically as the region required in *cis* to the DNA that is to be transferred. Conjugation-specific DNA replication is probably initiated within the *ori T* region. Conjugative DNA transfer requires the specific interaction of plasmid-encoded transfer factors at *ori T*. A 'transfer replicon' consists of a functional unit mediating conjugative transfer; it includes an *ori T* region and gene products that interact specifically at *ori T*. This is a constituent of the 'DNA transfer and replication system' (Dtr), as defined by Willetts (62). In addition to Dtr the plasmid-encoded 'mating pair formation system' (Mpf), responsible for the synthesis of pili and other surface proteins, is essential for conjugative DNA transfer.

It has been demonstrated that both strands of the RK2 *ori T* can function as origins for complementary strand synthesis in single-stranded M13 phage DNA replication in cells containing the RK2/RP4 DNA primase (for details see Section III E.). Molecular cloning of *ori T* into vector plasmids (preferably such vectors lacking any *ori T* sequence and transfer or mobilization functions) provides an *in vivo* assay for *ori T* activity (Figure 4). The assay measures the ability of a recombinant *ori T* plasmid to be mobilized by a compatible helper plasmid. Transconjugants confer a phenotype which is distinguishable from donor and recipient. To prevent any event depending on homologous recombination a *Rec A* background is preferred for mating experiments. The fact that most of the transconjugants in short mating processes (approximately 30

Figure 3. Genetic organization of IncP transfer region Tra1 studied on plasmids RP4 and R751. The section of RP4 between map coordinates 38.0–55.6 (above the dashed line) containing the Tra1 region is shown. Restriction sites are as in Figure 1 including *Not*I sites and one of several *Acc*I sites. Promoter sites are indicated P; sites were determined by RNA polymerase binding studies and confirmed by sequence comparison to the consensus sequence for *E. coli* promoters (47). The lines with arrowheads represent proposed transcripts. Genes are drawn as boxes with arrows at their 5'-ends. The numbers above the boxes refer to the apparent molecular masses in kilodaltons of the corresponding gene products as determined from SDS-polyacrylamide gels. The boxes drawn with dotted lines represent open reading frames deduced from nucleotide sequences. Data presented in this scheme are confirmed by nucleotide sequencing: *aphA* (46); the DNA primase region (L.Miele and E.L.); *Sst*II to P_{L2}(B.Strack, G.Ziegelin and E.L.:), $P_{L2}-Acc$I (G.Ziegelin and E.L.); *Acc*I–*Not*I (M.Kröger and E.L.) and *Not*I–*Sph*I (W.Schilf and E.L.) will be published elsewhere. The region of R751 containing Tra1 (below the dashed line) is defined by *Sal*I (38.7) and *Pst*I (53.0) according to the map in Figure 1. Two restriction sites, *Pvu*I and *Acc*I have been included. The maps of RP4 and R751 were aligned at an *Acc*I site which appears to be conserved in the *ori T* regions of both plasmids. The *ori T* regions are marked as black boxes.

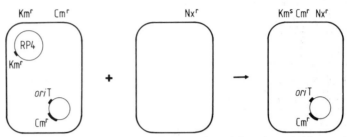

Figure 4. Scheme of the *in vivo* assay for *oriT* activity of recombinant plasmids. Donorcells harboring the helper and the *oriT* plasmid are mixed in a 1:10 ratio with recipient cells, filtered on a nitrocellulose membrane and incubated for a defined mating time on a nutrient agar plate. Mating of the cells is interrupted by vigorous vortexing and transconjugants are quantified by subsequent plating on selective medium (23).

min) contain only the recombinant *oriT* plasmid demonstrates that transfer by cointegration does not interfere with the assay. Isolation of functional RK2/RP4 *oriT* fragments which lack structural genes has led to the conclusion that all IncP transfer factors function in *trans* (25,13,15).

Dissection of recombinant *oriT* plasmids of RK2 (25,19,20) and RP4 (13,15) and nucleotide sequencing revealed that *oriT* maps at 51 kb on the standard RP4 map in an intercistronic region comprising about 350 bp (Figure 5). Deletions from the right end reduce the transfer frequencies by about 100–200 fold, indicating that this part of the *oriT* sequence plays an important role for efficient transfer, probably as recognition sites for transfer factors. The sequence of a rudimentary *oriT* fragment (Figure 5) has been published (25). It contains an almost perfect 38 bp inverted repeat. It has been proposed that this region might form a secondary structure which comprises the site of specific interaction for relaxation proteins. A 24 bp deletion into the right hand arm of the repeat completely eliminates *oriT* activity indicating that the intact repeat is required for activity and that the direct repeat does not function as a transfer origin by itself. A comparison of the RK2 sequence with other *oriT* sequences reveals that all contain regions of dyad symmetry (63). Although partially homologous sequences have been described between RK2, ColE1, and RSF1010 *oriTs* (11), their functional significance is not obvious and remains to be demonstrated.

C. Relaxosomes

A number of conjugative and mobilizable plasmids can be isolated in the form of a supercoiled DNA-protein relaxation complex (relaxosome) by the gentle cell lysis technique. Upon treatment *in vitro* with SDS, pronase, or certain intercalating agents, the supercoiled DNA is relaxed by nicking and converted to the open circular form (Figure 6; for review see references 63,20). It has been shown for RK2 by alkaline sucrose gradient centrifugation that nicking occurs near or at the origin of transfer. The relaxation site (*rlx*), also called nick site (*nic*), was localized approximately 20 kb away from the vegetative origin of replication (*oriV*) of RK2 (23). Further analysis indicated that the transfer origin is contained in a 760 bp *Hae*II fragment that maps in the same region as the single-stranded nick observed in the RK2 relaxosome (25,19,20). Another approach to visualize the specific cleavage of a defined plasmid single strand created by a relaxation process, was to separate ssDNA fragments from linearized plasmid DNA by alkaline gel electrophoresis (15). The three distinct bands observed on the gel (full length linear and two smaller fragments of different size) demonstrate that the cleavage must have occurred at a unique site. Calculation of fragment sizes located the cleavage site within *oriT*. Relaxation of a recombinant plasmid containing the *oriT* region in the absence of a helper plasmid is not possible. This indicates that the specific cleavage process requires plasmid-encoded proteins. The coding regions for these functions map closely to *oriT*, a region named Tra1 core.

D. The Tra1 core region

As described in Section II, the IncPβ plasmid R751 can mobilize an RK2 *oriT* only if additional RK2 information is provided. This indicates that part of the conjugative machinery is interchangeable between the two plasmids, and probably consists of the complete Mpf-system but only part of the Dtr system. It seems likely that the noninterchangeable sections are needed for high precision protein–DNA recognition at the *oriT* sequence. This elegant assay system has been applied to define the Tra1 core region as the minimal fragment size of RK2/RP4 which is transferred at maximum efficiency by the heterologous helper plasmid R751. Defined RP4 fragments inserted into pBR329, as well as pDG4Δ22 carrying Tn5 insertions, were assayed for mobilization in the presence of R751 as a helper plasmid (13,15, Guiney and Deiss, unpublished). The information needed for heterologous mobilization comprises less than 3 kb and is located on both sides of *oriT* (Figure 7). Nucleotide sequence determination of an extended part of the Tra1 core of both plasmids RP4 and R751 followed by computer analysis show regions of homology which are interspersed with nonhomologous sections (Figure 8). The diverging regions probably determine the *oriT* specificity between the RP4 and the R751 transfer system. Different related *oriT* sequences apparently demand specific structural alterations in the proteins so that the transfer origins can be recognized.

The systematic approach as outlined in Section III A facilitated the detailed analysis of the genetic organization of Tra1 core. Two operons are transcribed divergently from a promoter region within the transfer origin. Each of the operons is polycistronic encoding several genes on either side of *oriT* (Figure 7). It can be concluded from this study and results of experiments involving

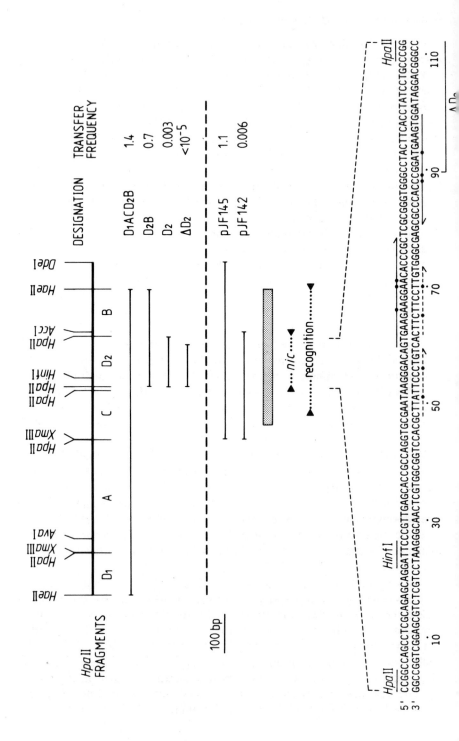

Tn5 insertions into pDG4Δ22, that efficient heterologous mobilization requires at least two genes to the left and one gene to the right side of *ori T*. Deletion of the left hand part from pMS226n (Figure 7; 49.1–50.3 *Not*I–*Bss*HII) reduces heterologous mobilization but does not abolish it. This indicates that the component encoded by the deleted part is partially interchangeable between the two systems (13). Heterologous mobilization could still be detected with the fragment containing only the genes for the 11 kDa and 15 kDa gene product. However, deletions at the 3' ends on either of these two genes abolish this property indicating that the products of the 11 kDa and 15 kDa genes possess *ori T* specificity. This result has been confirmed by the identification of Tn5 insertions in the 11 kDa and 15 kDa genes that abolish mobilization of pDG4Δ22 by R751.

The gel electrophoretic assay was used to detect specific relaxation at *ori T*, demonstrating that only the left hand portion of Tra1 is necessary for this reaction (15). The operon to the left hand side of *ori T*, carrying the functions mediating specific cleavage at *ori T*, is therefore called the relaxase operon. Predictions on the participation of host-encoded components in the relaxation process cannot yet be made.

Overproduction, isolation and purification of one of the Tra1 components—the 11 kDa protein—allowed its characterization as a dimeric DNA binding protein. In addition to its nonspecific DNA-binding capacity (15), it possesses a highly-specific affinity for an RP4 fragment containing *ori T* (Figure 9). An *ori T* fragment of R751 is not bound by this protein (data not shown) indicating that the 11 kDa protein binds exclusively to the RP4 *ori T*. The corresponding R751 *ori T* binding protein is the 12.5 kDa protein (Figure 3). Footprint analysis will be used to determine the site(s) of specific interaction, which might coincide with the large inverted repeat. Since the 11 kDa protein is encoded by the relaxase operon, the formation of the complex between the transfer origin and the 11 kDa protein might be the first step in the assembly of the relaxosome. Binding of 11 kDa protein to *ori T* might function to direct specific interactions of other proteins. A candidate for the second protein component of the relaxosome is the gene product encoded within the same operon adjacent to the gene for the 11 kDa product. Interestingly, an R751 component

---

Figure 5. The RK2/RP4 transfer origin. *Hpa*II subfragments of the 785 bp-*Hae*II fragment (25) are designated from A to D in the order of decreasing fragment size. The *oriT* activities of defined fragments are given in transconjugants/donor cell with RK2/RP4 as helper plasmid in the table on the right side. Data above the dashed line are taken from (25,19) and below from (13,15). The minimal fragment (ca. 350 bp) mediating full *oriT* activity is indicated as a shaded bar. The regions where the proposed relaxation nick (*nic*) must be localized and which is necessary for efficient mobilization are marked with arrowheads. The nucleotide sequence of the 112 bp *Hpa*II fragment D$_2$ is shown on the bottom (25). A 38 bp inverted repeat and a 12 bp direct repeat are designated by arrows. Bases which do not match are marked by dots. Nucleotides 90–112 are deleted of the *Hpa*II fragment D$_2$ to result in DD$_2$.

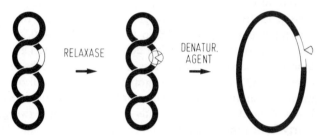

Figure 6. Assembly and *in vitro* assay of relaxosomes. Relaxation proteins ('relaxase') bind to the *oriT* region (marked as a white area) of the negatively supercoiled form of the plasmid. Relaxation induces strand- and site-specific cleavage of the DNA. The reaction *in vitro* is mediated by certain denaturing agents. In plasmid ColE1 the open circular product contains a free 3'-hydroxyl, but the 5'-end is covalently attached to one of the relaxation proteins (22).

appears to substitute for this RP4 gene product—at least partially—as detected by the heterologous mobilization assay. The role of 11 kDa protein could therefore be compared to replication proteins like O and P which operate to deliver host replication proteins to *oriV* in the initiating prepriming steps of phage lambda DNA replication (40).

What could be the role of the 15 kDa product encoded by the remaining operon of Tra1 core? Undoubtedly the product is essential for a step in the transfer process involving *oriT* specificity. Since specific relaxation appears to be covered by functions of the relaxase operon of the Tra1 core, the 15 kDa protein might be necessary to facilitate the precise restoration of the cleavage site at *oriT*.

E. DNA Primases of IncP Plasmids

Plasmid-specific DNA primase activity was detected by an assay for the conversion of single-stranded circular DNA of small phages to its duplex form *in vitro* (36). A survey of plasmids representing most of the known incompatibility groups (31) revealed that several conjugative plasmids specify this activity (Table I). For the IncI and IncP complexes it has been demonstrated that the primases are accessory transfer factors (31,43) involved in generation of RNA primers for transfer replication. All P-type plasmids studied to date encode DNA primases (33). The assay used for detecting priming activities measures the ability of crude extracts from cells harboring plasmids to mediate DNA synthesis on single-stranded phage DNA (Figure 10). Conversion of single-stranded circular templates to the duplex form occurs in a rifampicin-resistant reaction indicating that synthesis is independent of *E. coli* RNA polymerase. The reaction requires plasmid-encoded DNA primase and a DNA polymerase (the large fragment of DNA polymerase I, T7 DNA polymerase,

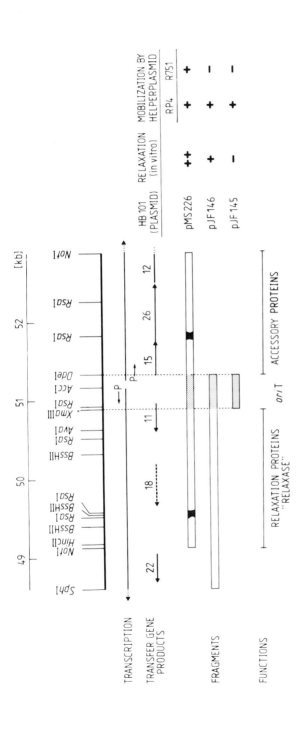

Figure 7. Functional map of Tra1 core encoding relaxosome formation. A physical map of the NotI-F fragment of Tra1 located between coordinates 48.6 and 52.9 (Figure 1) is shown. Only one of several XmaIII sites has been included. Directions of transcription and proposed promoter sites (P) are indicated by arrows. Gene products are identified by numbers referring to molecular masses in kilodaltons. Fragments inserted into vector pBR329 are represented by bars. The shaded part marks an intergenic region which functions as oriT. The black thick bars mark the region which is necessary for efficient heterologous mobilization as determined by Tn5 insertions. Relaxation and mobilization properties of RP4 oriT derivatives are summarized in the table.

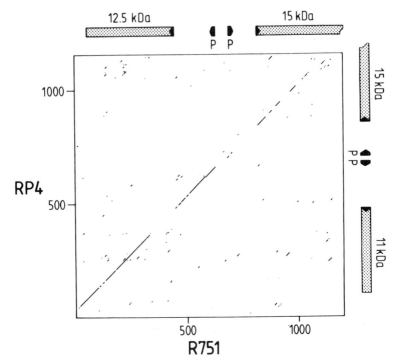

Figure 8. Sequence comparison of *ori T* regions of plasmids RP4 and R751. Nucleotide sequences of about 1200 bp of each of the two plasmids are compared via dot matrix analysis (39). Identical sequence coordinates are represented by dots (67% positional identity in a range of 21 bp). Direction of transcription starting from promoter sites (P) is indicated by black arrowheads. Genes on either side of the intercistronic *ori T* regions are drawn as shaded bars. The numbers above the bars refer to molecular masses of the gene products. Details will be published elsewhere (W.Pansegrau and E.L.).

or DNA polymerase III holoenzyme in the presence of single-strand DNA binding protein (Ssb)) as well as ribo- and deoxyribonucleoside triphosphates and magnesium or manganese (Mn^{2+}) ions. In addition, the reaction needs CMP or cytidine. A variety of single-stranded phage DNAs (fd, If1, PR64FS, Pf3, ϕX174 and G4) can serve as templates for the plasmid-specified primases with almost equal efficiencies (32), indicating that the enzyme is active in the general priming reaction (35,61).

Existence of a plasmid-encoded DNA primase (36) explains how Inc I plasmids can partially suppress the temperature-sensitive defect of the host primase—the Dna G protein (60). The plasmid specified enzyme can substitute for the host primase during discontinuous DNA replication *in vivo* allowing extensive colony formation at the high temperature in *dna*G mutants, thus providing a potent *in vivo* assay for this class of enzymes (Table I). Other

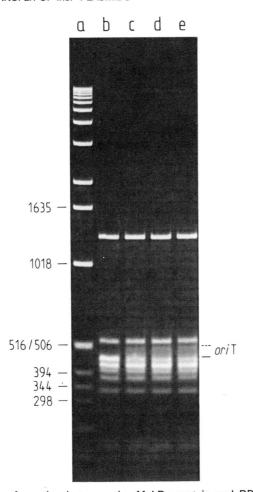

Figure 9. Complex formation between the 11 kDa protein and RP4 *oriT* DNA. A *Bam*HI/*Dde*I digest of pJF145 (see Figure 5) was incubated for 10 min at 37°C in 30 mM Tris-HCl (pH 7.6), 50 mM NaCl, 5 mM MgCl$_2$, and 10.1 μg/ml BSA with various amounts of purified 11 kDa protein. Samples were run on 2% (w/v) agarose gel in 40 mM Tris, 5 mM sodium acetate (adjusted with acetic acid to pH 7.9) at 3.5 V/cm for 7 hrs. A photograph of the ethidium bromide stained gel is shown. Lane a, DNA fragment size markers; lanes b–e, 0.75 μg of pJF145 fragments incubated in the presence of 11 kDa protein; b, 11 kDa protein omitted; c, 30 ng; d, 60 ng; e, 90 ng. The shift of the *oriT* containing fragment (450 bp) is indicated by the dashed line. The band consists of two fragments, the RP4 *oriT* and a similarly sized vector fragment.

plasmids may specify primases that are not able to function in general priming (e.g. the replicon specific priming enzyme of RSF1010) and are therefore not detectable in this assay (49).

The plasmid-encoded DNA primase synthesizes oligoribonucleotides *in vitro*

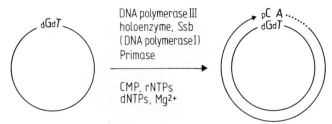

Figure 10. Initiation of complementary strand synthesis on single-stranded phage DNAs *in vitro*. Protein and substrate requirements in the presence of plasmid encoded DNA primase are shown.

which can be extended by DNA polymerases. The primers range in size from 2 to 10 nucleotides containing cytidine or cytidine 5'-monophosphate at the 5'-terminus with AMP as the second nucleotide (32). Oligoribonucleotide synthesis by plasmid primase on short oligodeoxynucleotide templates of defined sequences indicated that the enzyme can recognize $dTdG$ sequences in $(dTdG)_n$, $n = 1-9$, and that the products are complementary to the template DNA (Figure 11). The enzyme recognizes a 3'-$dGdT...-5'$ sequence located either at the end (Figure 12C) or within a polydeoxynucleotide chain (Figure 12A and B). This ability of the plasmid-encoded DNA primases to synthesize oligoribonucleotides with the sequence 5'-(p)CpA... by recognizing the dinucleotide sequence 3'-$dGdT...5'$ without the assistance of additional protein components makes the enzyme clearly distinct from the DnaG protein. Despite these enzymatic differences, the plasmid DNA primase can apparently replace the DnaG protein in chromosomal, phage, and plasmid replication as demonstrated by suppression studies. The plasmid primases differ from other primases or priming systems (e.g. the phage T7 DNA primase, the phage T4 priming system, the DnaG protein requiring systems, or the priming of fd/M13 complementary strand synthesis by RNA polymerase) in their specific requirement for CMP or cytidine instead of a purine triphosphate. In addition, the property that these plasmid-encoded primases can initiate primer synthesis from a 3'-$dGdT...5'$ sequence at the termini of single-stranded DNA, may be significant for conjugative DNA replication.

All IncP plasmids examined contain genes specifying DNA primases (*pri*) (33). The IncP primase locus belongs to the Tra1 region—as shown for the Birmingham plasmids RK2, R18, R68, RP1 and RP4 (Figure 1). It codes for two polypeptides of 118 and 80 kDa, which arise from an in-phase overlapping gene arrangement (35). The promoter region for the *pri* region maps about 5 kb upstream from the start codon of *pri*. Data for RNA polymerase binding sites, deletion analysis and the sequence comparison with the *E. coli* promoter consensus sequence are in good agreement. The promoter sequence contains direct and inverted repeats suggesting that transcription of the *pri* region is regulated (G.Ziegelin, and E.L. to be published elsewhere). Both

TABLE I. *Plasmids specifying DNA primase activity and their ability to suppress the* Escherichia coli dnaG3 *mutation*

Plasmid(s)	Inc group	Primase activity	Suppression of dnaG3
R16, R864a	B	+(13,70,71)	+(70,71)
R40a	C	+(72)	+(72)
ColIb, R64, R144	Iα	+(13,35,37)	+(38,73)
R621A	Iγ	+(13,35,37)	+(38,73)
TP114	Iδ	+(13)	n.t.
R805a	Iζ	+(72)	+(71)
R391	J	−(13)	+(72,73)
R387	K	+(13)	+(72)
R831b, R446b	M	+(13)	−(72)
RP4 (RK2, R18, R68, RP1)	Pα	+(13)	+(13,20,74)
R1033	Pα	+(16)	+(16)
R751, R772, R906	Pβ	+(16)	+(16)
RA3	U	+(72)	+(72)

n.t. = not tested; reference numbers are indicated in parentheses.

purified *pri* polypeptides have identical specific activities, demonstrating that a large portion of the N-terminus of the 118 kDa primase is dispensable for the priming function *in vitro*. This portion might serve another as yet unknown function (32). Both polypeptides bind strongly to single- and double-stranded DNA. The 118 kDa Pri polypeptide functions as a monomer and has an anisometric shape as concluded from glycerol gradient centrifugation and gel filtration studies (31). The Pri proteins are the most abundant gene products of RP4 with about 4000 copies of the 118 kDa protein per cell. This also suggests that the primases are multifunctional proteins, because the priming reaction *in vitro* requires only catalytic amounts of enzyme (31,42).

The *pri* genes of the α and β subgroups are related to each other as judged from Southern blot hybridization and immunological data (33). Extensive DNA and protein sequence homologies have been detected, although the gene products of the α and β P-subgroups exhibit substantial differences in size. IncPβ plasmids encode three Pri polypeptides of 180, 135 and 70 kDa, respectively, which cross react with the anti IncPα primase serum directed against the RP4 118 kDa polypeptide. Cross reaction with I-type primase or DnaG protein has not been observed. Expression vector cloning of the R751 *Sph*I-C fragment (45) facilitated overproduction of the three Pri polypeptides plus two additional proteins of 72 and 25 kDa (W.Pansegrau, B.Strack and E.L. to be published elsewhere). The latter proteins do not react with IncPα primase antiserum. Because the coding capacity of the *Sph*I-C fragment would only permit the synthesis of a polypeptide chain of a total length of 270 kDa, the gene organization of the IncPβ *pri* locus can only be explained by an overlapping gene arrangement similar to that of the IncPα *pri* gene (35).

RIBO – PRODUCT ------------- DEOXY – TEMPLATE	SUBSTRATES
$5'$ pCA $3'$ GTGTGTGTGTGTGT $5'$	pC, ppp*A
$5'$ pCACACACACA $3'$ GTGTGTGTGTGTGT $5'$	pC, ppp*A, pppC
$5'$ pCA $3'$ GT $5'$	pC, ppp*A, pppC
- - $3'$ TG $5'$	pC, ppp*A pA, ppp*C
$5'$ pCAAAAAAAA $3'$ GTTTTTTTT $5'$	pC, ppp*A
- - - - - - - - - - $3'$ TTTTTTTTTT $5'$	pC, pA, ppp*A

Figure 11. Template dependent synthesis of oligoribonucleotides by plasmid-encoded DNA primase. Various synthetic oligodeoxynucleotides were applied as templates for the synthesis of oligoribonucleotides by plasmid-encoded primase. Reaction conditions have been described previously (32). The asterisk designates the ^{32}P-labeled phosphate of the nucleoside triphosphate used as a substrate. Products of the reaction were analyzed by electrophoresis on 20% (w/v) polyacrylamide–7M urea slab gels.

The plasmid-encoded primase has undoubtedly a role in bacterial conjugation. However, effects of mutations in *pri* on vegetative replication and plasmid stability are more likely to be pleiotropic (31,42). RP4 *pri* mutants (RP4::Tn7 *pri*) transfer with a lower efficiency into some bacterial species, such as *S. typhimurium*. Back-transfer experiments showed that this effect is recipient-specific (31). A similar recipient-specific conjugation defect was observed when Pri$^-$ mutants of the Birmingham plasmid R18 were tested for conjugation proficiency (30). Thus the primase apparently contributes to the broad-host-range of IncP plasmids by facilitating the conjugation metabolism of these DNA elements. Merryweather *et al.* (42) have further investigated the physiological role of the RP4-specified primase in bacterial conjugation. Complementation tests involving recombinant plasmids carrying cloned fragments

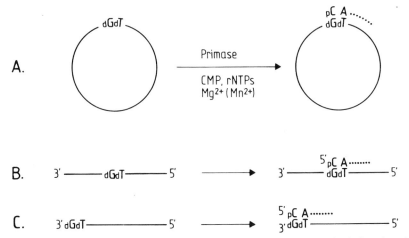

Figure 12. DNA directed oligoribonucleotide synthesis. Primer synthesis by plasmid-specified DNA primases uncoupled from DNA synthesis on circular and linear DNA templates.

of RP4 indicated that the primase acts to promote some post-transfer event in the recipient cell which can be satisfied by plasmid primase made in the donor cell. It is proposed that the enzyme, its products, or both are transmitted to the recipient cell during conjugation. Specificity of plasmid primases was assessed with derivatives of RP4 and the IncI$_1$ plasmid ColIb-P9. The latter is known to encode a DNA primase active in conjugation (43). When supplied in the donor cell, neither of the primases encoded by these plasmids substituted effectively in the nonhomologous system. Since ColIb primase provided in the recipient cell acts weakly on the transferred RP4 DNA, it is suggested that the specificity of these enzymes reflects their inability to be transmitted via the conjugation apparatus of the nonhomologous plasmid. A 'transfer replicon' specific interaction with plasmid specified primases has also been observed by Yakobson & Guiney (67). M13 phages lacking the primary origin for complementary strand synthesis are still viable but grow poorly and form very small plaques (29). Inserting part of the RK2 *ori T* sequence into the defective phage DNA specifically enhances phage replication in cells containing the IncPα DNA primase and restored phage production almost to the wild-type level. The IncPβ plasmid R751 was as effective as the IncPα plasmid(s) in promoting replication of M13ΔE101 *ori T* phage, despite the differences between α and β primase. However, R64drd11(IncIα) which also encodes a DNA primase (*sog*) did not enhance replication of the M13ΔE101 *ori T* phage. This indicates a highly specific interaction of plasmid-encoded primase with the *ori T* sequence and probably other transfer factors.

Further subcloning of fragments from the RK2 *ori T* region into the origin-defective vector M13ΔE *lac* has shown that both strands of *ori T* will function as efficient initiation sites for complementary strand synthesis in the presence of the RK2 DNA primase (66). Since several subfragments from *ori T* will promote efficient M13ΔE *lac* replication, it is concluded that the *ori T* region contains multiple sites for primase-directed initiation of DNA synthesis, a finding consistent with the known *in vitro* properties of the enzyme.

IV. USE OF THE IncP CONJUGATION SYSTEM FOR THE GENETIC MANIPULATION OF GRAM-NEGATIVE BACTERIA

The IncP conjugation system represents an extremely versatile tool for genetic studies, since IncP plasmids can mediate gene transfer into virtually any Gram-negative organism. The power of this system is that it facilitates the application of *E. coli* recombinant DNA technology to a wide variety of species. In principle, any set of genes can be cloned and manipulated in *E. coli*, then returned to the native organism by IncP-mediated conjugation for functional studies. In addition, derivatives of P-plasmids have proven valuable in establishing genetic transfer systems within species by the mobilization of chromosome markers.

The use of Hfr strains that mediate oriented, high-frequency transfer of chromosomal genes, has greatly facilitated *E. coli* genetic studies. Unfortunately, endogenous plasmids that mediate chromosome mobilization —analogous to the F-plasmid—have not been found in most other Gram-negative bacteria (27). Therefore, considerable interest has centered on the use of P-plasmids to mediate gene transfer in organisms unrelated to *E. coli*. However, native IncP plasmids mobilize chromosomal genes at low or undetectable frequencies (27). A number of P-plasmid derivatives that are more efficient in chromosomal transfer have been isolated or constructed. These systems, described in detail in the Chapter 7, are based on one of two strategies for exploiting the IncP conjugation system: 1) integration of the entire plasmid into the chromosome is enhanced by transposable elements on the plasmid, or regions of homology between the plasmid and chromosome, and 2) the transfer origin (*ori T*), present on a transposon, is integrated into the chromosome and then mobilized *in trans* by the conjugation system present on a helper plasmid.

Detailed studies on the transfer of genetic markers by integrated IncP plasmids have determined that the genes are transferred in a polarized fashion from a fixed point of integration, leading to the determination of the direction of transfer from *ori T* as shown in Figure 1 (17,1). These results conform closely to transfer mediated by the F-plasmid of *E. coli*, indicating that the general mechanism of transfer by these plasmids is probably similar. Recently, pULB113 (RP4::mini Mu) has been shown to mediate a novel retrotransfer of chromosomal markers from the recipient into the donor during matings (41). The mechanisms for this retro- or shuttle-transfer is unknown, but may be a specific property of IncP plasmids.

Based upon knowledge of the *ori T* region of RK2, a second approach to chromosome transfer was devised using the 760 bp *ori T* fragment cloned into Tn5 (66). The resulting hybrid transposon, Tn5-*ori T*, could be used to randomly and stably insert *ori T* into the chromosome. Using a transfer-proficient RK2 helper plasmid to mobilize the *ori T* sequence, polarized transfer of chromosomal markers was demonstrated in *E. coli* and *Rhizobium meliloti*. These results showed that mobilization of the RK2 *ori T* sequence *in trans* by the plasmid conjugation system occurs by the same mechanism of oriented transfer as in the native plasmid. The Tn5-*ori T* system is applicable in principle to virtually any Gram-negative organism in which Tn5 can transpose.

In addition to chromosome transfer, IncP plasmids are widely used to mobilize other plasmids in broad-host-range transfer applications. This plasmid mobilization occurs by one of two mechanisms: 1) use of the endogenous transfer origin and mobilization gene products of the plasmid to be transferred, or; 2) insertion of the RK2/RP4 *ori T* sequence into the target plasmid for use *in trans* by the helper plasmid.

Naturally-occurring non-conjugative plasmids usually contain a mobilization system consisting of a transfer origin and *mob* gene products capable of interacting with the transfer systems of self-transmissible plasmids. The specificity of this interaction varies considerably between different plasmids and is a poorly understood process. However, the IncP conjugation systems appear to be extremely versatile in promoting mobilization of a wide variety of plasmids from different hosts, including RSF1010 (IncQ broad-host-range group), ColE1, pSC101, pWD2 (*Neisseria gonorrhoea* β-lactamase plasmid), and pCP1 (*Bacteroides fragilis* R-plasmid) (62,24,51). For several of these plasmids, mobilization by F (RSF1010, pSC101, pWD2) or R64*drd11* (pCP1) occurred at much lower frequencies. In the cases that have been examined in detail, mobilization requires distinct regions on the target plasmid and does not appear to involve cointegrate formation. Therefore, the IncP conjugation system seems particularly efficient at mobilizing plasmids from diverse species. It is especially interesting that RK2 can mobilize the *Bacteroides* plasmid pCP1, while RK2 cannot be maintained in *Bacteroides* and pCP1 cannot replicate in *E. coli* (21). Thus, it is unlikely that the pCP1 *mob* system ever encounters the RK2 conjugation system in nature, and the efficient mobilization of pCP1 and other unrelated plasmids may reflect a certain lack of specificity in the broad-host-range transfer system.

Plasmids that lack a mobilization system can be efficiently transferred by insertion of the RK2 *ori T* sequence, analogous to chromosome mobilization discussed previously. This approach facilitates the construction of shuttle vectors for use in organisms such as *Bacteroides*, in which even IncP plasmids cannot be maintained. Hybrid vectors consisting of a native *Bacteroides* plasmid, an *E. coli* cloning vector, and the *ori T* of RK2 or the *mob* region of RSF1010, could be transferred from *E. coli* into *B. fragilis* using either RK2 or R751 as a helper plasmid (21,52). The IncP plasmid was never found alone in the *B. fragilis* transconjugants, indicating that P-plasmids cannot replicate in *Bacteroides*. These studies demonstrate that the host range of the IncP conjugation system exceeds that of the replication system. This approach will be useful for extending the range of P-plasmid-mediated genetic manipulations to organisms in which the usual broad-host-range plasmids cannot be maintained. Very recently it has been shown that plasmids carrying the RK2 transfer origin can be mobilized from *E. coli* donors to several Gram-positive organisms at frequencies of 2×10^{-8} to 5×10^{-7}. The plasmids are shuttle type vectors containing a pBR- and a pAMβ1 origin for vegetative replication in Gram-negatives as well as in Gram-positives (57).

The Tn5-*oriT* transposon has been used to insert the *oriT* sequence into other plasmids to promote mobilization by P-plasmids. This system proved particularly helpful in the analysis of large 'cryptic' plasmids in *Salmonella* species (5). These plasmids were identified as essential virulence factors by the combination of Tn5-*oriT* insertion mutagenesis and mobilization into plasmid-free strains (5). A similar Tn5 derivative with the *oriT* region from RP4 has been described (53).

V. CONCLUSIONS

The detailed analysis of the RK2/RP4 conjugation system has yielded considerable insight into the mechanism of plasmid DNA transfer. Based on the present information, the IncP transfer system conforms to the basic model of conjugation proposed for F- and related plasmids. RK2/RP4 transfer proceeds in a polarized manner from a fixed site, *oriT*, that is required *in cis*. Transfer is oriented so that the major portion of Tra1 enters the recipient last, analogous to the F-plasmid. *In vitro* nicking at *oriT* by the RK2/RP4 relaxosome is believed to be the correlate of the *in vivo* single strand nick that initiates the transfer process. Interactions between the *oriT* sequence and the conjugation system are highly specific and require at least three plasmid-encoded gene products, including the 11 kDa protein that binds specifically to *oriT*. The strand specificity and orientation of strand transfer (5' to 3') have not been confirmed *in vivo* but relaxosome nicking *in vitro* is strand-specific. The detailed mechanism of conjugative DNA replication remains unclear. The IncP plasmid DNA primases are products of the Tra1 region and clearly function in DNA transfer. Current evidence favors transfer of the RK2/RP4 primase to the recipient during conjugation, to function in priming of the DNA strand complementary to the transferred strand. Both DNA strands from the *oriT* region can serve as initiation sites for complementary strand synthesis mediated by the RK2/RP4 primase, implying that these enzymes could prime conjugative DNA synthesis in both the donor and recipient. An alternative mechanism, based on the rolling circle model for DNA transfer, dictates that the free 3' end of the transferred strand acts as a primer for DNA synthesis in the donor, with a second nicking event required to transfer a unit length strand (16). In addition to ongoing DNA synthesis, the intrinsic property of this model involves the action of a DNA helicase, like the *E. coli* Rep protein, which is needed to unwind the transferred DNA strand. No evidence for or against this model is available in the IncP system, although some indirect results with F- and IncI plasmids indicate that RNA priming is required for conjugative DNA synthesis in the donor (63).

The control of conjugative transfer gene expression in IncP plasmids remains unclear. The high transfer frequencies of native IncP plasmids indicate that the transfer phenotype is constitutive. However, interactions of other conjugative plasmids with the IncP group indicate that control of transfer may be complex. RP1 inhibits the transfer of certain IncW plasmids (68). This Fi$^+$ (IncW) property maps in at least two regions of RP1: one comprising the 6.5 kb *Pst*I-C fragment and another in an unspecified location. The latter function inhibits IncW pilus formation while the *Pst*I-C fragment inhibits fertility by an unknown mechanism. The IncPβ plasmids R906 and R751 lack the Fi$^+$ (IncW) property. Several plasmids inhibit IncP transfer, with F, R100, and the IncN plasmid pKM101 being the best studied (44,55,56,64). All of these plasmids contain well-defined DNA regions which decrease P-plasmid transfer when cloned *in trans*. The F-plasmid gene has been identified as *pifC*, located in the replication origin region, and an analogous gene, *tir*, of R100 has been identified (56,64). The evidence indicates that RP4 contains a binding site for the *pifC* gene product. The pKM101 Fi$^+$ (IncN) property is encoded in the transfer region of the plasmid (64). None of these transacting factors abolish P-pilus synthesis, and the mechanism of their Fi$^+$ action remains unknown.

The molecular basis for the exceedingly broad-host-range transfer capabilities of P-plasmids remains a fascinating enigma. Although the basic mechanism of P-plasmid transfer resembles the F-plasmid system, a direct comparison of the host ranges of RK2 and F showed that RK2 was 10^4 times more efficient at promoting plasmid transfer from *E. coli* to *P. aeruginosa* (18). This study indicated that the IncP conjugation system possesses specific properties of broad-host-range transfer not shared by the F-plasmid. In addition, the host range of the P-plasmid conjugation system exceeds that of the replication apparatus, since IncP plasmids can mediate gene transfer to *Bacteroides* species but cannot be maintained in these organisms (21). The broad-host-range capability of the P-plasmids must reflect specific adaptations at the level of both mating pair formation (Mpf) and DNA transfer and replication (Dtr).

There is virtually no information available on the Mpf system of IncP plasmids, although P-pili appear to be required. The fact that purified F-pili bind to *E. coli* but not to *P. aeruginosa* cells may partially account for the narrow host range of F transfer (26), and, by analog, P-pili are likely to be crucial for broad-host-range transfer. The requirements for the Mpf system are daunting; it is difficult to conceive of a common structure on the surface of all Gram-negative bacteria that could form a recognition site for the IncP Mpf proteins.

Even lipopolysaccharide, which has a highly conserved core structure in many Gram-negatives, is substantially different in *Bacteroides*, yet P-plasmid transfer readily occurs to these organisms (21).

The Inc P plasmid primases represent a well characterized broad host range adaptation of the Dtr system. Although host primases can substitute in many bacteria, P-plasmid transfer into certain species is markedly diminished without the plasmid primase. However, DNA primases are not a unique feature of broad-host-range plasmids, having been well characterized in Inc I plasmids and also found in at least five other incompatibility groups (Table I). Future work will be needed to identify the additional specific features of both the Mpf and Dtr systems that endow the Inc P plasmids with such a remarkable transfer capability.

ACKNOWLEDGEMENTS

This work was supported by Grant GM 28924 from the National Institute of General Medical Sciences to D.G.G., who is grateful for the expert technical assistance of Cornelia Deiss.

E.L. thanks Heinz Schuster for generous support and encouragement and gratefully acknowledges the expert assistance of Marianne Schlicht. We appreciate the critical reading of this manuscript by Jens Peter Fürste and Werner Pansegrau.

VI. REFERENCES

1. Al-Doori Z, Watson M, Scaife J (1982) The orientation of transfer of the plasmid RP4. *Genet Res Camb* **39**, 99–103
2. Barth PT (1979) RP4 and R300B as wide-host-range plasmid cloning vehicles. *In*: Plasmids of Medical, Environmental and Commercial Importance (Timmis KN, Pühler A, eds) Elsevier/North-Holland Biomedical Press. pp.399–410
3. Barth PT, Grinter NJ (1977) Map of plasmid RP4 derived by insertion of transposon C. *J Mol Biol* **113**, 455–74.
4. Barth PT, Grinter NJ, Bradley DE (1978) Conjugal transfer system of plasmid RP4: Analysis by transposon 7 insertion. *J Bacteriol* **133**, 43–52
5. Chikami GK, Fierer J, Guiney DG (1985) Plasmid-mediated virulence in *Salmonella dublin* demonstrated by use of a Tn5-*ori T* construct. *Infect Immun* **50**, 420–2
6. Cowan P, Krishnapillai V (1982) Tn7 insertion mutations affecting the host range of the promiscuous Inc P-1 plasmid R18. *Plasmid* **8**, 164–74
7. Dalrymple BP (1982) Plasmid encoded DNA primases. Ph.D. thesis, University of Leicester, England.
8. Dalrymple BP, Boulnois GJ, Wilkins BM, Orr E, Williams PH (1982)

Evidence for two genetically distinct DNA primase activities specified by plasmids of the B and I incompatibility groups. *J Bacteriol* **151**, 1–7

9. Dalrymple BP and Williams PH (1982) Detection of primase specified by IncB plasmid R864a. *J Bacteriol* **152**, 901–3

10. de Boer HA, Comstock LJ and Vasser M (1983) The *tac* promoter: a functional hybrid derived from the *trp* and *lac* promoters. *Proc Natl Acad Sci USA* **80**, 21–5

11. Derbyshire KM, Willetts NS (1987) Mobilization of the nonconjugative plasmid RSF1010: A genetic analysis of its origin of transfer. *Mol Gen Genet* **206**, 154–60

12. Figurski D, Meyer R, Miller DS, Helinski DR (1976) Generation *in vitro* of deletions in the broad-host-range plasmid RK2 using phage Mu insertions and a restriction endonuclease. *Gene* **1**, 107–19

13. Fürste JP (1986) Konjugativer Transfer des promiskuitiven Plasmids RP4: Wechselwirkung von plasmidkodierten Faktoren mit dem Transferorigin. Ph.D. thesis, Freie Universität Berlin, Berlin, Germany.

14. Fürste JP, Pansegrau W, Frank R, Blöcker H, Scholz P, Bagdasarian M, Lanka E (1987) Molecular cloning of the plasmid RP4 primase region in a multi-host-range *tacP* expression vector. *Gene* **48**, 119–31

15. Fürste JP, Ziegelin G, Pansegrau W, Lanka E (1987) Conjugative transfer of promiscuous plasmid RP4: plasmid specified functions essential for formation of relaxosomes. *In*: DNA Replication and Recombination, (McMacken R, Kelly TJ, eds) *UCLA Symp Mol Cell Biol, New Series*, Vol. 47, Alan R. Liss, Inc. pp.553–64.

16. Gilbert W, Dressler D (1966) DNA replication: the rolling circle model. *Cold Spring Harbor Symp Quant Biol* **33**, 474–84

17. Grinter NJ (1981) Analysis of chromosome mobilization using hybrids between plasmid RP4 and a fragment of bacteriophage lambda carrying IS*1*. *Plasmid* **5**, 267–76

18. Guiney DG (1982) Host range of conjugation and replication functions of the *Escherichia coli* sex plasmid F'*lac*: comparison with the broad-host-range plasmid RK2. *J Mol Biol* **162**, 699–703

19. Guiney Jr DG (1984) Promiscuous transfer of drug resistance in Gram-negative bacteria. *J Infect Diseases* **149**, 320–29

20. Guiney DG, Chikami G, Deiss C, Yakobson E (1985) The origin of plasmid DNA transfer during bacterial conjugation. *In*: Plasmids in Bacteria. Basic Life Sciences, Vol. 30, (Hollaender A, ed) Plenum Press, New York and London. pp.521–3

21. Guiney DG, Hasegawa P, Davis CE (1984) Plasmid transfer from *Escherichia coli* to *Bacteroides fragilis*: differential expression of antibiotic resistance phenotypes. *Proc Natl Acad Sci USA* **81**, 7203–6

22. Guiney DG, Helinski DR (1975) Relaxation complexes of plasmid DNA and protein. III. Association of protein with the 5′ terminus of the broken

DNA strand in the relaxed complex of plasmid ColE1. *J Biol Chem* **250**, 8796-803

23. Guiney DG, Helinski DR (1979) The DNA–protein relaxation complex of the plasmid RK2: Location of the site-specific nick in the region of the proposed origin of transfer. *Mol Gen Genet* **176**, 183-9

24. Guiney DG, Ito J (1982) Transfer of the gonococcal penicillinase plasmid: mobilization in *E. coli* by P group plasmids and isolation as a DNA–protein relaxation complex. *J Bacteriol* **150**, 298-302

25. Guiney DG and Yakobson E (1983) Location and nucleotide sequence of the transfer origin of the broad-host-range plasmid RK2. *Proc Natl Acad Sci USA* **80**, 3595-8

26. Helmuth R, Achtman M (1978) Cell–cell interactions in conjugating *E. coli*: purification of F-pili with biological activity. *Proc Natl Acad Sci USA* **75**, 1237-41

27. Holloway BW (1979) Plasmids that mobilize bacterial chromosome. *Plasmid* **2**, 1-19

28. Ippen-Ihler KA, Minkley EG (1986) The conjugation system of F, the fertility factor of *Escherichia coli*. *Ann Rev Genet* **20**, 593-624

29. Kim MH, Hines JC, Ray DS (1981) Viable deletions of the M13 complementary strand origin. *Proc Natl Acad Sci USA* **78**, 6784-8

30. Krishnapillai V, Nash J, Lanka E (1984) Insertion mutations in the promiscuous IncP-1 plasmid R18 which affect its host range between *Pseudomonas* species. *Plasmid* **12**, 170-80

31. Lanka E, Barth PT (1981) Plasmid RP4 specifies a deoxyribonucleic acid primase involved in its conjugal transfer and maintenance. *J Bacteriol* **148**, 769-81

32. Lanka E, Fürste JP (1984) Function and properties of RP4 DNA primase. *In*: Proteins Involved in DNA Replication (Hübscher U, Spadari S, eds) *Advances in Experimental Medicine and Biology*, Vol. **179**, Plenum Publ Corp, pp.265-80

33. Lanka E, Fürste JP, Yakobson E, Guiney DG (1985) Conserved regions at the DNA primase locus of IncPα and IncPβ plasmids. *Plasmid* **14**, 217-23

34. Lanka E, Lurz R and Fürste JP (1983) Molecular cloning and mapping of *Sph*I restriction fragments of plasmid RP4. *Plasmid* **10**, 303-7.

35. Lanka E, Lurz R, Kröger M, Fürste JP (1984) Plasmid RP4 encodes two forms of a DNA primase. *Mol Gen Genet* **194**, 65-72

36. Lanka E, Scherzinger E, Günther E, Schuster H (1979) A DNA primase specified by I-like plasmids. *Proc Natl Acad Sci USA* **76**, 3632-6

37. Lanka E, Schuster H (1983) The *dnaC* protein of *Escherichia coli*. Purification, physical properties and interaction with *dnaB* protein. *Nucl Acids Res* **11**, 987-97

38. Ludwig RA, Johansen E (1980) DnaG-suppressing variants of R68.45

with enhanced chromosome donating ability in *Rhizobium. Plasmid* 3, 359–61
39. Maizel Jr JV, Lenk RP (1981) Enhanced graphic matrix analysis of nucleic acid and protein sequences. *Proc Natl Acad Sci USA* **78**, 7665–9
40. McMacken R, Alfano C, Gomes B, LeBowitz, JH, Mensa-Wilmot K, Roberts JD, Wold M (1987) Biochemical mechanisms in the initiation of bacteriophage lambda DNA replication. *In*: DNA Replication and Recombination (McMacken R, TJ Kelly, eds), *UCLA Symp Mol Cell Biol, New Series*, **47**, pp.227–45. Alan R. Liss, Inc.
41. Mergeay M, Lejeune P, Sadouk A, Gerits J, Fabry L (1987) Shuttle transfer (or retrotransfer) of chromosomal markers mediated by plasmid pULB113. *Mol Gen Genet* **209**, 61–70
42. Merryweather A, Barth PT, Wilkins BM (1986) Role and specificity of plasmid RP4-encoded DNA primase in bacterial conjugation. *J Bacteriol* **167**, 12–7
43. Merryweather A, Rees CED, Smith NM, Wilkins BM (1986) Role of *sog* polypeptides specified by plasmid ColIb-P9 and their transfer between conjugating bacteria. *EMBO J* **5**, 3007–12
44. Miller J, Lanka E, Malamy M (1985) F-factor inhibition of conjugal transfer of broad-host-range plasmid RP4: requirement for the protein product of *pif* operon regulatory gene *pifC. J Bacteriol* **163**, 1067–73
45. Pansegrau W, Lanka E (1987) Conservation of a common 'backbone' in the genetic organization of the IncP plasmids RP4 and R751. *Nucl Acids Res* **15**, 2385
46. Pansegrau W, Miele L, Lurz R, Lanka E (1987) Nucleotide sequence of the kanamycin resistance determinant of plasmid RP4: Homology to aminoglycoside-3'-phosphotransferases. *Plasmid* **18**, 193–204
47. Rosenberg M, Court D (1979) Regulatory sequences involved in the promotion and termination of RNA transcription. *Ann Rev Genet* **13**, 319–53
48. Sasakawa C, Yoshikawa M (1978) Requirements for suppression of a *dnaG* mutation by an I-type plasmid. *J Bacteriol* **133**, 485–91
49. Scherzinger E, Bagdasarian MM, Scholz P, Lurz R, Rückert B, Bagdasarian M (1984) Replication of the broad host range plasmid RSF1010: Requirement for three plasmid-encoded proteins. *Proc Natl Acad Sci USA* **81**, 654–8
50. Schilf W, Krishnapillai V (1986) Genetic analysis of insertion mutations of the promiscuous IncP-1 plasmid R18 mapping near *oriT* which affect its host range. *Plasmid* **15**, 48–56
51. Shoemaker NB, Getty C, Guthrie EP, Salyers AA (1986) Regions in *Bacteroides* plasmids pBFTM10 and pB8–51 that allow *Escherichia coli–Bacteroides* shuttle vectors to be mobilized by IncP plasmids and a conjugative *Bacteroides* tetracycline resistance element. *J Bacteriol* **166**, 959–65
52. Shoemaker N, Guthrie E, Salyers A and Gardener J (1985) Evidence that the clindamycin–erythromycin resistance gene of *Bacteroides* plasmid pBF4 is

on a transposable element. *J Bacteriol* **162**, 626–32

53. Simon R, O'Connell M, Labes M, Pühler, A (1986) Plasmid vectors for the genetic analysis and manipulation of *Rhizobia* and other Gram-negative bacteria. *Meth Enzymol* **118**, 640–59

54. Smith CA, Thomas CM (1987) Comparison of the organization of the genomes of phenotypically diverse plasmids of incompatibility group P: members of the IncPβ subgroup are closely related. *Mol Gen Genet* **206**, 419–27

55. Tanimoto K, Iino T (1983) Transfer inhibition of RP4 by F-factor. *Mol Gen Genet* **192**, 104–9

56. Tanimoto K, Iino T, Ohtsubo H, Ohtsubo, E (1985) Identification of a gene, *tir*, of R100, functionally homologous to the F3 gene of F in the inhibition of RP4 transfer. *Mol Gen Genet* **198**, 356–7.

57. Trieu-Cuot P, Carlier C, Martin P, Courvalin P (1987) Plasmid transfer by conjugation from *Escherichia coli* to Gram-positive bacteria. *FEMS Micrbiol Lett* **48**, 289–94

58. Ubben D, Schmitt R (1986) Tn*1721* derivatives for transposon mutagenesis, restriction mapping and nucleotide sequence analysis. Gene **41**, 145–52.

59. Watson J, Schmidt L, Willetts N (1980) Cloning the Tra1 region of RP1. *Plasmid* **4**, 175–83.

60. Wilkins BM (1975) Partial suppression of the phenotype of *Escherichia coli* K12 *dnaG* mutants by some I-like conjugative plasmids. *J Bacteriol* **122**, 899–904

61. Wilkins BM, Boulnois GJ, Lanka E (1981) A plasmid DNA primase active in discontinuous bacterial DNA replication. *Nature* **290**, 217–21

62. Willetts N (1981) Sites and systems for conjugal DNA transfer in bacteria. *In*: Molecular Biology, Pathogenicity, and Ecology of Bacterial Plasmids (Levy SB, Clowes RC, Koenig EL, eds) Plenum Press, New York. pp.207–15

63. Willetts N, Wilkins B (1984) Processing of plasmid DNA during bacterial conjugation. *Microbiol Rev* **48**, 24–41

64. Winans S, Walker G (1985) Fertility inhibition of RP1 by IncN plasmid pKM101. *J Bacteriol* **161**, 425–27

65. Yakobson E, Guiney G (1983) Homology in the transfer origins of broad-host-range IncP plasmids: Definition of two subgroups of P-plasmids. *Mol Gen Genet* **192**, 436–38

66. Yakobson E and Guiney DG Conjugal (1984) Transfer of bacterial chromosomes mediated by the RK2 plasmid transfer origin cloned into transposon Tn*5*. *J Bacteriol* **160**, 451–3

67. Yakobson E and Guiney DG (1988) Initiation of DNA synthesis in the transfer origin region of RK2 by the plasmid-encoded primase: Detection using a defective M13 phage. (in press)

68. Yusoff K, Stanisich V (1984) Location of a function on RP1 that fertility inhibits IncW plasmids. *Plasmid* **11**, 178–81

CHAPTER 3
RELATIONSHIPS AND EVOLUTION OF Inc P PLASMIDS

Christopher A. Smith and Christopher M. Thomas

I. TYPICAL Inc P PLASMIDS

Typical plasmids belonging to *Escherichia coli* incompatibility group P (= com4; classified as Inc P-1 in *Pseudomonas*) have a very broad host range, being capable of self transfer to and maintenance within members of most Gram-negative genera so far tested (Chapter 1). Such plasmids are today widespread, and have been isolated from diverse Gram-negative bacterial species collected at diverse geographical locations (Table I). All of these plasmids were isolated as agents conferring antibiotic resistance on their bacterial hosts, often from clinical specimens, with the exception of pJP4 which was isolated because it confers on *Alcaligenes* the ability to catabolize the pesticide 2,4-dichlorophenoxyacetic acid. Most of these plasmids confer resistance to more than one antimicrobial agent, some determining resistance to as many as seven commonly used antibiotics as well as to mercurials (Table I).

II. THE Inc Pα/Inc Pβ DIVISION

The plasmids listed in Table I have been divided into two subgroups, Inc Pα and Inc Pβ, which appear to represent major evolutionary branches of the Inc P group. This division of the Inc P plasmids into two subgroups was proposed because Southern blotting *Hae*II digests of ten Inc P plasmids with a probe carrying the *ori T* region of Inc P plasmid RK2 revealed only two patterns of fragment sizes, which correlated with the results of complementation studies on this locus (75). Studies using probes derived from the *trf A* and *ori V* regions

Promiscuous Plasmids of Gram-Negative Bacteria   © 1989 Academic Press Limited
ISBN 0-12-688480-3   All rights of reproduction in any form reserved

TABLE I. *Naturally-occurring typical IncP plasmids*

	Resistance markers[a]										Other markers[b]				
Plasmid	Km	Tc	Pn	Sm	Su	Hg	Cm	Gm	Tp	*bla*	[Te]	Fi	Ae	Tn*1*	IS*21*
Assigned to IncPα:															
Birmingham isolates[c]															
RK2	+	+	+	−	−	−	−	−	−	TEM	+	?	?	+	+
RP1/R18	+	+	+	−	−	−	−	−	−	TEM−2	?	+	+	+	+
RP4	+	+	+	−	−	−	−	−	−	TEM−2	+	?	?	+	+
R68	+	+	+	−	−	−	−	−	−	TEM−2	?	?	+	+	+
Other isolates															
R26	+	+	+	+	+	+	+	+	−	?	+	+	+	+	+
R702	+	+	−	+	+	+	−	−	−	−	?	+	?	−	−
R839	+	+	+	+	+	+	−	−	−	TEM−1	?	+	?	+	−
R934	+	+	+	−	−	+	−	−	−	TEM−1	?	?	?	−	+
R938	+	+	+	+	+	+	+	−	−	TEM−1	?	+	?	−	+
R995	+	+	−	−	−	−	−	−	−	−	?	?	?	−	−
R1033	+	+	+	+	+	+	+	+	−	TEM−1	?	?	?	+	+
pUZ8	+	+	−	−	−	+	−	−	−	−	+	?	?	−	−
Provisionally assigned to IncPα:															
R$_{GN}$823	+	+	+	+	+	?	+	−	−	TEM	?	?	+	?	?
Rm16b	+	+	+	−	−	−	−	−	−	TEM−2	?	?	+	?	?
RP8	+	+	+	−	−	−	−	−	−	?	?	?	?	?	?
RP638	+	+	+	−	−	−	−	−	−	TEM−2	?	?	+	?	?
R527	+	+	+	+	+	+	+	+	−	TEM−1	?	?	+	?	?
R690	+	+	+	−	−	−	−	−	−	?	?	?	?	?	?
R842	+	+	+	−	−	−	−	−	−	TEM−1	?	?	?	?	?
R940	+	+	+	+	+	?	−	−	−	?	?	?	?	?	?
pMG22	+	+	−	−	+	+	−	+	−	−	?	?	?	−	?
pVS9	+	+	+	+	+	+	+	+	−	?	?	?	+	?	?
Assigned to IncPβ:															
R751	−	−	−	−	−	−	−	−	+	−	−	−	−	−	−
R772	+	−	−	−	−	−	−	−	−	−	−	?	?	−	−
R906	−	−	+	+	+	+	−	−	−	OXA−2	?	−	−	−	−
pJP4[d]	−	−	−	−	−	+	−	−	−	−	?	?	?	−	−

[a] Km, kanamycin; Tc, tetracycline; Pn, penicillin; Sm, streptomycin; Su, sulphonamide; Hg, mercuric salts; Cm, chloramphenicol; Gm, gentamicin; Tp, trimethoprim; +, resistance conferred;−, resistance not conferred; ?, not known to have been tested. *bla*: classification of β-lactamase, if present (36,45).

[b] [Te], cryptic tellurite resistance (2); Fi, inhibition of the fertility of coresident IncW plasmids (76); Ae, insensitivity to aeruginosin AP41 (37,64); Tn*1*, presence of a Tn*1*-like element at the same position as Tn*1* (70); IS*21*, presence of an IS*21*-like element at the same position as IS*21* (70).

[c] R18 and R68 are also designated R1882 and R6886, respectively (5,32). Further isolates of IncP plasmids confering Km[r] Tc[r] & Pn[r] from Birmingham (UK) clinical specimens, which are probably closely related or identical to these plasmids are: RK1, RK3, R74/R7475, from *Klebsiella aerogenes* (5,33); R436a, R437a, R438–R440, from *Proteus mirabilis/vulgaris* (21); R708–R710, R711a, R712a, R713 from Providence (20); RP2/R3425, RP9/R91/R91a/R9169, R30, R88 (5,32), from *Pseudomonas aeruginosa*.

[d] Plasmid pJP4 also confers (in *Alcaligenes*) the ability to utilize 2,4-dichlorophenoxyacetic acid and 3-chlorobenzoic acid.

Plasmid	Species of origin	Country of origin	Size (kb)
Assigned to IncPα:			
Birmingham isolates			
RK2	*Klebsiella aerogenes*	UK	60
RP1/R18	*Pseudomonas aeruginosa*	UK	60
RP4	*Pseudomonas aeruginosa*	UK	60
R68	*Pseudomonas aeruginosa*	UK	60
Other isolates			
R26	*Serratia marcescens*	Spain	72
R702	*Proteus mirabilis*	USA	77
R839	*Serratia marcescens*	UK	87
R934	*Serratia marcescens*	France	72
R938	*Serratia marcescens*	France	84
R995	*Proteus mirabilis*	Hong Kong	57
R1033	*Pseudomonas aeruginosa*	Spain	75
pUZ8	*Pseudomonas aeruginosa*	Spain	58
Provisionally assigned to IncPα:			
$R_{GN}823$	*Klebsiella pneumoniae*	Japan	—
Rm16b	*Pseudomonas aeruginosa*	Japan	58
RP8	*Pseudomonas aeruginosa*	UK	93
RP638	*Pseudomonas aeruginosa*	S. Africa	66
R527	*Pseudomonas aeruginosa*	Spain	72
R690	Providence	UK	—
R842	*Proteus mirabilis*	UK	—
R940	*Proteus mirabilis/vulgaris*	USA	—
pMG22	*Pseudomonas aeruginosa*	USA	—
pVS9	*Pseudomonas aeruginosa*	W. Germany	78
Assigned to IncPβ:			
R751	*Klebsiella aerogenes*	UK	53
R772	*Proteus mirabilis*	USA	61
R906	*Bordetella bronchiseptica*	Japan	58
pJP4	*Alcaligenes eutrophus*	Australia	80
R772	*Proteus mirabilis*	USA	61
R906	*Bordetella bronchiseptica*	Japan	58
pJP4	*Alcaligenes eutrophus*	Australia	80

TABLE I. *continued*

Plasmid	Reference
Assigned to IncPα:	
Birmingham isolates	
RK2	33
RP1/R18	5,32
RP4	11
R68	32
Other isolates	
R26	64
R702	22
R839	22
R934	26
R938	27
R995	70
R1033	62
pUZ8	23
Provisionally assigned to IncPα:	
$R_{GN}823$	55
Rm16b	36
RP8	32
RP638	52
R527	64
R690	20
R842	45
R940	21
pMG22	35
pVS9	74
Assigned to IncPβ:	
R751	38
R772	8
R906	27
pJP4	13

of RK2 (7,60) and the *pri* locus of RP4 (44) confirmed the validity of this subgrouping, which is also supported by the results of heteroduplex mapping (70,53,54). Although at present the number of known members of the IncPβ subgroup is much smaller than the number of plasmids assigned to the IncPα subgroup, both subgroups show considerable diversity in the geographic locations and the range of host species from which they have been isolated (Table I), consistent with a relatively ancient evolutionary separation.

III. ATYPICAL IncP PLASMIDS

Table II lists a number of plasmids which show group P-incompatibility but do not appear to belong to either of the subgroups shown in Table I. Of these

R446a has not been investigated beyond being assigned to the IncP group, while the others are known to be atypical, lacking the broad host range of typical IncP plasmids and showing asymetric incompatiblity, displacing but not being displaced by other IncP plasmids (19,30,48,49). Such unidirectional incompatibility is also shown by plasmids constructed *in vitro* which carry group P-incompatibility determinants, but possess a replication system belonging to another incompatibility group (Chapter 1). This suggests that these narrow-host-range plasmids may be the products of replicon fusions followed by deletions, which have resulted in the presence of P-group incompatibility determinants on plasmids whose replication systems belong to some other (narrow host range) incompatibility group. Plasmid pAV1 is confined to *Acinetobacter calcoaceticus* and has only been tested for incompatibility against group P (30). Plasmid pHH502 is incompatible with both IncIα and IncP plasmids and determines IncIα-type pili; it gives rise spontaneously to segregants such as pHH502–1 which have lost all IncIα characteristics (49). Plasmid pHH502–1 determines rigid pili which confer sensitivity to bacteriophages PR4 and PRD1 (which infect bacteria carrying IncN-, IncP-and IncW-type pili), but not to the IncP specific bacteriophage PRR1, and which are immunologically distinct from those of IncM, IncN, IncP and IncW plasmids (49). Plasmid pHH502–1 expresses unidirectional incompatibility against IncP plasmids, and has been shown to have homology to the *oriV* region of IncP plasmids but no homology to the *trfA* gene, which is essential for replication from *oriV* (Chapter 1; 49,61). Thus pHH502–1 is presumably a plasmid of an unknown incompatibility group which carries relicts of an IncP plasmid, including P-group incompatibility determinant(s) as the result of a replicon fusion followed by deletion(s). Plasmid pTM89 is incompatible with members of the FII and P-groups, and gives rise both to segregants incompatible only with group FII and to segregants which are compatible with all known groups (48). Plasmids pMU700-pMU707 are incompatible both with IncIα and IncP plasmids (19). It is likely that the narrow host range of all of these atypical IncP plasmids results from the lack of a complete IncP replication system, and they will not be discussed further.

IV. THE IncPα SUBGROUP

The known natural members of the IncPα subgroup all confer resistance both to tetracycline and to kanamycin, while none of the known IncPβ plasmids confers resistance to tetracycline and only R772 confers kanamycin resistance (Table I). A number of IncP plasmids which confer resistance to both of these antibiotics, but have not been examined by hybridization pattern or heteroduplex analysis, have been included in Table I as provisionally assigned to the IncPα subgroup. The plasmids definitely assigned to the IncPα subgroup have been divided into two sections. The first section consists of plasmids which

TABLE II. *Other plasmids which show P group incompatibility*

Plasmid	Resistance markers[a]										Other Inc groups[b]
	Km	Tc	Pn	Sm	Su	Hg	Cm	Gm	Tp	*bla*	
R446a	−	−	+	−	−	−	−	−	−	TEM	−
pAV1	−	−	−	−	+	−	−	−	−	−	?
pHH502	−	+	−	−	[+]	+	+	−	+	−	Iα, Unc
pHH502−1	−	−	−	−	[+]	+	−	−	+	−	Unc
pTM89	+	+	+	+	+	−	+	−	−	?	FII, Unc
pMU700	−	−	−	−	+	+	−	−	−	−	Iα
pMU707											−

[a] Km, kanamycin; Tc, tetracycline; Pn, penicillin; Sm, streptomycin; Su, sulphonamide; Hg, mercuric salts; Cm, chloramphenicol; Gm, gentamicin; Tp, trimethoprim: +, resistance conferred; [+] low level resistance conferred (49);−, resistance not conferred; ?, not known to have been tested. *bla*: classification of β-lactamase, if present (45).

[b] Unc, evidence for the presence of a replication system not belonging to a known incompatibility group; ?, not tested for other incompatibility groups.

Plasmid	Species of origin	Country of origin	Size (kb)
R446a	*Proteus morganii*	S. Africa	−
pAV1	*Acinetobacter calcoaceticus*	UK	−
pHH502	*Escherichia coli*	UK	71
pHH502−1	*Escherichia coli*	UK	57
pTM89	*Escherichia coli*	Italy	−
pMU700−	*Escherichia coli*	Australia	−
pMU707			

Plasmid	Reference
R446a	24
pAV1	30
pHH502	49
pHH502−1	49
pTM89	48
pMU700−	19
pMU707	

were isolated from specimens collected during 1969 in a single hospital at Birmingham (U.K.) and were used to define the incompatibility groups P of *Escherichia coli* and P-1 of *Pseudomonas aeruginosa* (33,34). These plasmids, which confer resistance to penicillin/carbenicillin as well as to tetracycline and kanamycin, are the most thoroughly studied members of the group. The five Birmingham plasmids are indistinguishable by gross restriction map (9,65) and by heteroduplex analysis (4), although minor differences have been reported between RP4 and R68 in such traits as plating of IncP plasmid-specific phages such as PRR1 and their effect on virulence of *P. aeruginosa* (40,73,74) and a few base changes have been observed between the Tc^r regions of RP1 and RK2 (M. Pinkney, unpublished). Figure 1 shows the organization of these plasmids. An obvious feature of this map is the clustering of restriction sites in the regions which contain the resistance determinants, with relatively few sites in the regions involved in plasmid maintenance and conjugal transfer. Similar clustering of restriction sites in the regions carrying the resistance determinants is seen in the maps of plasmids belonging to incompatibility groups N, Q and W. One feature which is not shared by plasmids of other groups is the interruption by these regions of the segments carrying the plasmid maintenance and conjugal transfer functions: the Tc^r determinant lies between the essential replication loci *trfA* and *oriV*, and a region including IS*21* and the Km^r determinant lies between the transfer regions Tra1 and Tra2.

Heteroduplex analysis has shown that the other IncPα plasmids are very similar in organization to the Birmingham isolates, showing high homology to these plasmids throughout the regions encoding functions involved in replication, maintenance and conjugal transfer and the determinants for Tc^r and Km^r (70). These plasmids differ mainly by the insertion or deletion of small numbers of segments of DNA, most of which are known or putative transposable elements and many of which carry resistance determinants (70).

The tetracycline resistance genes *tetA* and *tetR* are not transposable, but differ from those of Tn*1721* by only a few base changes (72), suggesting a very close evolutionary relationship. However, sequence analysis has shown that there are no sequences related to Tn*1722* (the minor transposon within Tn*1721* which carries the transposition functions; 56) associated with the Tc^r genes in these plasmids (72; M. Pinkney, unpublished); therefore the nature of the relationship remains obscure. The *aph* gene confering resistance to kanamycin is related to that of Tn*903*, showing about 58% nucleotide sequence homology (51). Although this gene is not transposable, the defective insertion element IS*21* (73; also designate IS*8*; 12), is adjacent to it in most of the IncPα plasmids (Figure 1; Table I) and the *aph* gene is apparently transcribed from a promoter lying partly within IS*21* (51; N. S. Willetts, personal communication), which suggests that the two may once have formed part of a Km^r transposon. The four IncPα plasmids known to lack IS*21* show deletions of three diferent sizes adjacent to its position, suggesting that it was present in

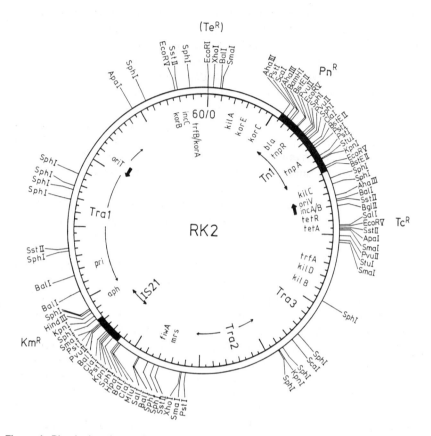

Figure 1. Physical and genetic map of IncPα plasmids RK2/RP1/RP4/R18/R68. The 60 kb restriction map is based on coordinates for RP4 (43; E. Lanka, personal communication). Pn^R, Km^R and Tc^R indicate the ampicillin, kanamycin and tetracyline resistance determinants; (Te^R) indicates a cryptic tellurite resistance determinant (66). Functions associated with plasmid maintenance (Chapter 1): inc, incompatbility towards IncP plasmids; kil, host lethal or plasmid inhibitory function; kor, suppression of the effect of kil gene(s); oriV, origin of vegetative replication (arrow indicates direction of replication); trf, trans-acting replication function (trfB, korA and korD are the same locus). Functions involved in conjugal transfer (Chapter 2): oriT, origin of conjugal DNA transfer (arrow indicates direction of transfer); pri, primase; Tra, block of transfer loci. Other genetic loci are defined as follows: bla, β-lactamase; tnp, transposition function (28); tet, tetracycline resistance (72); aph, aminoglycoside phosphotransferase (51); fiw, fertility inhibition towards IncW plasmids (76); mrs, multimer resolution system (N. J. Grinter, G. Brewster & P. T. Barth, personal communication). The DNA sequence of one of the Birmingham isolates has been determined for coordinates 0.0 to 5.5 (N. Wang & C. M. Thomas, unpublished); 10.5 to 19.2 (58,63,72; M. Pinkney, C. A. Smith, J. P. Ibbotson & C. M. Thomas, unpublished); 36.2 to 38.3 (IS21; R. Moore & N. Willetts, personal communication); 38.0 to 55.5 (E. Lanka, personal communication); 55.5 to 0.0 (1,39,67,68; C. A. Smith & C. M. Thomas, unpublished).

their common ancestor, as expected if it was involved in the acquisition of the Km^r determinant, but has been lost by three separate imprecise deletion events (70). Presumably these deletions have coupled transcription of the *aph* gene to a new promoter.

There are three other markers which appear to be shared by the IncPα plasmids, but to be absent from the IncPβ plasmids, although in each case only representatives of each subgroup have been tested, as shown in Table I. Firstly IncPα but not IncPβ plasmids have been shown to confer on *P. aeruginosa* insensitivity to aeruginosin AP41 (64). Secondly IncPα but not IncPβ plasmids inhibit the fertility of coresident IncW plasmids (76), and the IncPβ plasmids tested lacked homology to the *Pst*I fragment of RP1 which carries the locus *fiwA*, which is partly responsible for this effect (Figure 1). Thirdly IncPα but not IncPβ plasmids have been shown to possess a cryptic tellurite resistance locus (Figure 1), active variants of which can easily be selected (2) and which has been reported to be transposable (3).

Many of the IncPα plasmids also share an homologous Pn^r region, carried by Tn*1* (also designated Tn*801*, Tn*401* and Tn*A*), which is closely related to Tn*3* (22,28), but confers a higher level of resistance to ampicillin due to a point mutation which creates two new *bla* promoters, P$_a$ and P$_b$, which are stronger than the Tn*3* *bla* promoter P$_3$(6) and shows some differences in restriction map (Figure 1; Table I) which reflect other minor sequence variations between Tn*1* and Tn*3* (6a). Since Tn*1* is an active transposon it is not clear whether those IncPα plasmids which lack a Tn*1*-like element have lost it in one or more precise deletion events or are descended from a common ancestor of the IncPα subgroup which had not yet acquired it. There has been minor divergence between the β-lactamases (*bla*) carried by the Tn*1*-like elements of the IncPα plasmids, some being classified as TEM-1 and others as TEM-2 (Table I; 45). Two of the plasmids which lack Tn*1* carry Pn^r determinants located elsewhere.

Although substantial regions of representatives of the Birmingham IncPα isolates have been sequenced, there is no sequence information available from other members of this subgroup for comparison. However, the results of Southern blotting and heteroduplex experiments (7,44,60,70,75) suggest that the degree of homology is high and therefore that the members of the IncPα subgroup are derived from a relatively recent common ancestor.

V. THE IncPβ SUBGROUP

In contrast to the members of the IncPα subgroup, all of which confer Tc^r and Km^r, the plasmids designated IncPβ do not all share any phenotypic marker (Table I). Although R906 and pJP4 both confer resistance to mercuric salts and carry the same Hg^r determinant, which is also present (in an inactive form) on R772, R751 shows no trace of this Hg^r determinant (60).

Figure 2 shows the organization of R751, which is the most thoroughly studied of the IncPβ plasmids, and Figure 3 shows that of R906. The restriction maps of both IncPβ plasmids show clustering of the restriction sites similar to that in the map of the IncPα plasmid RK2. Although in each of the IncPβ plasmids one cluster of restriction sites separates the replication loci *oriV* and *trfA*, and the other separates Tra1 from Tra2, there is no similarity between the clusters of sites in the two IncPβ plasmids.

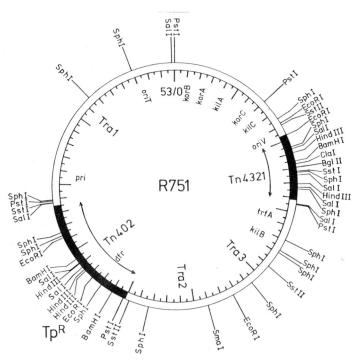

Figure 2. Physical and genetic map of R751. Restriction maps of R751 published previously show some conflicts and it appears that there is some restriction site polymorphism amongst laboratory strains (31,47,57,60,71); the 53 kb restriction map shown is based on that of W. Pansegrau and E. Lanka (50; E. Lanka, personal communication) with minor additions (31; C.A. Smith, unpublished). The genetic map is derived from the results of several studies (47,53,54,60,70; C.A. Smith, unpublished). Tp^r indicates the position of the trimethoprim resistance determinant. The genetic loci are as defined for RK2 (Figure 1), except for *dfr*, trimethoprim-resistant dihydrofolate reductase (17). The end points indicated for Tn402 (47) are approximate only.

Heteroduplex analysis of the relationship between R751 and R906 (53,54) showed that the homology between them is extensive, but is limited to the regions encoding functions involved in replication, maintenance and conjugal transfer. The restriction maps of the two plasmids show some similarities in these regions, although the map of R906 is less detailed than that of R751 (Figures 2 & 3). The restriction map of R772 (29,60) also shows clustering of restriction sites in two regions, one of which contains the Km^r determinant and the transposable element IS70; these do not show similarity in restriction map to IS21 and the Km^r determinant of the IncPα plasmids. Between these clusters of sites the restriction map of R772 is similar to the maps of R751 and R906 (60). Southern blotting has confirmed that the segments between these two clusters are homologous to the corresponding segments of R751 and R906 and revealed homology to the Hg^r determinant of R906 at the end of the cluster of restriction sites containing the Km^r determinant (60). However, the relative orientation of these two segments in R772 is opposite to that in R751 and R906, and to that of the corresponding segments of IncPα plasmid RK2, implying that a large inversion has occurred in the lineage of R772 (60). The map available for pJP4 is much less detailed than those for the other IncPβ plasmids, showing cleavage sites only for the restriction enzymes EcoRI, HindIII and BamHI (14). However, this map does show that most of the restriction sites lie in one half of the plasmid, which carries the Hg^r determinant and the genes which confer (in suitable hosts) the ability to catabolize 2,4-dichlorophenoxyacetic acid and 3-chlorobenzoic acid (15,18). No transposable elements have been demonstrated on pJP4, but renaturation studies revealed the presence of 1.8 kb inverted repeats flanking a 1 kb unique region and Southern hybridization experiments indicated the presence of further repeats within this half of the plasmid (18). These sequences may be involved in the specific rearrangements of pJP4 which are involved in the efficient expression of the 3-chlorobenzoic acid catabolism genes (18). The results of Southern hybridization experiments indicate that this region containing the phenotypic markers of pJP4 and the majority of the restriction sites separates the replication loci trfA and oriV, and that the end of this region adjacent to trfA, which carries the Hg^r determinant is homologous to the corresponding region of R906 (60).

In the case of R751 (Figure 2), one cluster of restriction sites contains Tn402 (47) which carries the dfr locus which confers Tp^r, the only known phenotypic marker of the plasmid. This is very closely related to that carried by the IncW plasmid R338, the two trimethoprim-resistant dihydrofolate reductases differing by only 11 out of 78 amino acids (17). The other cluster is not associated with any phenotype and has not been shown to transpose; however, it has been shown by EM studies to consist of a unique region flanked by 1.4 kb inverted repeats and has been designated Tn4321 as a putative cryptic transposon (53,54). Comparison of the nucleotide sequence of the oriV region of R751 (59) with that of the proximal end of the trfA region of R751 (C. A. Smith, unpublished) has revealed that the outer ends of the inverted repeats of Tn4321 are homologous to one end of Tn501, but Southern blotting revealed no homology to the transposition region of this transposon (60). Comparison of these sequences with those for the $oriV$ (59) and $trfA$ (58) regions of IncPα plasmid RK2/RP1 and the $oriV$ region of R906 (C. A. Smith, unpublished) revealed that a region which is adjacent to $oriV$ in the latter two plasmids is adjacent to $trfA$ in R751 (Figure 4). This finding is consistent with the results of heteroduplex formation between R751 and R906 (53,54) and between R751 and a derivative of IncPα plasmid RP4 (70).

No transposable elements have been reported on R906. One cluster of restriction sites (Figure 3) contains the Su^r determinant; this cluster is very similar to the restriction map of the Su^r region of Tn21 and related transposons (44). The second cluster contains the Pn^r, Sm^r and Hg^r determinants. The end of this cluster adjacent to the $trfA$ locus shows similarities to the restriction map of the Hg^r region of Tn501 and Southern blotting showed that the region indicated on the map by an arrow is homologous to Tn501, but did not reveal any region homologous to the transposition region of this transposon (60). DNA sequence analysis has shown that one of the two terminal repeats of Tn501 is present at the end of the Hg^r determinants of R906, R772 and pJP4 adjacent to the $trfA$ locus, and that the sequences of the three plasmids are identical in this region (60). Comparison of these sequences with the sequence of the end of the $trfA$ region of R751 (C. A. Smith, unpublished) reveals that the Tn501-like terminal repeat is inserted 80 bp beyond the termination codon of the $trfA$ gene in these plasmids (Figure 4). This suggests that the Hg^r region was originally acquired by a common ancestor of these three plasmids by transposition.

Comparison of the nucleotide sequences of the $oriV$ regions of R751 and R906 (59; C. A. Smith, unpublished) has revealed that although both plasmids are assigned to the IncPβ subgroup there is only 76% homology between them. In contrast partial sequence data for the $oriV$ regions of R772 and pJP4 show complete homology to the R906 sequence (C. A. Smith, unpublished). This suggests that the latter three plasmids have a much more recent common ancestor than their most recent common ancestor with R751.

VI. RELATIONSHIP BETWEEN THE SUBGROUPS

Comparison of the maps of the IncPα plasmid RK2 (Figure 1) and the IncPβ plasmid R751 shows that the arrangement of the loci involved in plasmid maintenance and conjugal transfer is very similar, but that there is no correspondance between the restriction maps of the two plasmids in these regions. This is consistent with the results of heteroduplex formation between an IncPα test-plasmid derived from RP4 and the IncPβ plasmids R751 and R906 (70). These showed extensive regions of heteroduplex between the plasmids, within which symmetrical unpaired regions occurred at variable positions in different heteroduplices, indicating that the plasmids belonging to the two subgroups possessed homologous colinear backbone regions, but that substantial sequence

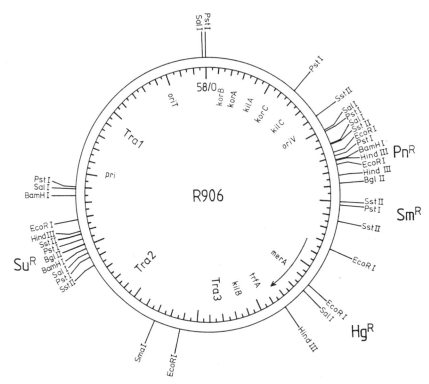

Figure 3. Physical and genetic map of R906. The 53 kb restriction map is based on published maps (60,71) with modifications (C. A. Smith, unpublished). The genetic map is based on the results of several studies (53,54,60,70; C. A. Smith, unpublished). Pn^r, Sm^r, Hg^r and Su^r indicate the positions of the determinants for resistance to penicillin, streptomycin, mercuric salts and sulphonamide. The genetic loci are as defined for RK2 (Figure 1), except for *merA*, mercuric reductase. R906 is not known to carry any transposable elements, but the region from coordinates 18.3 kb to 23.5 kb is homologous to one end of Tn*501* (60).

divergence had occurred between them. The Southern blotting studies used to define the subgroups also imply a substantial degree of sequence divergence between the homologous regions of the Inc Pα and Inc Pβ plasmids (7,44,75).

There are three regions of the Inc P plasmids for which nucleotide sequences are available from members of both the Inc Pα and Inc Pβ subgroups: the *ori V* region; the *kilC* region; and the *ori T* region.

The *ori V* region contains the origin of vegetative replication and major determinants of group P-incompatibility. Comparison of the sequences of the *ori V* regions of the Inc Pα plasmid RK2 (63,72), and the Inc Pβ plasmids R751 (59; C. A. Smith, unpublished) and R906 (C. A. Smith, unpublished) reveals about 65% homology between the sequences from the two subgroups. Despite this

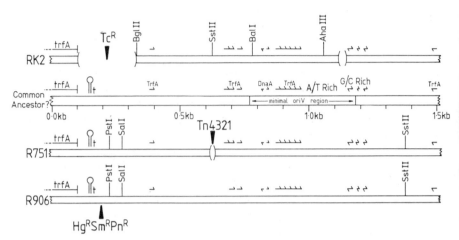

Figure 4. Organization of the *trfA/oriV* region. Brackets indicate deletions in the IncP backbone of the *trfA/oriV* region and vertical arrows indicate the positions of insertions. Features conserved between two or more plasmids are indicated above the reconstruction of the probable arrangement of the *trfA/oriV* region of the common ancestor of the IncPα and IncPβ plasmids as follows: 3' end of the *trfA* coding region; potential transcriptional terminator for rightward transcription; single copy and three direct copies of repeat to which TrfA (the product of the *trfA* gene) binds; two inverted copies of the consensus sequence for DnaA (the product of the *E. coli dnaA* gene); five direct copies of the repeat to which TrfA binds; an A/T-rich region; three G/C-rich inverted repeats; a single copy of the repeat to which TrfA binds inverted with respect to the other copies. The minimal *oriV* region indicated is the shortest region of RK2 shown to retain a functional *oriV* in *E. coli* (Chapter 1).

substantial degree of sequence divergence, these essential replication loci have remained interchangable, as the *ori V* loci of both R751 (59) and R906 (7) will function if the *trfA* gene of RK2 is provided in *trans*. Consistent with this functional interchangability is the conservation between the three sequences of various features, including a series of direct repeats which are binding sites for the product(s) of the essential replication gene *trfA* (Figure 4; Chapter 1). Other features which are conserved include a pair of putative binding sites for the product of the *E. coli dnaA* gene and an A/T rich region and three G/C rich inverted repeats; all of these appear to be of functional importance in at least some species (Chapter 1).

The *kilC* region of RK2 contains a locus which is lethal to the host cell unless it is regulated by the *korC* and *trfB/korA/korD* loci (Chapter 1). Comparison of the nucleotide sequences of the *kilC* regions of RK2 (J. P. Ibbotson, unpublished) and R751 (C. A. Smith, unpublished) reveals the presence of an open reading frame in both plasmids, preceded by a putative promoter region with high homology to the consensus for *E. coli* Eσ[70] promoters. The coding regions of these putative *kilC* genes show 80% nucleotide sequence homology and 70% amino acid sequence homology between their predicted products. The putative *kilC* promoter regions of the two plasmids show the conservation of two inverted repeats, one of which is also present in the *kilA*, *trfA* and *trfB/korA/korD* promoter regions of RK2 and has been strongly implicated as an operator site for repression of these promoters by the product of the *trfB/korA/korD* gene (Chapter 1). The second conserved repeat is likely to be an operator site for control of the promoter by the product of the *korC* gene. The putative *kilC* promoter region of R751 also shows a third inverted repeat, which is absent from the corresponding region of RK2, but is present in the *kilA*, *trfA*, *trfB/korA/korD* and putative *kilB* promoter regions of RK2 and has been implicated as an operator site for repression by the product of the *korB* gene (Chapter 1). This conservation of probable operator sites for the products of *kor* genes between IncPα and IncPβ plasmids is consistent with the finding that IncP plasmids of both subgroups carry functions able to supress the lethality of the *kilA* and *kilB* loci of RK2 (16).

The *oriT* region contains the origin of conjugal transfer and the site of nicking in the relaxation complex (Chapter 2). Comparison of the nucleotide sequences of the *oriT* regions of IncPα plasmid RP4 and IncPβ plasmid R751 revealed the conservation of several features including divergent promoters and a 38 bp inverted repeat, one end of which is the site of the relaxation nick (W. Pansegrau & E. Lanka, personal communication). The product of one of the adjacent genes in RP4 was found to bind the the inverted repeat region in RP4, but not to the (different) inverted repeat in R751 (G. Ziegelin, W. Pansegrau & E. Lanka, personal communication). Complementation studies have shown that the *oriT*/Tra1 regions of IncPα plasmid RK2 and IncPβ plasmid R751 can each be complemented by the Tra2/Tra3 region of the other plasmid to form a functional conjugal transfer system (47). Similarly a plasmid carrying a segment of RK2 including both *oriT* and the adjacent genes necessary for the formation of the relaxation complex can be mobilized by the complete conjugal transfer systems of the IncPβ plasmids R751 or R772 provided in *trans* (75). However, consistent with the differing binding specificities observed at *oriT*, a plasmid carrying the *oriT* locus of RK2 alone can be mobilized by IncPα plasmids, but not by IncPβ plasmids (75).

The substantial degree of sequence divergence between homologous regions of the IncPα and IncPβ plasmids suggests that the subgrouping reflects an evolutionary branching of some antiquity.

VII. ORGANIZATION AND ORIGIN OF THE ANCESTRAL IncP PLASMIDS

The results of the comparisons of the nucleotide sequences from the *oriV* regions and from the proximal ends of the *trfA* regions (Figure 4) suggest that these essential replication loci were adjacent in the most recent common ancestor of the IncPβ plasmids R751 and R906 and also in the common ancestor of the IncPα and IncPβ subgroups. Thus the IncP plasmids are probably derived from an ancestor in which the replication functions were as tightly clustered as they are in members of other incompatibility groups. It appears that these loci have been separated by insertions into the IncP backbone on at least three separate occasions; by the insertion of the Tc^r region in the IncPα lineage, by the insertion of Tn4321 in R751 and by the insertion of a Hg^r transposon (followed by further insertion/deletion events) in the lineage of the IncPβ plasmids R906, R772 and pJP4.

The segments inserted into the IncP backbone between the transfer regions oriT/Tra1/pri and Tra2/Tra3 in the IncPα plasmids and in IncPβ plasmids R751 and R906 are unrelated (Figures 1, 2 & 3; 53,54,70). It therefore appears probable that the conjugal transfer functions formed a continuous segment in the common ancestor of the IncP plasmids. It is possible that these regions remain adjacent in pJP4, as there is no cluster of restriction sites in the present map of this plasmid within the segment carrying the transfer and maintenance functions, and the length of this segment is close to that of the backbone regions of the other IncP plasmids after the removal of known insertions (14,60).

Present day IncP plasmids are geographically widespread, and have been isolated from diverse Gram-negative bacteria including several enteric species (Table I). However, a study of the plasmids present in isolates of enteric species collected before the clinical use of antibiotics revealed no IncP plasmids, although members of other present-day incompatibility groups were recovered (10). Since the large degree of nucleotide sequence divergence between members of the two subgroups suggests that the IncP plasmids have a relatively ancient origin, it is probable that they originated in other genera and have spread into the enteric bacteria comparatively recently, as a result of their promiscuity combined with the inadvertent application of powerful selection for the intergeneric spread of antibiotic resistance. Comparison of present day IncP plasmids has not revealed any phenotypic marker (apart from fertility) likely to have been carried by the common ancestor of the group, and it is possible that this was a cryptic plasmid. The members of the IncPα subgroup appear to share a cryptic tellurite resistance determinant, and three of the four members of the IncPβ subgroup share a mercury resistance determinant (inactive in the case of R772). This suggests that before the therapeutic use of antibiotics the usual hosts of the ancestors of both subgroups may have been bacteria from environments such as water or soil where resistance to the toxic effects of heavy metals conferred a potential selective advantage.

VIII. REFERENCES

1. Bechhofer DH, Figurski DH (1983) Map location and nucleotide sequence of korA, a key regulatory gene of promiscuous plasmid RK2. *Nucl Acids Res*

11, 7453–69

2. Bradley DE (1985) Detection of tellurite-resistance determinants in IncP plasmids. *J Gen Microbiol* **131**, 3135–7

3. Bradley DE, Taylor DE (1987) Transposition from RP4 to other replicons of a tellurite-resistance determinant not normally expressed by IncPα plasmids. *FEMS Microbiol Letts* **41**, 237–40

4. Burkardt HJ, Riess G, Pühler A (1979) Relationship of group P1 plasmids revealed by heteroduplex experiments: RP1, RP4, R68 and RK2 are identical. *J Gen Microbiol* **114**, 341–8

5. Chandler PM, Krishnapillai V (1974) Phenotypic properties of R-factors of *Pseudomonas aeruginosa*: R-factors readily transferable between *Pseudomonas* and the Enterobacteriaceae. *Genet Res Camb* **23**, 239–50

6. Chen S-T. Clowes RC (1984) Two improved promoter sequences for β-lactamase expression arising from a single base substitution. *Nucl Acids Res* **12**, 3219–34

6a.Chen S-T. Clowes RC (1987) Variations between the nucleotide sequences of Tn*1*, Tn*2* and Tn*3* and expression of β-lactamase in *Pseudomonas aeruginosa* and *Escherichia coli*. *J Bacteriol* **169**, 913–6

7. Chikami GK, Guiney DG, Schmidhauser TJ, Helinski DR 1985. Comparison of ten IncP plasmids: homology in the regions involved in plasmid replication. *J Bacteriol* **162**, 656–60

8. Coetzee JN (1978) Mobilization of the *Proteus mirabilis* chromosome by R-plasmid R772. *J Gen Microbiol* **108**, 103–9

9. Currier TC, Morgan MK (1981) Restriction endonuclease analyses of the incompatibility group P-1 plasmids RK2, RP1, RP4, R68 and R68.45. *Current Microbiol* **5**, 323–7

10. Datta N, Hughes V (1983) Plasmids of the same Inc groups in Enterobacteria before and after the medical use of antibiotics. *Nature* **306**, 616–7

11. Datta N, Hedges RW, Shaw EJ, Sykes RB, Richmond MH (1971) Properties of an R-factor from *Pseudomonas aeruginosa*. *J Bacteriol* **108**, 1244–9

12. Depicker A, De Block M, Inze D, Van Montagu M, Schell J (1980) IS-like element IS*8* in RP4 plasmid and its involvement in cointegration. *Gene* **10**, 329–38

13. Don RH, Pemberton JM (1981) Properties of six pesticide degradation plasmids isolated from *Alcaligenes paradoxus* and *Alcaligenes eutrophus*. *J Bacteriol* **145**, 681–6

14. Don RH, Pemberton JM (1985) Genetic and physical map of the 2,4-dichlorophenoxyacetic acid-degradative plasmid pJP4. *J Bacteriol* **161**, 466–8

15. Don RH, Weightman AJ, Knackmuss H-J, Timmis KN (1985) Transposon mutagenesis and cloning analysis of the pathways for degrada-

tion of 2,4-dichlorophenoxyacetic acid and 3-chlorobenzoate in *Alcaligenes eutrophus* JM134 (pJP4). *J Bacteriol* **161**, 85–90

16. Figurski DH, Pohlman RF, Bechhofer DH, Prince AS, Kelton CA (1982) The broad-host-range plasmid RK2 encodes multiple *kil* genes potentially lethal to *Escherichia coli* host cells. *Proc Natl Acad Sci USA* **79**, 1935–9

17. Flensburg J, Steen R (1986) Nucleotide sequence analysis of the trimethoprim resistant dihydrofolate reductase encoded by R-plasmid R751. *Nucl Acids Res* **14**, 5933

18. Ghosal D, You I–S, Chatterjee DK, Chakrabarty AM (1985) Genes specifying degradation of 3-chlorobenzoic acid in plasmids pAC27 and pJP4. *Proc Natl Acad Sci USA* **82**, 1638–42

19. Grant AJ, Bird PI, Pittard J (1980) Naturally-occurring plasmids exhibitting incompatibility with members of incompatibility groups I and P. *J Bacteriol* **144**, 758–65

20. Hedges RW (1974) R-factors from Providence. *J Gen Microbiol* **81**, 171–81

21. Hedges RW (1975) R-factors from *Proteus mirabilis* and *P. vulgaris*. *J Gen Microbiol* **87**, 301–11

22. Hedges RW, Jacob A (1974) Transposition of ampicillin resistance from RP4 to other replicons. *Mol Gen Genet*. **132**, 31–40

23. Hedges RW, Matthew M (1979) Acquisition by *Escherichia coli* of plasmid-born β-lactamases normally confined to *Pseudomonas* spp. *Plasmid* **2**, 269–78

24. Hedges RW, Datta N, Coetzee JN, Dennison S (1973) R-factors from *Proteus morganii*. *J Gen Microbiol* **77**, 249–59

25. Hedges RW, Jacob A, Smith JT (1974) Properties of an R-factor from *Bordetella bronchiseptica*. *J Gen Microbiol* **84**, 199–204

26. Hedges RW, Rodriguez-Lemoine L, Datta N (1975) R-factors from *Serratia marcescens*. *J Gen Microbiol* **86**, 88–92

27. Hedges RW, Matthew M, Smith DI, Cresswell JM, Jacob AE (1977) Properties of a transposon conferring resistance to penicillins and streptomycin. *Gene* **1**, 241–53

28. Heffron F, McCarthy BJ, Ohtsubo H, Ohtsubo E (1979) DNA sequence analysis of the transposon Tn*3*: three genes and three sites involved in the transposition of Tn*3*. *Cell* **18**, 1153–63

29. Hille J, van Kan J, Klasen I, Schilperoort R (1983) Site-directed mutagenesis in *Escherichia coli* of a stable R772::Ti cointegrate plasmid from *Agrobacterium tumefaciens*. *J Bacteriol* **154**, 693–701

30. Hinchliffe E, Vivian A (1980) Naturally-occurring plasmids in *Acinetobacter calcoaceticus*: a P-class R-factor of restricted host range. *J Gen Microbiol* **116**, 75–80

31. Hirsch PR, Beringer JE (1984) A physical map of pPH1JI and pPH4JI. *Plasmid* **12**, 139–41

32. Holloway BW, Richmond MH (1973) R-Factors used for genetic studies in strains of *Pseudomonas aeruginosa* and their origin. *Genet Res Camb* 21, 103–5

33. Ingram LC, Richmond MH, Sykes RB (1973) Molecular characterization of the R-factors implicated in the carbenicillin resistance of a sequence of *Pseudomonas aeruginosa* strains isolated from burns. *Antimicrob Ag Chemother* 3, 279–88

34. Jacoby GA (1977) Classification of plasmids in *Pseudomonas aeruginosa*. *In* Microbiology–1977 (Schlessinger D, ed) Washington DC: American Society for Microbiology, pp. 119–26.

35. Jacoby GA (1980) Plasmid determined resistance to carbenicillin and gentamicin in *Pseudomonas aeruginosa*. *In* Plasmids and transposons, environmental effects and maintenance mechanisms, (Stuttard C, Rozee KR, eds) New York: Academic Press, pp 83–93

36. Jacoby GA, Matthew M (1979) The distribution of β-lactamase genes on plasmids found in *Pseudomonas*. *Plasmid* 2, 41–7

37. Jacoby GA, Shapiro JA (1977) Plasmids studied in *Pseudomonas aeruginosa* and other Pseudomonads. *In* DNA Insertion Elements, Plasmids, and Episomes, (Bukhari AI, Shapiro JA, Adhya S, eds) Cold Spring Harbor, NY: Cold Spring Harbor Laboratory, pp 639–56.

38. Jobanputra RS, Datta N (1974) Trimethoprim R-factors in Enterobacteria from clinical specimens. *J Med Microbiol* 7, 169–77

39. Kornacki JA, Balderes PJ, Figurski DH (1987) Nucleotide sequence of *korB*, a replication control gene of broad-host-range plasmid RK2. *J Mol Biol* 198, 211–22

40. Krishnapillai V (1977) Superinfection inhibition by prophage B3 of some R-plasmids in *P. aeruginosa*. *Genet Res* 29, 47–54

41. Kratz J, Schmidt F, Wiedemann B (1983) Characterization of Tn*2411* and Tn*2410*, two transposons derived from R-plasmid R1767 and related to Tn*2603* and Tn*21*. *J Bacteriol* 155, 1333–42

42. Lanka E, Barth PT (1981) Plasmid RP4 specifies a deoxyribonucleic acid primase involved in its conjugal transfer and maintenance. *J Bacteriol* 148, 769–81

43. Lanka E, Lurz R, Fürste JP (1983) Molecular cloning and mapping of *Sph*I restriction fragments of plasmid RP4. *Plasmid* 10, 303–7

44. Lanka E, Fürste JP, Yakobson E, Guiney DG (1985) Conserved regions at the DNA primase locus of the IncPα and *Inc*Pβ plasmids. *Plasmid* 14, 217–23

45. Matthew M, Hedges RW (1976) Analytical isoelectric focusing of R-factor-determined β-lactamases: correlation with plasmid compatibility. *J Bacteriol* 125, 713–8

46. Meyer R, Helinski DR (1977) Unidirectional replication of the P-group plasmid RK2. *Biochim Biophys Acta* 487, 109–13

47. Meyer RJ, Shapiro JA (1980) Genetic organization of the broad-host-range IncP-1 plasmid R751. *J Bacteriol* **143**, 1362–73

48. Monti-Bragadin C, Samer L (1975) Compatibility of pTM89, a new F-like R-factor, and of derivative plasmids. *J Bacteriol* **124**, 1132–6

49. Nugent ME, Ellis K, Datta DE (1982) pHH502, a plasmid with IncP and IncIα characters, loses the latter by a specific recA-independent deletion event. *J Gen Microbiol* **128**, 2781–90

50. Pansegrau W, Lanka E (1987) Conservation of a common 'backbone' in the genetic organization of the IncP plasmids RP4 and R751. *Nucl Acids Res* **15**, 2385

51. Pansegrau W, Miele L, Lurz R, Lanka E (1987) Nucleotide sequence of the kanamycin resistance determinant of plasmid RP4: homology to aminoglycoside-3'-phosphotransferases. *Plasmid* **18**, 193–204

52. van Rensburg AJ, de Kock MJ (1974) A new R-factor from *Pseudomonas aeruginosa*. *J Gen Microbiol* **82**, 207–8

53. Sakanyan VA, Krupenko MA, Alikhanyan SI (1983) Homology of broad-host-range plasmids. *Genetika* **19**, 1409–18

54. Sakanyan VA, Azaryan NG, Krupenko MA (1985) Molecular organization of the plasmid R906. *Molekulyarnaya Biologiya* **4**, 964–73.

55. Sawai T, Takahashi K, Yamagishi S, Mitsuhashi S (1970) Variant of penicillinase mediated by an R-factor in *Escherichia coli*. *J Bacteriol* **104**, 620–9

56. Schmitt R, Altenbuchner J, Wiebauer K, Arnold W, Pühler A, Schöffl F (1981) Basis of transposition and gene amplification by Tn*1721*, and related tetracycline-resistance transposons. *Cold Spring Harbour Symp Quant Biol* **45**, 59–65

57. Shoemaker NB, Getty C, Gardener JF, Salyers AA (1986) Tn*4351* transposes in *Bacteroides* spp. and mediates the integration of plasmid R751 into the *Bacteroides* chromosome. *J Bacteriol* **165**, 929–36

58. Smith CA, Thomas CM (1984) Nucleotide sequence of the trfA gene of broad-host-range plasmid RK2. *J Mol Biol* **175**, 251–62

59. Smith CA, Thomas CM (1985) Comparison of the nucleotide sequences of the vegetative replication origins of broad-host-range IncP plasmids R751 and RK2 reveals conserved features of probable functional importance. *Nucl Acids Res* **13**, 557–72

60. Smith CA, Thomas CM (1987) Comparison of the organization of the genomes of phenotypically diverse plasmids of incompatibility group P: Members of the IncPβ subgroup are closely related. *Mol Gen Genet* **206**, 419–27

61. Smith CA, Thomas CM (1987) Narrow-host-range IncP plasmid pHH502-1 lacks a complete IncP replicon. *J Gen Microbiol* **133**, 2247–52

62. Smith DI, Gomez LR, Rubio CM, Datta N, Jacob AE, Hedges RW (1975) Third type of plasmid conferring gentamicin resistance in *Pseudomonas*

aeruginosa. Antimicrob Ag Chemother **8**, 227–30

63. Stalker DM, Thomas CM, Helinski DR (1981) Nucleotide sequence of the region of the origin of replication of the broad-host-range plasmid RK2. *Mol Gen Genet* **181**, 8–12

64. Stanisich VA, Ortiz JM (1976) Similarities between plasmids of the P-incompatibility group derived from different bacterial genera. *J Gen Microbiol* **94**, 281–89

65. Stokes HW, Moore RJ, Krishnapillai V (1981) Complementation analysis in *Pseudomonas aeruginosa* of the transfer genes of the wide-host-range R-plasmid R18. *Plasmid* **5**, 202–12

66. Taylor DE, Bradley DE (1987) Location on RP4 of a tellurite-resistance determinant not normally expressed in Inc Pα plasmids. *Antimicrob Ag Chemother* **31**, 823–5

67. Theophilus BDM, Thomas CM (1987) Nucleotide sequence of the transcriptional repressor gene *korB* which plays a key role in regulation of the copy number of broad-host-range plasmid RK2. *Nucl Acids Res* **15**, 7443–50

68. Thomas CM, Smith CA (1986) The *trfB* region of broad-host-range plasmid RK2: The nucleotide sequence reveals *incC* and key regulatory gene *trfB/korA/korD* as overlapping genes. *Nucl Acids Res* **14**, 4453–69

69. Thomas CM, Stalker DM, Helinski DR (1981) Replication and incompatibility properties of the origin region of replication of broad-host-range plasmid RK2. *Mol Gen Genet* **181**, 1–7

70. Villarroel R, Hedges RW, Maenhaut R, Leemans J, Engler G, van Montagu M, Schell J (1983) Heteroduplex analysis of P-plasmid evolution: the role of insertion and deletion of transposable elements. *Mol Gen Genet* **189**, 390–99

71. Ward JM, Grinsted J (1982) Analysis of the Inc P-1 group plasmids R906 and R751 and their relationship to RP1. *Plasmid* **8**, 244–52

72. Waters SH, Grinsted J, Rogowsky P, Altenbuchner J, Schmitt R (1983) The tetracycline resistance determinants of RP1 and Tn*1721*: nucleotide sequence analysis. *Nucl Acids Res* **11**, 6089–105

73. Willetts NS, Crowther C, Holloway BW (1981) The insertion sequence IS*21* of R68.45 and the molecular basis for mobilization of the bacterial chromosome. *Plasmid* **6**, 30–52

74. Wretlind B, Becker K, Haas D (1985) Inc P-1 plasmids decrease the serum resistance and the virulence of *Pseudomonas aeruginosa*. *J Gen Microbiol* **131**, 2701–4

75. Yakobson E, Guiney D (1983) Homology in the transfer origins of broad-host-range Inc P plasmids: definition of two subgroups of P-plasmids. *Mol Gen Genet* **192**, 436–8

76. Yusoff K, Stanisich VA (1984) Location of a function on RP1 that fertility inhibits Inc W plasmids. *Plasmid* **11**, 178–81

CHAPTER 4A

THE MOLECULAR BIOLOGY OF Inc Q PLASMIDS

Joachim Frey and Michael Bagdasarian

I. INTRODUCTION

The incompatibility group Inc Q (sometimes referred to as Inc P4) is a distinct group of plasmids characterized by their relatively small size, medium range copy number and an extremely wide-host-range. As expected, they were isolated from various species including *E. coli*, *Salmonella* spp., *Proteus* spp., and *Pseudomonas* spp. They mainly confer resistance to streptomycin and sulfonamide. However, some plasmids with other resistances were reported. Properties of the representative Inc Q plasmids known to date are listed in Table I. A common feature of Inc Q plasmids is their ability to be be mobilized even among different bacterial species if transfer functions are provided in *trans* by a conjugative 'helper' plasmid (2,41,15).

Recently, replicons of the Inc Q plasmids and in particular the replicon of RSF1010 have been used to develop vectors, useful for the manipulation of various Gram-negative bacteria unaccessible to the 'classic' *E. coli* cloning systems based on Col E1 type replicons or *E. coli* specific bacteriophages. These broad-host-range vectors have largely contributed to the genetic analysis of the species of *Pseudomonas* and are reviewed in (4,36; and Franklin and Spooner, Chapter 10). In addition it has been shown that the origin of transfer and the mobilization functions of RSF1010 promote plasmid transfer to plants (7), a finding which enhances considerably the usefulness of RSF1010 and its derivative vectors.

In addition to being of great practical value Inc Q replicons have attracted considerable attention in the attempts to answer the fundamental questions

Promiscuous Plasmids of Gram-Negative Bacteria
ISBN 0-12-688480-3
© 1989 Academic Press Limited
All rights of reproduction in any form reserved

TABLE I. *Plasmids of the IncQ group*

Name	Size (kb)	Phenotype	Copies/ chromos.	Origin	Ref
R300B	8.61	Sm Su Tra$^-$	11	*Salmonella typhi*	5
RSF1010	8.685	Sm Su Mob$^+$	12	*Escherichia coli*	16
R1162	8.3	Sm Su Tra$^-$	11	*Pseudomonas aeruginosa*	5
R305c	8.6	Sm Su Tra$^-$	9.3	*Salmonella typhi*	5
R310	8.6	Sm Su Tra$^-$	9.2	*Salmonella typhi*	5
R450B	8.6	Sm Su Tra$^-$	8.6	*Proteus morganii*	5
R464C	8.6	Sm Su Tra$^-$	11.2	*Proteus morganii*	5
R676	8.6	Sm Su Tra$^-$	8.5	*Salmonella senftenburg*	5
R678	14.0	Sm Su Tra$^-$	10.8	*Salmonella dublin*	5
R750	8.9	Sm Su Tra$^-$	11.5	*Providencia*	17a
R682	8.6	Sm Su Tra$^-$	7.9	*Proteus mirabilis*	5
R684	9.5	Sm Su Tra$^-$	10.5	*Proteus mirabilis*	5
PB165	11.9	Sm Su Tra$^-$	4.1	*Escherichia coli*	5
R89S	8.18	Sm		*Escherichia coli*	33
pFM739	9.45	Sm Su Ap Mob$^+$		*Neisseria sicca*	32

concerning the molecular basis of the broad-host-range property. This has prompted studies to determine whether the broad-host-range of the IncQ plasmids was due to plasmid stability functions such as partitioning, to the presence of particular broad-host-range 'compatibility' functions, or to a particular mode of replication. Recent studies conducted mainly on RSF1010 show that it is a special mode of replication rather than an individual broad host range function which allows RSF1010 as well as probably other IncQ plasmids to propagate in nearly all Gram-negative bacteria. Table II lists bacterial species in which RSF1010 or derivatives of it were shown to replicate.

Three plasmids of the IncQ group, R300B (5), R1162 (5) and RSF1010 (16), which are very similar to or identical with one another have been studied to date and have provided an insight into the molecular mechanisms of IncQ plasmid replication. We will therefore discuss in this review results which concern these three plasmids, and in particular RSF1010 whose physical structure and biochemical and genetic functions have been studied most extensively.

II. PHYSICAL AND GENETIC MAP

RSF1010 has a size of 8685 bp (37; Scholz and Bagdasarian, unpublished). Very detailed genetic and physical maps have been generated for it. Figure 1 shows a map which is based on restriction analysis (4), *in vivo* and *in vitro* complementation studies (35,17; Haring and Scherzinger, Chapter 4B), and nucleotide sequence analysis (38,37; Scholz and Bagdasarian, unpublished). RSF1010 contains two drug resistance genes, Sm^R (*aphC$^+$*) conferring a high level resistance to streptomycin, and Su^R, the sulfonamide resistance gene. They are located between coordinates 7.8 kb and approximately 1.0 kb and are cotranscribed from a promoter at 7.9 kb (18,1). The other major genetic units on RSF1010 are the origin of vegetative replication, *ori V*, at 2.5 kb (17); the essential replication genes, *rep*, extending from 4.4 kb to 7.6 kb (35); as well as the determinants for mobilization located between 3.0 and 4.3 kb (28,1,10).

TABLE II. *Host range of RSF1010—derived cloning vectors.*
A list of species in which RSF1010 or RSF1010 derived vectors have been shown to replicate. This list is not exhaustive.

Species	Ref
Acetobacter xylinum	
Actinobacillus pleuropneumoniae	J. Perrin (pers. comm.)
Aerobacter aerogenes	25
Aeromonas hydrophila	15a
Agrobacterium tumefaciens	19
Alcaligenes eutrophus	25
Azotobacter vinelandii	9
Caulobacter crescentus	42
Erwinia carotovora	25
Erwinia chrysanthemi	15a
Escherichia coli	16
Gluconobacter sp.	42
Hypomicrobium sp.	25
Klebsiella pneumoniae	25
Methylophilus methylotrophus	42
Moraxella sp.	25
Paracoccus denitrificans	15a
Proteus mirabilis	25
Pseudomonas aeruginosa	29
Pseudomonas putida	29
Pseudomonas sp.	15
Pseudomonas testosteroni	25
Rhizobium leguminosarum	42
Rhizobium meliloti	8
Rhodopseudomonas sphaeroides	34
Serratia marcescens	25
Thiobacillus ferooxidans	15
Vibrio cholerae	34
Xanthomonas campestris	25
Xanthomonas maltophila	15b
Yersinia enterocolitica	34

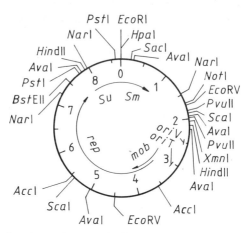

Figure 1. A physical and genetic map of RSF1010 based on the sequencing data (36; Scholz and Bagdasarian, unpublished).

Comparison of the available restriction maps, location of the antibiotic resistance genes and *oriV* and *oriT* of the plasmids R300B (6) and R1162 (26) shows the identity or close similarity of these plasmids with RSF1010.

III. REPLICATION

Replication of RSF1010 was dissected by cloning, expression and purification of gene products, *in vitro* reconstitution of the DNA replication process and *in vivo* and *in vitro* complementation studies. Thus it has been shown that a recircularized 2.1 kb DNA fragment of RSF1010 containing *oriV*, pMMB12, will replicate in *E. coli*, if the essential *rep* gene products, expressed from the DNA segment between coordinates 4.2– 8.9 kb, are supplied in *trans* (35). The location of *oriV* on pMMB12 was determined by electron microscopy. Subsequent analysis of RSF1010 DNA fragments, cloned into M13 vectors, in an *in vitro* replication system allowed precise mapping of the active origin to a fragment approximately 400 bp. These results also showed that the replication of RSF1010 may proceed in both directions (38; Haring and Scherzinger, Chapter 4B).

Deletion and complementation experiments revealed that each of the three *rep* genes, *repA*, *repB* and *repC* is essential for replication (35,38). These genes are located at a considerable distance from *oriV* and separated from it by regions which are not required for replication. Moreover, each of the three *rep* gene products of RSF1010 is also essential for initiation of plasmid DNA replication *in vitro* (12,35; Haring and Scherzinger, Chapter 4B). The same studies have established the exact locations of each *rep* gene on the physical map of RSF1010 (Figure 2). Biochemical functions have been determined for three of the *rep* gene products. They are summarized in Table III.

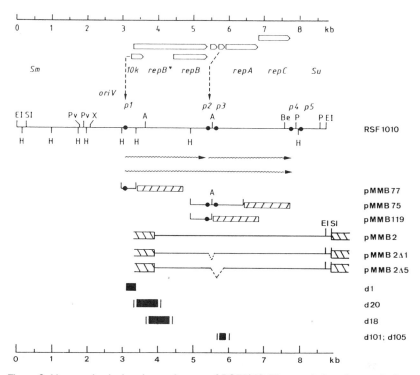

Figure 2. Upper: physical and genetic map of RSF1010. The restriction sites are indicated as follows: A, *Acc*I; Be, *Bst*EII; EI, *Eco*RI; H, *Hinf*I; P, *Pst*I; Pv, *Pvu*II; SI, *Sst*I; X, *Xmn*I. *E. coli* RNA polymerase binding sites are indicated as filled circles and corresponding promoters marked as p1 through p5. Open boxes with arrowheads represent reading frames in the *rep* region of the plasmid, the arrowheads show the transcriptional orientation. (*oriV*) origin of vegetative replication; (Sm) gene(s) conferring resistance to streptomycin; (Su) sulfonamides resistance gene. Wavy lines represent transcripts originating from the *rep* region with the arrows pointing in the direction of transcription. Middle: DNA fragments of RSF1010, containing the promoters, cloned upstream of a promoterless galactokinase gene (hatched box), used to determine the strength and regulation of promoters p1 and p2. Lower: Positions of the deletions d1, d18, d20, d101 and d105 and part of RSF1010 sequences contained in pMMB12, pMMB2, pMMB2Δ1 and pMMB2Δ5. The hatched boxes represent parts of the vectors, drawn lines represent RSF1010 DNA.

DNA sequence analysis indicated that two molecular forms of the protein RepB are made. A smaller form, RepB (MW = 38 kDa), purified to homogeneity, was found essential for replication of RSF1010 (35). The bigger form RepB* (MW = 70 kDa), translated in the same reading frame as

TABLE III. *Regulation of the RSF1010 rep genes at promoter p1.*
The p1 promoted galactokinase activity was measured in cells containing plasmid pMMB77 in the presence or absence of RSF1010 or an RSF1010 derivative plasmid.

Coresident plasmid	copy nr. of coresident	equivalent	Galacto-kinase Units	Replication in *P. putida*
–	–	–	121.1	
RSF1010	12.0 ± 1	1	45.1	+
d1	32.4 ± 5	2.6	128.5	–
d18	13.1 ± 2	1.1	31.4	+
d20	37.4 ± 6	3	110.2	–

RepB (Figure 2), has been detected on Western blots by crossreaction with antibodies raised against RepB (37; Scherzinger, personal communication) Deletion experiments (1) and *in vitro* complementation (Haring and Scherzinger, Chapter 4B) have shown, that RepB* is not essential for autonomous replication of RSF1010 in *E. coli*. It is still unclear, however, whether this protein is required for conjugational transfer of the plasmid. Overlapping of the mobilization genes with the *repB** frame make genetic analysis of both *mob* and *rep* functions difficult.

R1162 has been shown to contain 2 regions with genes essential for plasmid replication: *repI* and *repII* (27,21). Their map location corresponds with *repAC* and *repB* respectively. It was shown that *repI* encodes 2 polypeptides of 29 kDa and 31 kDa which seem to be similar or identical with RSF1010 repC and RepA proteins, respectively (21; Haring and Scherzinger, Chapter 4B).

To determine the influence of different *rep* genes on the copy number of an RSF1010 replicon, we have used a binary plasmid system based on the RSF1010 derivative pKT210 (2) and a controlled expression vector containing either genes *repA*, *repB*, or *repAC* under the control of the *tac* promoter. The replication rate, and hence the change in copy number of the plasmid pKT210 was measured using a DNA:DNA hybridization technique (14) before and after induction of the cloned genes *repB*, *repA*, and *repAC* (17). The results illustrated in Figure 3 show that the replication rate of the RSF1010 replicon increased six fold upon induction of the gene products from the cloned *repAC* fragment, while only a slight induction was registered if *repB* was overproduced in the cell. Induction of *repA* or the presence of the cloning vector alone did not affect the replication rate of RSF1010 (17).

DNA-protein binding studies showed that purified RepC protein binds specifically to a 200 bp *Dde*I fragment of the origin region of RSF1010 which contains three perfectly conserved and one partial 20 bp directly repeated sequence units (17). These results strongly suggest that the RepC protein binds specifically to *oriV* sequences, that its concentration in the cell is limiting and that it positively regulates the frequency of RSF1010 replication. In other words RepC protein acts as specific initiator in RSF1010 replication.

These studies, described more fully in Chapter 4B, showed that RSF1010 encodes three components of the primosome, a protein complex essential for initiation and propagation of DNA replication (22). It was postulated therefore that replication of broad-host-range plasmids is independent of many host replication functions normally used by the narrow-host-range replicons. The independence of RSF1010 DNA replication of *dnaB*, *dnaC* and *dnaG* functions (38) as well as of *dnaA* (M.Yarmolinsky, personal communication) and *rpoB* (12,35) was confirmed experimentally in *E. coli*. The only factor known to date to be required from the host for replication of RSF1010 is DNA polymerase III (38). The elucidation of replication requirements in other bacterial species must await the isolation of appropriate *dna* mutants.

IV. REGULATION OF REPLICATION (INCOMPATIBILITY)

Plasmids are maintained in host cells at a number of copies characteristic for each individual or a group of replicons. This results from a functional balance between the positive regulation, exerted by initiation factors, such as RepC protein in case of RSF1010, and negative regulatory elements. The effect of incompatibility between plasmids is the result of identity or close similarity in the negative regulatory mechanisms of their replicons. Negative regulation is of primary importance in replication and no naturally-occurring replicons known to date can function properly without negative control (30).

The origin region of the IncQ plasmids has several regulatory sites. The 20 bp direct repeats at *oriV* regions of RSF1010 and R1162 are identical (23,24,17). They are essential for origin function (36,24) and if cloned onto a multicopy vector they exert IncQ incompatibility in *trans* (24,31). This effect originally ascribed to a hypothetical product, expressed by the origin fragment, can now be explained as resulting from titration of the RepC initiator protein that binds to the direct repeats (17). Replication genes themselves are under feedback regulatory control. Figure 2 shows the location of promoters of RSF1010, marked as p1 through p5. Identified originally as RNA polymerase binding sites (2) their positions were confirmed later by DNA sequence analysis (36). Mapping of transcripts from RSF1010 *rep* region by S1 resistance of DNA-RNA hybrids has suggested that *rep* genes are expressed as two operons, a long one, initiated at p1 and extending through *repB**, *repB*, *repA* and *repC* and a short one starting at p2 and extending through *repA* and *repC* (see Figure 2). Functions of the promoters p1 and p2 were tested by inserting appropriate DNA fragments of RSF1010 upstream of a promoterless galactokinase gene to give recombinant plasmids pMMB75, pMMB77 and pMMB119 as shown in Figure 2. Measurements of galactokinase activities in cells carrying these plasmids

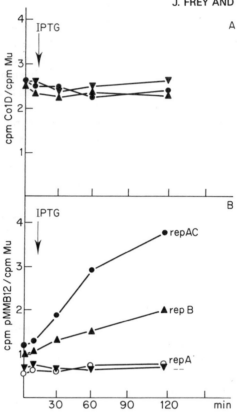

Figure 3. Replication of plasmids after induction of RSF1010 *rep* genes. Cultures of a
E.coli strain containing one copy of prophage Mu at *ilv* (14) and the plasmids pKT210,
pVH1 (a ColD derived vector containing the *lacIQ* gene) and an expression vector with
one of the RSF1010 *rep* genes inserted under *tac* promoter control were grown exponen-
tially in M9 supplemented medium containing methyl-^{3}H-thymine. 1.0 mM IPTG was
added at time indicated by the arrow. Samples equivalent to 1.0 ml culture of an OD
450 nm = 1.0 were withdrawn at times indicated, lysed in 0.5 M NaOH for 10 min,
neutralized and incubated with membrane filters containing 2 μg non-labelled
bacteriophage Mu, pMMB12, or pKT101 DNA for hybridization as described
previously(14). After the hybridization, the filters were washed and the methyl-^{3}H-
thymine cpm counted in a Scintillation counter. The ratios cpm− plasmid/cpm−Mu,
reported on this figure, reflect the evolution of the copy number of the plasmids after
induction of the various *rep* genes. Filled circles, *repAC*; filled triangles upwards, *repB*;
filled triangles downwards *repA*; open circles cloning vector alone. A: measurements
of the copy number variations of the ColD vector, pKT101, as control. B: measurements
of the copy number variations of the RSF1010 derivative pKT210.

showed not only that the promoters p1 and p2 can express downstream genes,
but also that they are regulated by genes present on RSF1010. As shown in Table
III, promoter p1 is repressed by the wild-type RSF1010 and by the deletion

derivative of RSF1010, d18, but not by deletion plasmids d1 or d20 (Figure 2).
Deletion d18 does not extend beyond AccI site at coordinate 3578 whereas d20
extends over this AccI site. Deletion d1 has removed 127 bp directly downstream
of p1 promoter (3). From these results it was concluded that a repressor gene
is located immediately downstream of the p1 promoter.

DNA sequence analysis indicated an open reading frame for a 10 k Da protein in this region. Both d1 and d20 deletions affect this reading frame whereas
d18 does not. The 10 k Da protein is a good candidate for the repressor of
p1 promoter. It must be pointed out, however, that although genetic evidence
clearly indicates the existence of a regulatory element expressed from the region
of p1 promoter, final proof must await its isolation and *in vitro* testing. In
accordance with the hypothesis that p1 promoter controls all three *rep* genes,
including the *repC* initiator gene, found to have the most pronounced effect
on replication rate (see Figure 3), it was shown that deletion derivatives d1
and d20 have a strongly increased copy number in *E. coli* (Table IV; see also
Bagdasarian *et al*, 1986).

TABLE IV. (a). *Regulation of the RSF1010* rep *genes at promoter p2 and p3.*
p2 and p3 promoted galactokinase activity.

galK plasmid[a]	RSF1010 promo-ter	Coresident plasmid	Galactokinase Units
pMMB75	p2, p3	−	54.2
pMMB75	p2, p3	RSF1010	42.0
pMMB119	p2	−	133.5
pMMB119	p2	pMMB2	17.3
pMMB119	p2	pMMB2Δ1	31.8
pMMB119	p2	pMMB2Δ5	120.3

[a] see Figure 2.

TABLE IVB. *Copy numbers of RSF1010 and deletion derivatives in* E. coli *and* P. putida.

| Plasmid | E. coli host | | P. putida host | |
	copy Nr	RSF1010 equiv.	copy Nr.	RSF1010 equiv.
RSF1010	12.1 ± 0.3	1	13.8 ± 2.5	1
d101	37.0 ± 3.8	3	19.5 ± 2.8	1.4
d105	15.3 ± 3.8	2.8	15.3 ± 2	1.2

Regulation at promoter p2 shows marked similarity to the regulation at p1.
Thus p2 is repressed in the presence of plasmid pMMB2 (35) which contains
sequences downstream of p2 (Figure 2) and by plasmid pMMB2Δ1 carrying
a small deletion around AccI site at coordinate 5.5 kb. However, plasmid

pMMB2Δ5, having a bigger deletion at this site no longer represses p2 (Table V). Recently we have isolated small deletions at this *Acc*I site in the wild-type RSF1010. As expected these derivatives, called d101 and d105, are maintained in *E. coli* cells at elevated copy numbers similarly to deletion plasmids d1 and d20 (Table IV). These results are consistent with the proposal that the second of the two 7 kDa reading frames, located immediately downstream of p2 as predicted by the sequencing data, encodes the repressor of the *repAC* operon. This polypeptide has been purified and shown to bind to p2 DNA sequences (Haring and Scherzinger, Chapter 4B).

In the case of R1162 it was found that its copy number is directly dependent on the concentration of two proteins encoded at the region *repI* (21). These proteins may have similar functions to the RepA and RepC proteins of RSF1010. However, a 75 bp RNA molecule complementary to the *repI* mRNA was found which regulates negatively the expression of the *rep* genes (21). It remains to be established whether the regulation of R1162 and RSF1010 are indeed different or whether two different mechanisms of the same regulatory system, common to all IncQ plasmids are being studied.

V. TRANSCRIPTIONAL REGULATION AFFECTS THE HOST RANGE

Attempts to study the regulation of copy number by the repressors of p1 or of p2 in other bacterial species have revealed that all copy number mutants obtained by deleting the repressor of p1 (including d1 and d20) have lost the broad-host-range properties. They could no longer be introduced into *Pseudomonas putida* (Table III) or *Pseudomonas aeruginosa*, although they replicate and are stably maintained in *E. coli* at a higher copy number. The results obtained with the mutation d1 suggests that the loss of the broad-host-range property is associated with the damage to the 10 kDa reading frame or other small repressor encoded in this region, rather than to the *repB** protein, since both forms of the primase, the *repB* as well as the *repB**, were detected in the cells carrying the d1 mutant (Haring and Scherzinger, Chapter 4B). In addition, plasmid carrying the deletion d18 within the *repB** reading frame could replicate in *Pseudomonas*.

Copy number mutants carrying deletions in the regulatory region for the p2 repressor, show an increased copy number in *E. coli*, as well as in *P. putida*. However, the copy number elevation in *P. putida* is distinctly lower than in *E. coli*. We therefore believe that the two promoters governing the *rep* region (p1 and p2) have different significance in different Gram-negative bacteria. It is possible that regulation at the promoter p1 plays a decisive role in copy number control in *P. putida*, and that a deletion of the repressor of p1 would lead in *P. putida* to a 'runaway replication' observed previously in other replicons (40). Regulation at p2 seems to be less critical in *P. putida*, since deletions of the regulatory element of p2 does not prevent the mutant plasmids from being stably maintained in this host. In *E. coli*, regulation at p1 and at p2 seems to be of equal importance; deletion of the regulatory elements at p1 or p2 results in equal increase of copy number but does not prevent replication. It should be noted that a given promoter can have different strength in different bacterial hosts (25). This can explain the differential behaviour of the deletions d1 and d20 (which delete the regulatory element at p1) in *E. coli* and in *P. putida* and shows the importance of these negative control elements in the expression of *rep* genes in broad-host-range plasmids in particular.

VI. MOBILIZATION

RSF1010 is not self-transferable by conjugation. It may be mobilized, however, by certain coexisting conjugative plasmids. Relatively high mobilization frequencies were obtained with 'helper' plasmids from the incompatibility groups IncP, IncIα, IncM and IncX (41). Mobilization of RSF1010 derived cloning vectors was used to transfer the vectors into hosts where transformation gave only poor results (2) or to mobilize directly entire gene banks made with RSF1010 derived cosmids from *E. coli* into another hosts for screening (15). Several genes are required by a mobilizable plasmid to be transferred in conjugation if conjugative functions are supplied in *trans*: an origin of transfer, *oriT*, site (previously known as *nic* site) and plasmid-specific mobilization genes (*mob*) which encode proteins of as yet unknown functions. The origin of transfer has been mapped to the 800 bp *Hae*II fragment of RSF1010 between coordinates 2.7 and 3.5 kb. It was identified as the so-called 'relaxation nick site' (28). An 80–88 bp fragment of RSF1010, which allowed mobilization at 10% efficiency was localized (11). This fragment stretches from coordinate 3080 to 3068 bp (on our map Figure 1) and includes a 10 bp inverted repeat which may be involved in recognition and/or nicking by the mobilization proteins (11). Inversion of the *oriT* containing fragment with respect to *oriV* showed that *oriT* is functionaly independent of *oriV*. A segment of 1.8 kb containing the entire region required for mobilization of RSF1010 was sequenced (10). Using insertion mutagenesis and complementation the authors have identified three genes required for mobilization. A gene with a reading frame of >65 kDa (which presumably is identical to the *repB** gene in Figure 2), a gene overlapping with the 65 kDa reading frame which encodes a 16 kDa reading frame, and a third gene encoding a 9 kDa frame which is transcribed in the opposite direction (10). However, the precise function of these proteins and their role in conjugative transfer remain to be clarified.

VII. CONCLUDING REMARKS

A concerted action of several essential genes is required for replication of Inc Q plasmids. In RSF1010 the concentration of *repC* gene product is limiting and therefore it positively regulates plasmid replication. The replication genes *repA* and *repB* provide plasmid specific functions of a helicase and a primase, respectively. RSF1010 thus encodes its own proteins essential for primosome formation. This property is believed to render the replication of the plasmid independent of the host primosomal functions and to allow its maintenance in a wide range of hosts.

Replication genes of RSF1010 are clustered in two regulatory units, one driven by promoter p1 transcribes all 3 *rep* genes, the second driven by promoter p2 transcribes only *repA* and *repC*. Both promotors seem to be negatively regulated by gene products encoded downstream from each promoter. Regulation at both promotors controls the number of plasmid copies in the cell. Regulation at the promoter p1 is essential for the ability of RSF1010 to replicate in *P. putida* and hence essential for the broad host range property of RSF1010.

Inc Q plasmids are not self-transferable in conjugation, but may be mobilized into a variety of different species, including plant cells. In RSF1010 the origin of transfer is a site different from the origin of vegetative replication. However, mobilization genes and certain replication control genes (repressor of p1, *mob* genes) are located on the same DNA segment and their reading frames overlap. The same is true for genetically identified loci that change the host range. Further work is needed to elucidate the interdependence of these functions.

It is perhaps not surprising that RSF1010, a relatively small replicon, carries a rather complex and elaborate replicative and regulatory apparatus. Although reminiscent of the bacteriophage lambda regulatory circuit of the *cro*, O and P-genes, RSF1010 system contains two autoregulatory loops, one at p1 promoter and one at p2. It must be remembered, however, that RSF1010 has evolved to accommodate to many very different cellular environments ranging from Enterobacteriaceae to soil organisms and from human to plant pathogens.

Other plasmids, notably the representatives of Inc P group, such as RK2, also exhibit a very wide-host-range. Although in RK2 only one protein, Trf A, seems to be essential for replication whereas its primase encoding genes are not required in *E. coli*, the control mechanisms, leading to its regulated replication are extremely complex (see Chapter 1). In this respect the two broad-host-range plasmids RK2 and RSF1010 share the common property of requiring several plasmid encoded replication factor for their normal maintenance.

ACKNOWLEDGEMENTS

This work was supported by the Swiss National Science foundation Grant No.3.594.084 and No.3.627.087 to J.F. and by Research Excellence and

Economic Development Grant from the State of Michigan to M.B. A short term fellowship (ASTF 5031) from the European Molecular Biology Organization to J.F. is gratefully acknowledged.

VIII. REFERENCES

1. Bagdasarian M, Lurz R, Rückert B, Franklin FCH, Bagdasarian MM, Frey J, Timmis KN (1981) Specific-purpose plasmid cloning vectors. II. Broad host range, high copy number, RSF1010-derived vectors, and a host-vector system for gene cloning in *Pseudomonas*. *Gene* **16**, 237–47

2. Bagdasarian M, Timmis KN (1982) Host vector systems for gene cloning in *Pseudomonas*. *Curr Topics Microb Immun* **96**, 47–67

3. Bagdasarian M, Bagdasarian MM, Lurz R, Nordheim A, Frey J, Timmis KN (1982) Molecular and functional analysis of the broad-host-range plasmid RSF1010 and construction of vectors for gene cloning in Gram-negative bacteria. *In* Drug resistance in bacteria, Genetics, Biochemistry, and Molecular Biology. (Mitsuhashi S, ed) Tokyo, New York. pp 183–97

4. Bagdasarian MM, Scholz P, Frey J, Bagdasarian M (1986) Regulation of the *rep* operon expression in the broad-host-range plasmid RSF1010. *Banbury Report* **24**, 209–23

5. Barth P, Grinter NJ (1974) Comparison of the deoxyribonucleic acid molecular weights and homologies of plasmids conferring linked resistance to streptomycin and sulfonamides. *J Bacteriol* **120**, 618–30

6. Barth PT, Tobin L, Sharpe GS (1981) Development of broad-host-range plasmid vectors. *In* Molecular Biology, pathogenicity, and ecology of bacterial plasmids. (Levy SB, Clowes RC, Koenig EL, eds) Plenum. pp 439–48

7. Buchanan-Wollaston V, Passiatore JE, Cannon F (1987) The *mob* and *oriT* mobilization functions of bacterial plasmid promote its transfer to plants. *Nature* **328**, 172–5

8. David M, Tronchet M, Denarié JL (1981) Transformation of *Azotobacter vinelandii* with DNA of plasmids RP4 (Inc-P-1 group) and RSF1010 (IncQ group). *J Bacteriol* **146**, 1154–7

9. David M, Vielma M, Julliot JS (1983) Introduction of IncQ plasmids into *Rhizobium meliloti*. Isolation of a host range mutant of RSF1010 plasmid. *FEMS Microbiol Lett* **16**, 2–3

10. Derbyshire KM, Hatfull G, Willetts N (1987) Mobilization of the nonconjugative plasmid RSF1010: A genetic and DNA sequence analysis of the mobilization region. *Mol Gen Genet* **206**, 161–8

11. Derbyshire KM, Willetts NS (1987) Mobilization of the nonconjugative plasmid RSF1010: A genetic analysis of its origin of transfer. *Mol Gen Genet* **206**, 154–60

12. Diaz R, Staudenbauer WL (1982) Replication of the broad-host-range plasmid RSF1010 in cell-free extracts of *Escherichia coli* and *Pseudomonas*

aeruginosa. Nucl Acid Res **10**, 4687–702

14. Frey J, Chandler M, Caro L (1979) The effects of an *Escherichia coli* *dna*Ats mutation on the replication of the plasmids ColE1, pSC101, R100.1 and RTF.TC. *Mol Gen Genet* **174**, 117–26

15. Frey J, Bagdasarian M, Feiss D, Franklin FCH, Deshusses J (1983) Stable cosmid vectors that enable the introduction of cloned fragments into a wide range of Gram-negative bacteria. *Gene* **24**, 299–308

15a. Frey J, Krisch HM (1985) Ω mutagenesis in Gram-negative bacteria: a selectable interposon which is strongly polar in a wide range of bacterial species. *Gene* **36**, 143–50

15b. Frey J, Mudd EA, Krisch HM (1988) A bacteriophage T_4 expression cassette that functions efficiently in a wide range of Gram-negative bacteria. *Gene*, **62**, 237–47

16. Guerry P, van Embden J, and Falkow S (1974) Molecular nature of two nonconjugative plasmids carrying drug resistance genes. *J Bacteriol* **117**, 619–39

17. Haring V, Scholz P, Scherzinger E, Frey J, Derbyshire K, Hatfull G, Willetts NS, Bagdasarian M (1985) Protein RepC is involved in copy number control of the broad-host-range plasmid RSF1010. *Proc Natl Acad Sci USA* **82**, 6090–4

17a. Hedges RW (1974) R-factors from Providence. *J Gen Microbiol* **81**, 171–82

18. Heffron F, Rubens C, Falkow S (1975) Translocation of a plasmid DNA sequence which mediates ampicillin resistance: molecular nature and specificity of insertion. *Proc Natl Acad Sci USA* **72**, 3623–7

19. Hille J, Schilperoort R (1981) Behaviour of IncQ plasmids in *Agrobacterium tumefaciens. Plasmid* **6**, 360–2

20. Kim K, Meyer RJ (1985) Copy number of the broad-host-range plasmid R1162 is determined by the amounts of essential plasmid-encoded proteins. *J Mol Biol* **185**, 755–67

21. Kim K, Meyer RJ (1986) Copy-number of broad-host-range plasmid R1162 is regulated by a small RNA. *Nucl Acids Res* **14**, 8027–8046

22. Kornberg A (1980) DNA Replication. WH Freeman and Co, San Francisco.

23. Lin LS, Meyer RJ (1984) Nucleotide sequence and functional properites of DNA encoding incompatibility in the broad-host-range plasmid R1162. *Mol Gen Genet* **194**, 423–31

24. Lin LS, Meyer RJ (1986) Directly repeated, 20 bp sequence of plasmid R1162 DNA is required for replication, expression of incompatibility and copy-number control. *Plasmid* **15**, 35–47

25. Mermod N, Ramos JL, Lehrbach PR, Timmis KN (1986) Vector for regulated expression of cloned genes in a wide range of Gram-negative bacteria. *J Bacteriol* **167**, 447–54

26. Meyer R, Laux R, Bach G, Hinds M, Bayly R, Shapiro J (1982) Broad-host-range IncP-4 plasmid R1162: Effects of deletions and insertions on plasmid maintenance and host range. *J Bacteriol* **152**, 140–150

27. Meyer RJ, Lin LS, Kim D, Brasch M (1985) Broad-host-range plasmid R1162: replication, incompatibility and copy-number control. *In* Plasmids in Bacteria. (Helinski DR, Cohen SN, Clewell DB, Jackson DA, Hollaender A, eds) Plenum, New York/London, pp 173–88

28. Nordheim A, Hashimoto-Gotoh T, Timmis KN (1980) Location of two relaxation nick sites in R6K and single sites in pSC1010 and RSF1010 close to origins of vegetative replication: Implication for conjugal transfer of plasmid deoxynucleic acid. *J Bacteriol* **144**, 923–32

29. Nagahari K, Sakaguchi K (1978) RSF1010 plasmid as potentially useful vector in *Pseudomonas* species. *J Bacteriol* **133**, 1527–9

30. Nordström K (1985) Control of plasmid replication. Theoretical considerations and practical solutions. *In* Plasmids in Bacteria. (Helinski DR, Cohen SN, Clewell DB, Jackson DA, Hollaender A, eds) Plenum, New York/London, pp 189–214.

31. Persson C, Nordström K (1986) Control of replication of the broad-host-range plasmid RSF1010: The incompatibility determinant consists of directly repeated DNA sequences. *Mol Gen Genet* **203**, 189–92

32. Rotger R, Rubio F, Nombela C (1984) A multi-resistance plasmid isolated from commensal *Neisseria* species is closely related to the enterobacterial plasmid RSF1010. *J Gen Microbiol* **132**, 2491–6

33. Saano AD, Zinchenko VV (1987) A new IncQ plasmid R89S: properties and genetic organization. *Plasmid* **17**, 191–201

33. Sandkvist M, Hirst TR, Bagdasarian M (1987) Alterations at the carboxyl terminus change assembly and secretion properties of the B subunit of *Escherichia coli* heat-labile enterotoxin. *J Bacteriol* **169**, 4570–6.

34. Scherzinger E, Bagdasarian MM, Scholz P, Lurz R, Rückert B, Bagdasarian M (1984) Replication of the broad-host-range plasmid RSF1010: Requirement for three plasmid-encoded proteins. *Proc Natl Acad Sci USA* **81**, 654–8

35. Schmidhauser TJ, Ditta G, Helinski DR (1988). Broad-host-range plasmid cloning vectors for Gram-negative bacteria. *In* Vectors. A survey of molecular cloning vectors and their uses. (Rodriguez RL, Denhardt DT, eds), Butterworth Publishers, pp 287–382

37. Scholz P (1985) Strukturelle und Funktionelle Analyse der Replikationsdeterminanten von RSF1010- ein Plasmid mit extrem breitem Wirtsspektrum. (Thesis) Freie Universität Berlin, Fachbereich Biologie (23), Berlin

38. Scholz P, Haring V, Scherzinger E, Lurz R, Bagdasarian MM, Schuster H, Bagdasarian M (1985) Replication of the broad host range plasmid RSF1010. *In* Plasmids in Bacteria (Helinski DR, Cohen SN, Clewell DB,

Jackson DA, Hollaender A, eds) Plenum, New York/London, pp 243–59
40. Uhlin BE, Nordström K (1978) A runaway replication mutant of plasmid R1drd-19: Temperature-dependent loss of copy number control. *Mol Gen Genet* **165**, 167–72
41. Willetts NS, Crowther C (1981) Mobilization of the non-conjugative IncQ plasmid RSF1010. *Genet Res* **37**, 311–6
42. Windass JD, Worsey MJ, Pioli EM, Pioli D, Barth PT, Atherton KT, Dart EC, Byrom D, Powell K, Senior PJ (1980) Improved conversion of methanol to single-cell protein by *Methylophilus methylotrophus*. *Nature* **287**, 396–401

CHAPTER 4B

REPLICATION PROTEINS OF THE IncQ PLASMID RSF1010

Volker Haring and Eberhard Scherzinger

I. INTRODUCTION

Bacterial plasmids of the *E. coli* incompatibility group IncQ are noncon-jugative, multicopy replicons conferring resistance to streptomycin and sulfonamides (21). Although not self-transmissible, they are efficiently mobilized into many different species of Gram-negative bacteria by certain conjugative plasmids such as RP4 (5,43). Furthermore, it has been shown that the origin of transfer and the mobilization functions of an IncQ group plasmid can mediate the transfer of plasmid DNA from *Agrobacterium* into plants (6).

The best known representatives of IncQ group plasmids are RSF1010 (22), R300B (4) and R1162 (32). They all have a size of 8.7 kb and, though isolated from diverse bacterial backgrounds, they seem to be very similar or identical (4,25). The combination of small size, high copy number and broad host range has made them attractive for use as DNA cloning vehicles (as reviewed in ref. 37) and has been the motivating factor in studies of their mode of replication.

For RSF1010, replication in *Escherichia coli* and *Pseudomonas aeruginosa* starts at a unique site and proceeds via θ−form intermediate structures either bi- or unidirectionally from this origin (11,39). With cell-free systems for the study of plasmid replication in *E. coli* and *P. aeruginosa*, it has been shown that RSF1010 requires plasmid-encoded factors for its replication (14). Three essential replication genes, *repA, B* and *C*, were identified on RSF1010 which are situated at considerable distance from the origin of vegetative replication, *ori V* (Figure 1) (36). The copy number of RSF1010 which is 10–12 per chromosome in *E. coli* (4) was found to be dependent on the intracellular

Promiscuous Plasmids of Gram-Negative Bacteria   © 1989 Academic Press Limited
ISBN 0-12-688480-3   All rights of reproduction in any form reserved

concentration of the *rep*C gene product (24; Frey and Bagdasarian, Chapter 4A).

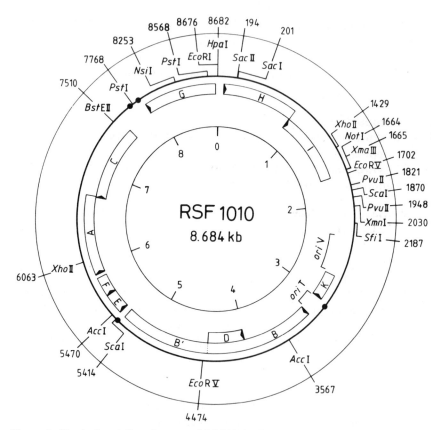

Figure 1. Physical and Genetic map of RSF1010. The map is based on the complete nucleotide sequence of RSF1010 (Scholz and Scherzinger, unpublished results). The numbering of the nucleotide sequence begins at the unique *Hpa*I site: the first A in the sequence ...GTTAAC... is designated as nucleotide 1. The map shows the positions of restriction sites for enzymes that cut the plasmid once or twice; the number refers to the coordinate of the first (5′) nucleotide in the recognition sequence. The position of the eleven known protein coding genes are shown as boxes; all coding sequences with the exception of K are transcribed clockwise as indicated by the arrow heads. The coding sequences labelled A, B/B′, and C correspond to the replication genes *repA*, *repB/B′*, and *repC* respectively. The map also includes the positions of the four major *E. coli* RNA polymerase binding sites (black circles), the origin of vegetative replication (*oriV*) and the origin of transfer (*oriT*).

The *repB* gene product, protein RepB, is found in two molecular forms (ES *et al*, unpublished). The cross reaction of the larger form (77.9 kDa) with antibodies raised against the smaller form (35.9 kDa) as well as DNA and amino acid sequence analysis indicate that both forms are translated from the same reading frame. The smaller protein is formed by reinitiation of translation in the middle of gene B (Figure 1), as is found for the bacteriophage ϕX174 proteins A and A′ (31). By analogy to these proteins we named the larger form RepB and the smaller one RepB′. The smaller protein bears all the functions required for RSF1010 replication *in vivo* and *in vitro* but it can be replaced by the larger protein in all our *in vitro* replication tests (ES *et al*, unpublished).

In this communication we will report the results of our studies on the mechanism of RSF1010 replication *in vitro*. We describe the overproduction and purification of the plasmid-encoded Rep proteins, an *in vitro* system that replicates RSF1010 DNA, and the role of the proteins RepA, B′, and C in the initiation of RSF1010 replication. Most of the results presented here are taken from refs. 23 and 38.

II. OVERPRODUCTION AND PURIFICATION OF THE RSF1010 REPLICATION PROTEINS

A. Expression systems for the RSF1010 replication genes

In order to prepare the RSF1010 replication proteins in large quantities we constructed *rep* expression plasmids (24) that carry either one or two of the *rep* genes under transcriptional control of the strong, hybrid *trp/lac* (*tac*) promoter (10). The physical maps of these plasmids are given in Figure 2.

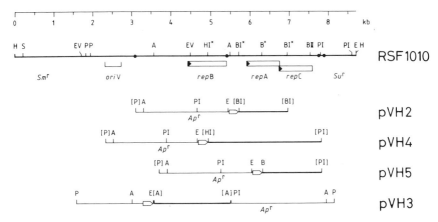

Figure 2. Simplified physical and genetic map of RSF1010 and diagrams of the *rep* expression plasmids used in this work. Restriction sites are shown as A, *Acc*I : B, *Bss*HII; BI, *Bal*I; BII, *Bst*EII; E, *Eco*RI; H, *Hpa*I; HI, *Hinf*I; P, *Pvu*II; and PI, *Pst*I. * Of the numerous *Bss*HII, *Bal*I, and *Hinf*I sites in RSF1010 DNA, only those relevant to plasmid constructions are indicated. (») direction of transcription from the *tac* promoter.

Since the *lac* operator site is an integral part of the *tac* promoter, transcription from it is blocked in the presence of the *lac* repressor and can be induced by addition of isopropyl-β-D-thiogalactoside to the growth medium. Therefore, to allow controlled expression of the *rep* genes the resulting expression plasmids were introduced into *E. coli* strain HB101 harboring pVH1 (24), a ColD-based multicopy plasmid that carries the *lac* repressor gene *lacI^Q*, and tested for their ability to direct IPTG-induced expression of the cloned genes. This was done by electrophoretic analysis of lysates of induced and uninduced cells harboring either of the *rep* expression plasmids.

The gel pattern of induced cells carrying pVH2 showed an additional polypeptide band of about 30 kDa, the *repA* gene product, compared to the pattern of uninduced cells (Figure 3; compare lanes a and b). Similarly, induction of pVH3-harboring cells resulted in the appearance of a polypeptide of ca. 39 kDa, the *repB'* gene product (Figure 3; compare lanes g and h). As expected, plasmid pVH4 directed the synthesis of two polypeptide species upon IPTG induction: one corresponding to the 30 kDa *repA* gene product and a second with an apparent molecular weight of 28 kDa, the *repC* gene product (Figure 3; compare lanes e and f).

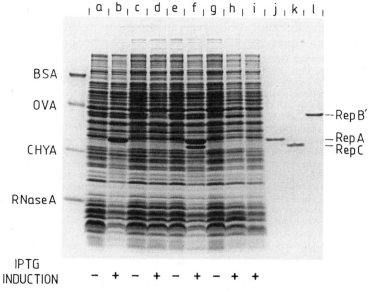

Figure 3. Induction of Rep protein synthesis directed expression plasmids pVH2, pVH3, pVH4 and pVH5. A Coomassie stained 15% polyacrylamide/SDS gel is shown. Whole cell lysates were prepared from HB101 cells harboring pVH1 plus one of the following plasmids: lanes a and b, pVH2; lanes c and d, pVH5; lanes e and f, pVH4; lanes g and h, pVH3; lane i, none; lanes j, k, and l, purified RepA, C, and B' proteins, respectively. Reference proteins are: bovine serum albumin (BSA), ovalbumin (OVA), chymotrypsinogen A (CHYA), and RNase H.

This 28 kDa RepC polypeptide band was not detectable in lysates of induced cells harboring pVH5 (Figure 3; compare lanes c and d), a plasmid that carries the 3'-end of the *repA* gene and the entire *repC* gene. The RepC protein could only be expressed in visible amounts when fragments were cloned which comprise the entire genes of *repA* and *repC* in their native adjacent positions. We propose that expression of RepC is closely coupled to that of RepA by a mechanism of translational interdependence as found in RNA phages (15). From our present knowledge of the RSF1010 genome organization we suppose that *repA* and *repC* are expressed from a polycistronic message. The sequence data of the *repA/C* region reveal that the coding sequences of both genes overlap by 14 nucleotides (Figure 4). The ribosome binding site and start codon of the *repC* gene are located in a region with dyad symmetry. At the symmetry region the mRNA can form a stem-loop structure that may repress synthesis of RepC by masking its initiation signals (Figure 4). Ribosomes translating the last codons of the RepA message will disrupt the secondary structure and the blocked sites become available to initiate RepC synthesis. Any alteration of the message that prevents translation across the secondary structure will also prevent effective *repC* expression.

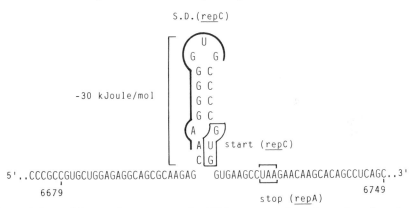

Figure 4. Potential secondary structure of mRNA sequences from the region of overlap between RSF1010 genes *repA* and C.

The leakiness of the 'repressed' structure in the absence of *repA* translation is not yet known. However, Meyer and co-workers (33,26) have shown by minicell experiments and complementation tests that a plasmid carrying a DNA fragment of the IncQ plasmid R1162 (homologous to RSF1010 sequences between coordinates 6.5 and 8.68) fused to a *tac* promoter can express RepC activity. This may be due to thermodynamic fluctuations in the secondary structure which make the RepC initiation signals temporarily accessible to ribosomes. The analyses of protein expression in minicells are unfortunately confused by the coexistence of the sulfonamide resistance gene of R1162 on the cloned fragments. This gene codes in plasmids R300B and RSF1010 for a 28.2 kDa polypeptide which has about the same electrophoretic mobility as RepC (5; ES *et al*, unpublished). Therefore, it is not possible to compare the rates of expression of *repA* and *repC* of RSF1010 to that of the analogous genes of R1162.

B. Purification of the RSF1010 Rep proteins

The proteins were purified from *E. coli* strains harboring pVH1 and one of the *rep* expression plasmids pVH2, pVH3 or pVH4. Extracts were prepared from cells after 3 hours of IPTG-induction and the purifications were monitored by both SDS-polyacrylamide gel electrophoresis and *in vitro* replication complementation assays (36)

The RepA protein was purified from induced HB101[pVH1,pVH2] cells by column chromatography using heparin-Sepharose, DEAE-Sephacel, phosphocellulose, phenyl-Sepharose, and Sephacryl S-200.

Starting with induced HB101[pVH1,pVH2] cells the RepB' protein was purified by chromatography through a heparin-Sepharose column, low salt precipitation, and three additional column chromatography steps using Sephacryl S-200, hydroxyapatite, and phosphocellulose.

The proteins RepA and C, contained in the extract of induced HB101[pVH1,pVH2] cells, were separated by chromatography through a heparin-Sepharose column. The RepA protein in the flow-through was purified as described above. The further purification of RepC was achieved by low salt precipitation and column chromatography using Sephadex G-100 and phosphocellulose. As expected from their different purification properties, no RepA activity was detectable in the final RepC preparation and vice versa.

Starting with 37 g of induced cells, these purification schemes yielded 86 mg RepA, 1 mg RepB', and 10 mg RepC. The purity of the proteins in the final fractions were electrophoretically estimated to be $\geqslant 95\%$ (Figure 3, lanes j, k, and l).

C. Physical properties of the purified Rep proteins

In order to identify the precise extent of the coding sequences for the Rep proteins in the nucleotide sequence we determined in collaboration with Drs B Wittman-Liebold and K Ashman the amino acid sequences of either terminus of the purified proteins. The NH_2-terminal sequences were determined by an automated Edman-degradation procedure (44). For each protein at least the first 18 amino acids identified were consistent with a run of 18 consecutive codons in an open reading frame assigned to the respective protein. For RepA and C, no initiator fMet could be found, presumably due to post-translational processing of these proteins *in vivo*. Moreover, these data reveal that *repC* starts with a GTG codon.

The COOH-terminal amino acid sequences were determined by incubation of the purified proteins with carboxypeptidase P for various times and identification of the released amino acids as phenylthiohydantoin derivatives. The first few amino acids released upon digestion of RepA and C match with the last codons of their predicted reading frames. In addition, these results confirm that the coding sequences of RepA and C overlap by 14 nucleotides. According to the nucleotide sequence the carboxy terminal sequence of RepB′ is Phe-Ser-Met-COOH. However, only Phe and Ser could be identified as final amino acids. It is not known whether the methionine is not incorporated as terminal amino acid or removed proteolytically to form the mature protein.

Under denaturing and reducing conditions each purified Rep protein migrates as a single polypeptide species in an sodium dodecyl sulphate-polyacrylamide gel (Figure 3; lanes j, k, and l). When compared to a set of standard proteins their molecular masses were estimated to be 29.7 kDA for RepA, 39.0 kDa for RepB′, and 27.8 kDa for RepC. Whereas for RepA the measured value is in good agreement with that calculated from the nucleotide sequence data (29.8 kDa), the estimated masses of RepB′ and C differ by about 10% from their predicted values (RepB′: 35.8 kDa, RepC: 30.9 kDa). These differences can be explained by anomalous binding affinities of SDS to these proteins or unusual structures of the SDS-protein complexes which cause an altered electrophoretic mobility (3). These and the following results are summarized in Table I.

TABLE I. *Summary of physical properties of proteins RepA, B′ and C*

Property	RepA	RepB′	RepC
Mass, DNA sequence, kDa	29.78	35.77	30.89
Mass, SDS-PAGE, kDa	29.7	39.0	27.8
Native mass, calculated [1], kDa	164.3	36.2	60.6
Sedimentation coefficient, S	7.9	2.8	4.1
Stokes radius, Å	48	31	34
Partial specific volume [2], cm³/g	0.74	0.73	0.74
Frictional coefficient [2]	1.28	1.42	1.29
Thermal stability [3], min	2	>30	10
NEM sensitivity	−	−	+
Subunit structure	hexamer	monomer	dimer

[1]calculated from sedimentation coefficient, Stokes radius, and partial specific volume
[2]calculated from amino acid sequence
[3]90 % inactivation at 80°C

The sedimentation coefficients of the native proteins were determined in 10 to 30% glycerol gradients. The gradients were fractionated and the fractions analysed by SDS-polyacrylamide gel electrophoresis as well as tested for activity of the Rep proteins by use of the *in vitro* replication system. Compared to standard proteins the sedimentation coefficients were calculated to be 7.9S (RepA), 2.8S (RepB'), and 4.1S (RepC). Under various conditions up to now no formation of a physical complex between the Rep proteins has been observed.

The Stokes radii of the Rep proteins were determined using the elution profiles of the gel filtration steps obtained in the course of the protein purification. The gel filtration columns were calibrated with a set of standard proteins. The estimated values are: 48A (RepA), 31A (RepB'), and 34A (RepC).

The molecular masses of the native proteins were determined from the sedimentation coefficients, the Stokes radii, and the apparent partial specific volumes (41). The apparent partial specific volumes (Table I) were calculated from the amino acid compositions as predicted by the nucleotide sequences (8). The native molecular masses of the Rep proteins were evaluated to be 164.3 kDa (RepA), 36.2 kDa (RepB'), and 60.6 kDa (RepC). These results indicate multimeric structures for RepA (hexamer) and RepC (dimer), whereas RepB' appears to be a monomer.

The multimer structures of the Rep proteins as predicted above were verified by incubation of the purified proteins with the crosslinking reagents dimethyl suberimidate (DMSI) or glutardialdehyde (GA) and analysing the reaction products by SDS-polyacrylamide gel electrophoresis. The electrophoretic mobility of RepB' was not altered significantly after incubation with either of the reagents. In Figure 5 the DMSI products of RepB' are shown. The product obtained after crosslinking of RepC with DMSI had a mobility corresponding to about twice the molecular mass as the unmodified monomer; greater products were not visible (Figure 5). The results obtained with GA were essentially the same, but the protein bands in the gel were more diffuse (not shown). The reaction of RepA with GA resulted in the appearance of six bands in the Coomassie blue-stained gel at positions consistent with molecular masses equal to integral multiplies of the RepA monomer mass (Figure 6). The main product of the reaction of RepA with DMSI had a size corresponding to a dimer; higher species were present only in small amounts even after prolonged incubation periods (not shown). This incomplete reaction of RepA with DMSI can be brought about by the alkaline reaction condition which can cause the dissociation of the multimeric form (9). These results confirm the multimer structures of the Rep proteins as predicted from their native molecular masses. Since low protein concentrations (50 µg/ml) were used in the crosslinking reactions and RepB' remained unaltered under the applied conditions, we can rule out a major contribution of unspecific intermolecular reactions to the observed results.

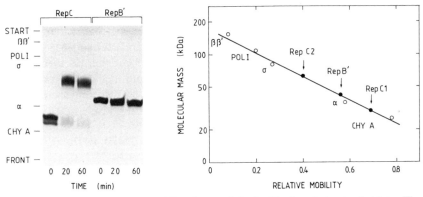

Figure 5. Analysis of RepB' and RepC crosslinked with dimethyl suberimidate. The reactions were carried out for various time periods as indicated and the reaction products analysed by SDS-polyacrylamide gel electrophoresis. Reference proteins are: the subunits of RNA polymerase β, β', σ, and α; DNA polymerase I (Pol I), and chymotrypsinogen A (CHYA).

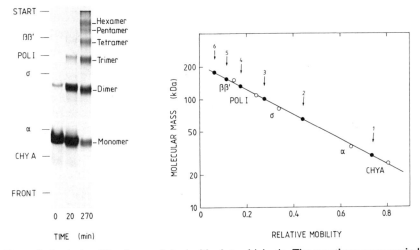

Figure 6. Analysis of RepA crosslinked with glutaraldehyde. The reactions were carried out for various time periods as indicated and the reaction products analysed by SDS-polyacrylamide gel electrophoresis. Reference proteins are as described in Figure 5.

Testing the proteins for functional sulfhydryl groups, DTE-free protein preparations were incubated with the SH-reagent N-ethylmaleimide (NEM). The reactions were stopped by addition of DTE, and the remaining activity of the proteins was tested in the *in vitro* replication system. Of the three Rep proteins only RepC was inactivated by the treatment with NEM.

The Rep proteins exhibited an extraordinary thermal stability. After 20 min at 60°C no significant decrease in their replication activity could be observed. At 80°C RepA was inactivated by 50% in less than 1 min, RepC in 3.5 min; 90% inactivation was achieved after 2 and 10 min, respectively. RepB' remained fully active for over 30 min at 80°C. The observed order of thermal inactivation may be explained by the multimer structures of the Rep proteins: with increasing complexity of the protein structure the thermal stability decreases.

III. *IN VITRO* REPLICATION OF RSF1010 PLASMID DNA

A. The *in vitro* replication system

An *in vitro* system based on a partially purified enzyme fraction prepared from plasmid free *E. coli* cells supported the replication of plasmid ColE1 but not of plasmid RSF1010 (14,36). However, upon supplementing the system with the three purified Rep proteins, extensive DNA synthesis was obtained with RSF1010 DNA as template (Figure 7). This reaction was absolutely dependent on the presence of each of the three Rep proteins (Table 2), which is consistent with the results of earlier studies using partially purified Rep protein fractions (36). As shown in Figure 7, DNA synthesis started after a short lag and continued for approximately 60 min. The uptake of radiolabelled precursor showed that DNA synthesis amounted to as much as 80% of the added RSF1010 DNA. Activity was maximal at Rep protein concentrations of 3–8 µg/ml RepA, 0.25–4 µg/ml RepB', and 3–8 µg/ml RepC; half maximal activity was achieved with 0.5 µg/ml RepA, 0.1 µg/ml RepB', and 0.75 µg/ml RepC which corresponds to Rep protein:DNA ratios of 1.2 RepA hexamers, 1.2 RepB' monomers, and 5.3 RepC dimers per RSF1010 molecule.

The requirements for the *in vitro* replication of RSF1010 DNA are summarized in Table II. In addition to a dependence on each of the three plasmid-encoded Rep proteins, replication was found to be strictly dependent upon the addition of closed circular DNA; linear RSF1010 DNA was inactive as template. Furthermore, the reaction was absolutely dependent on the presence of Mg^{2+} ions, ATP, deoxyribonucleotide triphosphates, and the *E. coli* protein fraction.

The products of the *in vitro* reaction after 60 min of incubation were analysed by gel electrophoresis (Figure 8) and velocity sedimentation in alkaline CsCl density gradients (not shown). In both types of experiments about 90% of the labelled product DNA banded in positions as were found for the supercoiled template DNA. Only a minor portion of the incorporated label was found in other positions and presumably represents replicative intermediates. This observation indicates that our *in vitro* system has the

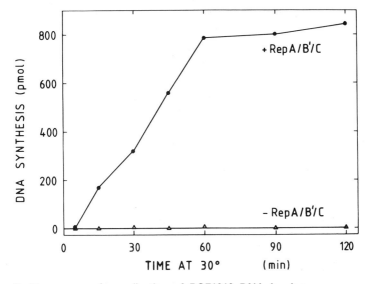

Figure 7. Time course for replication of RSF1010 DNA *in vitro*.

TABLE II. *Requirements for RSF1010 replication* in vitro

Reaction mixture	DNA synthesis (pmol)
complete	318
− RepA	< 3
− RepB′	< 3
− RepC	< 3
− *E. coli* extract	< 3
− RSF1010 DNA	< 3
complete[1]	3
complete[2]	268
− ATP	< 3
− rCTP, rGTP, rUTP	221
− dATP, dCTP, dGTP	4
− creatinephosphate, creatinekinase	42
− PEG 6000	64
+ Novobiocin (1 μg/ml)	10
+ Rifampicin (20 μg/ml)	284

[1]*Hpa*I-linearized RSF1010 DNA was used as template.

[2]The RSF1010 template DNA had been treated with *E. coli* topoisomerase I prior to its addition to the reaction mixture.

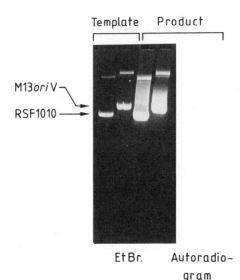

Figure 8. Analysis of the products of RSF1010 and M13*ori*V RFI replication *in vitro*. DNA syntheses were performed for 60 min at 30°C with [α–³²P]dCTP as radioactive label. The products were analysed in a neutral 1% agarose gel.

capacity to carry out at least one complete round of RSF1010 replication to produce product molecules which are indistinguishable by their topology from the template molecules.

In our *in vitro* system we tried to identify some of the host proteins participating in RSF1010 replication. This was done by testing the effects of antibiotics and antibodies directed against *E. coli* replication proteins. Furthermore, we assayed the activity of enzyme fractions prepared from temperature-sensitive *dna* mutants of *E. coli*. It was found that inhibition of DNA gyrase and DnaZ (gamma subunit of DNA polymerase III holoenzyme) reduced DNA synthesis to background levels. On the other hand, inactivation of the *E. coli* replication proteins DnaA, DnaB (helicase), DnaC, DnaG (primase), and DnaT (i protein) had little or no effect on RSF1010 replication. Furthermore, the activity of the RNA polymerase was also dispensable in the reaction (data not shown). These results are consistent with those previously described for the host protein requirements of RSF1010 replication *in vivo* (39) and *in vitro* (14,36).

The *in vitro* system has the capacity to replicate exogenously added RSF1010 and ColE1 DNA (Table III). But in contrast to observations made with RSF1010 DNA, the template activity of ColE1 DNA was inhibited in the presence of rifampicin. In addition, the RSF1010-encoded Rep proteins had no significant effect on ColE1 replication, neither in the presence nor in the absence of rifampicin. RFI DNA of bacteriophage M13mp9, which codes for its own initiator protein, was inert in the system. However, M13*ori*VΔ0, an *in vitro* recombinant consisting of a 2.5 kb *Pvu*II/*Pst*I fragment of plasmid pKT228 (RSF1010::Tn*3*; 1) cloned into vector M13mp9, was as effective as template as RSF1010. The inserted DNA comprises RSF1010 sequences between positions 1948 and 3982 in Figure 1 and includes both the origin of replication (*ori* V) and the origin of transfer (*ori* T) which have been mapped previously (11,12,34,39). Other DNA segments of RSF1010 showed no *ori* V activity in the *in vitro* system when inserted into M13mp9.

TABLE III. *Template Specificity of the* in vitro *Replication system*
The reactions were performed in the presence (column A) or in the absence (column B) of proteins RepA, B', and C.

Template	RSF1010 sequences present	DNA-Synthesis (pmol) A	B
RSF1010	1−8,684	332	<5
M13*ori*V RF I	1,948−3,982	498	<5
M13*rep*B RF I	3,571−5,473	<5	<5
M13*rep*AC RF I	5,472−7,773	<5	<5
M13mp9 RF I		<5	<5
Col E1		295	314
Col E1[1]		<5	≤5

[1]The reaction mixtures contained rifampicin.

B. Mapping of the *ori* V region

In an attempt to determine more precisely the sequences required to initiate RepA, B', and C dependent DNA synthesis we generated rightward and leftward deletions into the 2.5 kb *ori*VΔ0 fragment by *Bal*31 digestion. The resultant deletion fragments were cloned into M13mp8/9 vectors and the extents of the deletions were determined by dideoxy sequencing. The RFI DNAs of these chimeric phages were assayed for template activity in the *in vitro* system. Furthermore, the fragments were linked to a chloramphenicol resistance gene and tested for their ability to propagate as satellite plasmids in *E. coli* strain C600 harboring plasmid pMMB2, a ColD-based multicopy plasmid, that provides the Rep proteins in *trans* (36).

As summarized in Figure 9, we found that the sequences required for *ori* V function lie within a 395 bp segment between nucleotides 400 and 795 of the sequence shown in Figure 10. These *ori* V sequences are characterized by the presence of three perfectly conserved, 20 bp direct repeats and a large region (152 bp) with dyad symmetry which are separated by a 174 bp region containing a G+C-rich segment (28 bp) followed by an A+T-rich segment

(31 bp) and two small palindromic sequences. However, the origin function is not dependent on the integrity of this region. The orientation of the large symmetric region with respect to the *ori*-repeats can be changed without loss of activity. In addition, small insertions or deletions within the 174 bp spacer region did not detectably effect origin function *in vitro* (data not shown). These results are consistent with those obtained for the related plasmid R1162 (29,30,33).

Fragments Δ78 and Δ55, which have deletions in the large symmetric region (Figure 9), are still able to direct DNA synthesis *in vitro* but with only one fourth of the activity as found for the entire fragment Δ0 (data not shown). The *in vivo* assay with fragment Δ55 resulted only in few chloramphenicol resistant transformants. The analysis of the plasmid content of these transformants showed in addition to the helper plasmid pMMB2, the presence of a second plasmid with a size as expected for fragment Δ55 linked to the chloramphenicol resistant gene; this satellite plasmid was detectable only as very faint DNA band in the gel (Figure 11). We assume that sequences deleted in fragment Δ55 are essential for regular plasmid replication *in vivo*, but that the lack of these sequences in pPSΔ55 can be overcome under selective pressure to give rise to a satellite plasmid with an extremely low copy number. Such a by-pass mechanism is abolished by deletion of further sequences as in fragment Δ81.

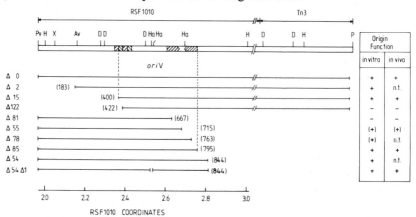

Figure 9. Mapping of *oriV*. Top: Structure of the 2495 bp *Pvu*II/*Pst*I-fragment (*oriV*Δ0) from pKT228 (RSF1010::Tn*3*). The four RSF1010 direct repeats (hatched boxes) and the region of 2-fold symmetry (hatched arrow) are indicated. Pv, *Pvu*II; H, *Hinf*I; X, *Xmn*I; Av, *Ava*I; D, *Dde*I; Ha, *Hae*II; P, *Pst*I. Bottom: Structure of various DNA-fragments used to define the essential sequences of the RSF1010 origin region. Numbers in parentheses indicate the nucleotide position at the deletion end points in the numbering system of Figure 10. Table: Summary of results obtained by testing the DNA sequences present in the various deletion mutants for template activity in the *in vitro* replication system and for ability to form satellite plasmids.

Figure 10. Nucleotide sequence of the *oriV*-containing 1043bp *Pvu*II/*Hin*fI fragment of RSF1010. Base pairs are numbered starting from the first nucleotide in the recognition sequence of *Pvu*II (position 1948 in the map of Fig.1). The box outlining the region from position 400 to 795 refers to the minimal DNA region with origin function as deduced from *in vitro* and *in vivo* replication studies. Direct repeats are boxed and palindromic sequences are indicated by arrows.

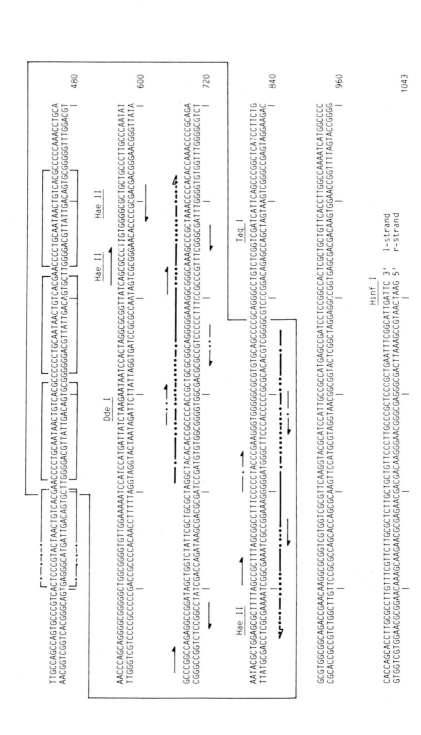

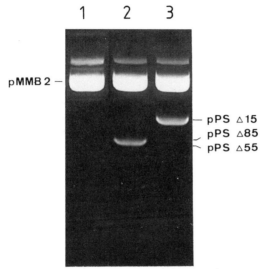

Figure 11. Agarose gel analysis of plasmid DNA extracted from *E. coli* C600 containing pMMB2 and one of the following satellite plasmids: 1, pPSΔ55; 2, pPSΔ85; 3, pPSΔ15.

IV. FUNCTIONAL PROPERTIES OF THE PLASMID-ENCODED REPLICATION PROTEINS

A. RepB acts as a DNA primase

Since the *in vitro* replication of RSF1010 was found to be independent of the host factors DnaG (primase) and RNA polymerase, it seemed plausible that RSF1010 codes for its own primase. To test this, the *in vitro* replication system was modified in that chimeric single-stranded phage DNA was used as template. Dependent on the vector (M13mp8 or 9), either orientation of the inserted fragment with respect to the M13 sequences can be obtained. This results in single-stranded, chimeric phage DNA containing either the l- (M13mp8) or r-strand (M13mp9) of the RSF1010 *ori V* region as labelled in Figure 10. To prevent RNA polymerase dependent initiation of M13 complementary strand synthesis rifampicin was added to the reaction mixture. In addition, the *E. coli* extract was prepared from strain PC22 (*dnaC*2, *polA*1) (19).

Of the templates tested, only those carrying the RSF1010 *ori V* sequences were active. The activity of single-stranded *ori V* DNA as template is no longer dependent on the RSF1010 proteins RepA and C, but it still depends on RepB' (Table IV). Furthermore, we found that in the *ori V* region each strand contains a sequence that can act as origin of the RepB'-dependent complementary strand synthesis. However, the templates containing the l-strand are more active than those with the complementary r-strand. This may reflect some variations in the primary or secondary structure of the RepB' recognition sites on the complementary strands.

TABLE IV. *REPB'-dependent replication of circular single-stranded* ori*V templates*

System	M13oriV(l)	M13oriV(r)	M13mp8
Complete mixture	110	41	<2
omit RepB'	<2	<2	<2
omit RepA	104	39	<2
omit RepC	113	44	<2
omit RepA+C	118	48	<2

The reactions were performed in the modified *in vitro* system (see text).

To determine the sequences that promote the RepB' dependent SS to RF DNA conversion, we used as templates in the modified *in vitro* system the single-stranded DNAs of the chimeric phages created to sequence and map the *ori V* region. As shown in Figure 12 these origins of complementary strand synthesis are located within the 152 bp palindromic sequence that forms part of the minimal *ori V* region. The origin of r-strand synthesis, which we propose to name *oriR*, was mapped to lie between nucleotides 749 and 796 in Figure 10 and the origin of l-strand synthesis, *oriL*, between nucleotides 649 and 716.

Recently we have found that the purified RepB' protein has the ability to synthesize a DNA primer *de novo*; this reaction does not require the presence of host factors or ribonucleotides. In addition, the three purified proteins RepB', *E. coli* single-strand binding protein, and T7 DNA polymerase are sufficient to convert single-stranded *ori V* DNA templates into double-stranded molecules in the presence of the four dNTPs (data not shown).

These results clearly show that the RepB' protein is directly responsible for the initiation of DNA synthesis on both strands of RSF1010 *ori V* templates, presumably by the synthesis of DNA primer. The ability of RepB' to initiate DNA synthesis in the absence of ribonucleotides can account for the observation that the presence of rCTP, rGTP, and rUTP in the *in vitro* system had only a stimulatory effect (Table II). The unusual specificity of the RepB' protein for IncQ *ori V* sequences explains why the homologous IncQ plasmid R300B was previously found not to code for a primase activity (27).

B. RepC is a site-specific DNA binding protein

Since several plasmids were found to code for a protein that initiates plasmid replication by binding to specific sequences at the respective origin of replication (reviewed in ref. 16), we tested the DNA-binding properties of the three purified RSF1010 Rep proteins. By use of a gel retardation assay (18) we found

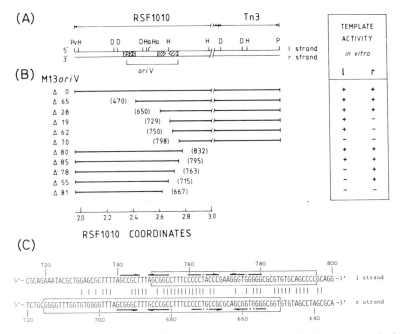

Figure 12. Mapping of the strand initiation determinants on the l- and r-strand of the RSF1010 *oriV* region. (A) Structure of the 2495bp *Pvu*II/*Pst*I fragment from pKT228 (RSF1010::Tn3). For symbols see Figure 9. (B) Structure of deletion derivatives cloned into M13mp8 and M13mp9 vector DNA. Numbers in parentheses indicate the nucleotide position at the deletion end points in the numbering system of Figure 10. The viral DNA of the hybrid phages were tested for template activity in *in vitro* reactions dependent on RepB' protein. The results are summarized in the table at the right. (C) Structural comparison of the l- and r-strands between position 715–800 and 635–720, respectively, in the sequence of Figure 10. The sequences identified as being essential for RepB'-dependent initiation of DNA synthesis on single-stranded DNA templates are boxed.

that RepC binds specifically to duplex DNA fragments carrying the direct repeats of the RSF1010 origin (Figure 13). The optimal protein:DNA ratio for this test was determined to be about 40 RepC dimers per plasmid molecule; at lower ratios significant amounts of the target fragments remain unbound, whereas at higher ratios a general band smearing occurs due to nonspecific binding of RepC to the other DNA fragments (data not shown).

The target fragments were not shifted in discrete steps to a defined position as reported for the lambda O protein (35), the R6K π protein (17) or the P1 RepA protein (7) but they appeared as a smear in the upper part of the gel. These results indicate a highly cooperative binding of RepC to the *ori* repeats resulting in structures larger than expected by a mere saturation of a definite number of binding sites.

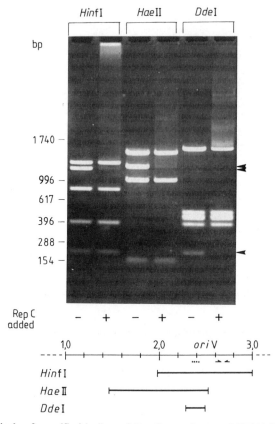

Figure 13. Analysis of specific binding of RepC protein to *ori*V DNA by agarose gel electrophoresis. Restriction fragments that are specifically converted to RepC-DNA complexes, manifested by their greatly reduced electrophoretic mobilities, are indicated by arrowheads. Their position on the RSF1010 map is shown in the bottom part of the figure. (....) 20bp direct repeats; (→ ←) 64bp inverted repeats.

To analyse the RepC-DNA complexes in more detail we prepared such complexes with the *ori*VΔO fragment and visualized them in the electron microscope. As shown in Figure 14, the observed structures can be divided into four groups: DNA molecules with apparently no bound protein
(a), DNA molecules with RepC bound in a dot-like structure
(b), loop-like structured DNA molecules with bound protein at the DNA junction
(c), and DNA clusters held together by RepC protein
(d). At a ratio of 30RepC dimers per DNA fragment most of the DNA molecules (60%) were found in the group of clusters; the other structures a, b, and c contained 30%, <1%, and 10% of the DNA molecules, respectively.

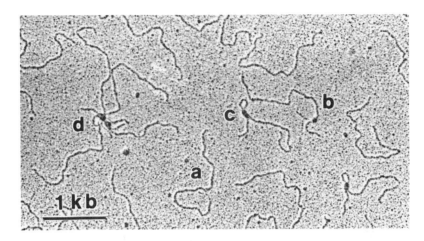

Figure 14. Electron micrograph of Rep C-DNA complexes. Purified Rep C protein was incubated for 15 min at 37°C with the 2.5 kb *oriV*Δ0 DNA fragment. The resulting Rep C-DNA complexes were fixed with glutaraldehyde and prepared for viewing in the electron microscope. Further details are explained in the text.

The electron micrograph depicted in Figure 14 demonstrate that Rep C possesses a strong tendency to link two DNA double-strands. This leads by an intra-DNA-molecular reaction to the loop structures and by an inter-DNA-molecular reaction to the observed clusters. The DNA concentration applied in the retardation assay probably favours the cluster formation resulting in the extreme and undefined shift of the target fragments. The coexistence of rotein-free DNA molecules and the clusters with large amounts of bound protein is a further indication that the binding of Rep C to DNA is highly cooperative.

In order to localize more precisely the sites of interaction of Rep C with the *ori V*Δ0 fragment, images of Rep C-DNA complexes were measured. The position of the dot-like Rep C complexes is about 19% of the distance from the left end of the DNA fragment, similar to the position found for the *ori*-repeats (data not shown). The position of one binding site in the loop-structured molecules coincides as well with the position of the *ori*-repeats (Figure 15); the second binding site is located about 500 bp apart in a region that has no apparent homology to the *ori*-repeats. The possibility of an artificial loop formation can largely be ruled out by the regularity of the observed loop sizes and the absence of such structures among RSF1010-DNA-RNA polymerase complexes prepared under identical conditions.

There are several indications that the interaction of Rep C with the secondary biding site is not specific. In the gel retardation assay only DNA

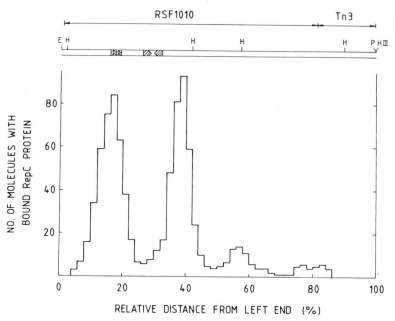

Figure 15. Analysis of Rep C binding to *oriV* DNA by electron microscopy. Rep C-DNA complexes were prepared as described in Figure 14. Loop structured (type a) molecules were randomly selected, photographed and measured at a magnification of 83, 200. The molecules were arranged so that the short free end of DNA is to the left. The structure of the *oriVΔ0* fragment is shown at the top. E, *Eco*RI; H, *Hin*fI; P, *Pst*I; HIII, *Hind*III; for other symbols see Figure 9.

fragments containing the *ori*—repeats were specifically converted to Rep C-DNA complexes with reduced electrophoretic mobility, but not those containing the secondary site. The dot-like structures were found predominantly at the position of the *ori*—repeats; the few others were randomly scattered along the entire DNA molecule. When complexes prepared with fragments lacking the second binding site were visualized in the electron microscope, loop structures with an apparently unaltered loop size were observed. The size of the loop and with that the second site of interaction seems to be dependent on the Mg^{2+} concentration. Moreover, using larger DNA fragments with the *ori*—repeats located more central, the observed loops have in common one binding site at a position as expected for the repeats, but the second sites are located either to the right or to the left of the primary site with about the same frequency (data not shown). With respect to these observations the loop-forming activity of Rep C differs significantly from that of the lambda cI repressor where loop formation occurs between homologous sequences (20).

C. RepA acts as a DNA helicase

In the course of the characterization of the purified proteins RepA was found to hydrolyse nucleoside triphosphates to nucleoside diphosphates and P_i; the best substrates are ATP and dATP followed by GTP and dGTP. This hydrolysing activity is stimulated by the presence of single-stranded DNA (Figure 16); no preference for RSF1010 *ori V* sequences was observed (Table V). To prove that the ATPase activity is not due to a contamination in the RepA preparation, a temperature-sensitive RepA mutant protein (RepA^ts) was purified as the wild-type protein. This mutant protein lost >90% of its ATPase activity within 1 min at 60°C whereas the wild-type protein remained active for over 10 min at the same temperature (data not shown).

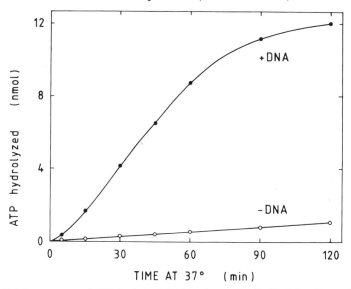

Figure 16. Time course of ATP hydrolysis by RepA protein. Purified RepA protein and [α−32P]ATP were incubated at 37°C either in the presence or absence of single-stranded M13*ori V*(I) DNA as indicated. After the times indicated the amount of [α−32P]ADP formed was measured.

The ss DNA-stimulated reaction started after a short lag and continued to hydrolyse >95% of the added ATP; in the absence of ss DNA no lag was detectable (Figure 16). This lag possibly reflects the time necessary to form an active RepA-DNA complex. The existence of such a complex could be demonstrated by gel filtration experiments in the presence of ATP. In the absence of ATP or with the non-hydrolysable ATP analog adenylyl-imidodiphosphate (AppNHp) no complex formation was observed (data not shown).

ATP dependence of the ATPase reaction does not follow the Michaelis-Menten equation. From kinetic data obtained in the absence of DNA a Hill-constant for ATP of 2 was determined (data not shown). We suspect that ATP itself can stimulate the ATPase activity of RepA by interaction with the DNA binding site of the protein: as a result, ATP could compete with DNA for the effector site. This idea is supported by the observation that with increasing ATP concentration the stimulatory effect of ssDNA on the ATPase activity decreases.

TABLE V. *ATPase activity of protein RepA in the presence of ssDNA*

	ATP hydrolysed DNA (nmol)
none	0.14
ssM13$oriV$(l)	3.81
ssM13$oriV$(r)	3.63
ssM13mp8	3.78
duplex RSF1010	0.18

The reaction mixture contained purified RepA protein, $[\alpha-^{32}]$ATP, and DNA as indicated. After 20 min at 37°C the amount of $[\alpha-^{32}]$ATP formed was measured.

Two facts prompted us to examine whether RepA also contains a DNA helicase activity. First, RSF1010 replication was found to be independent of the function of the *E. coli* replication protein DnaB which acts as helicase in the replication of the bacterial chromosome (28). Second, the two proteins RepA and DnaB are structurally related in that they both exist as hexamers and have sequence homologies with phage P22 gene 12 protein (Figure 17). However, a direct comparison of the DnaB and RepA sequences revealed only a weak homology.

As helicase substrate we hybridized a 789-nucleotide ^{32}P-labelled DNA fragment to an M13mp8 recombinant via a 591-nucleotide complementary sequence; the 3'- and 5'-termini of the annealed fragment remain single-stranded. As shown in Figure 18, RepA contains an activity that effects unwinding of extensive stretches of double-stranded DNA in a reaction requiring ATP hydrolysis. The activity was maximal at an ATP concentration of 0.75–1 mM; higher concentrations were inhibitory. This is consistent with the idea of competition of ATP and DNA for the DNA binding site of the protein.

V. CONCLUDING REMARKS

The overproduction and purification of the RSF1010-encoded proteins RepA, B' and C formed the basis for the *in vitro* studies of RSF1010 replication to yield some insight into the replication mechanism of RSF1010 and the func-

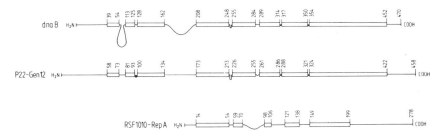

Figure 17. Amino acid homology of *E. coli* Dna B protein, bacteriophage P22 gene 12 protein, and RSF1010 RepA protein. The homologous regions of the three replication proteins are shown as boxes.

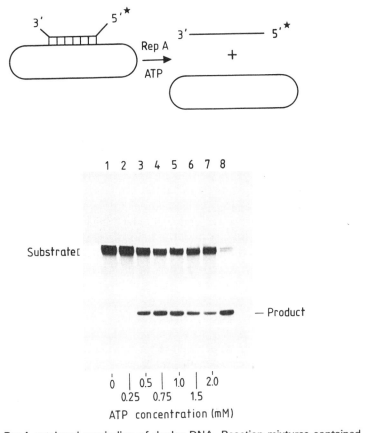

Figure 18. RepA-catalysed unwinding of duplex DNA. Reaction mixtures contained purified RepA protein, partially duplex DNA substrate, and various concentrations of ATP as indicated. After 30 min at 37°C, the product was analysed by electrophoresis through a 1% neutral agarose gel. Lane 8, 5 min, 100°C.

tional properties of the plasmid-encoded replication proteins.

By use of the purified Rep proteins and an ammonium sulfate fraction from an *E. coli* extract we have developed an *in vitro* system that efficiently replicates RSF1010 plasmid DNA. The properties of RSF1010 replication in this soluble enzyme system closely correspond to those characterized thus far *in vivo*. The system

(i) depends completely on exogenously supplied DNA,

(ii) specifically replicates supercoiled plasmid or phage RFI DNA that contains the unique RSF1010 origin region (*oriV*),

(iii) depends on each of the three RSF1010-encoded Rep proteins;

(iv) depends on host replication proteins such as DnaZ (gammasubunit of Pol III holoenzyme) but not on others [e. g. DnaA, DnaB (helicase), DnaC, DnaG (primase), DnaT];

(v) depends on DNA gyrase but does not require RNA polymerase,

(vi) initiates replication on both strands within the *oriV* region, and

(vii) continues replication to produce plasmid molecules identical in sequence and topology to the initial template.

From our present knowledge of the reactions catalysed by the purified Rep proteins we deduced a first model for RSF1010 replication that will now be presented and discussed.

The first step to initiate plasmid replication is the formation of the RepC-DNA complex. The primary recognition signals and binding sites for RepC are the three 20 bp, directly repeated sequences in the *oriV* region. The regular arrangement of these sites possibly mediates cooperativity of RepC binding. This causes by an unknown mechanism the bound protein to interact with a second site of the DNA, located about 500 bp apart. As a result, the DNA is folded into loop structures, containing the origins of complementary strand synthesis, *oriR* and *oriL*, within the looped segment.

In the second step of our model, the duplex DNA in the loop is unwound to enable RepB' to synthesize a DNA primer at *oriR* and *oriL*. A localized unwinding within the loop could be achieved by topological alterations of the superhelical template as a consequence of the loop formation. The RepA hexamer could bind to this stretch of single-stranded DNA and open the remaining duplex of the loop by its helicase activity. Although no interaction between RepA and RepC could be detected so far, the coordinate expression of these two proteins led us to speculate that the RepC-DNA complex directs the RepA helicase into the *oriV* region. Thus the formation of the loop structure could facilitate the entry of a helicase into both replication forks. Unwinding of the DNA in the *oriV* region by the action of gyrase is also conceivable. In this case, the loop structure could function additionally to stabilize the single-stranded segments in the *oriV* region.

The third step in the initiation process is the synthesis of primer by RepB'. The precise structure and starting positions of these primers are not known.

From our *in vitro* data we infer that RepB′ synthesizes nonsimultaneously on both strands, at *oriR* and *oriL*, a DNA primer. Subsequently, the primers are elongated by DNA polymerase III holoenzyme.

Since the DnaB helicase is dispensable for RSF1010 replication, we propose that RepA is required in the entire replication cycle and serves to unwind the DNA duplex in advance of the growing leading strand. A general priming system that mediates lagging strand synthesis seems not to be involved in RSF1010 replication. Such a system should still permit plasmid replication even in the absence of one of the complementary strand initiation sites. However, the sequences of fragment *oriVΔ55* that lacks *oriR* are not sufficient for replication of satellite plasmid pPSΔ55 *in vivo*; under selective pressure DNA synthesis on the displaced single-strand which lacks a RepB′ recognition sequence is possibly initiated by chance. These results suggest that RSF1010, unlike most other bacterial plasmids, replicates employing a strand displacement mechanism in which both daughter strands are synthesized unidirectionally from a single fixed strand origin,(i. e. *oriR* and *oriL*).

The picture emerging from our studies is that of a replicon encoding three essential replication genes that render plasmid replication independent of several vital replication functions of the host cell. We do not know of any other plasmid, even with a broad host range, that codes for a comparable set of proteins directly involved in plasmid replication. The strategy of IncQ plasmids to be maintained in many different species seems to be a high degree of independence from the host replication functions, rather than to recruit them for its own replication.

ACKNOWLEDGEMENTS

We thank H. Schuster for support and continuous encouragement. The expert assistance of Sabine Otto, Norbert Voll, and Marion Wassill is gratefully acknowledged.

REFERENCES

1. Bagdasarian M, Bagdasarian MM, Lurz R, Nordheim A, Frey J, Timmis KN (1982) Molecular and functyonal analysis of the broad-host-range plasmid RSF1010 and construction of vectors for gene cloning in Gram-negative bacteria. *In* Bacterial Drug Resistance (Mitsuhashi S, ed) Japan Scientific Society Press, Tokio, pp. 183–97
2. Bagdasarian M, Timmis KN (1982) Host:vector systems for gene cloning in *Pseudomonas. Curr Top Microbiol Immunol* **96** 47–67
3. Banker GA, Cotman CW (1972) Measurement of free electrophoretic mobility and retardation coefficient of protein-sodium dodecyl sulfate complexes by gel electrophoresis. *J Mol Biol* **247**, 5856–61

4. Barth PT, Grinter NJ (1974) Comparison of the deoxyribonucleic acid molecular weights and homologies of plasmids conferring linked resistance to streptomycin and sulfonamide. *J Bacteriol* **120**, 618–30

5. Barth PT, Tobin L, Sharpe GS (1981) Development of broad-host-range plasmid vectors. *In* Molecular Biology, Pathogenicity and Ecology of Bacterial Plasmids (Levy SB, Clowes RC, Koenig EL, eds) Plenum, New York, pp. 433–48

6. Buchanan-Wollaston V, Passiatore JE, Cannon F (1987) The *mob* and *oriT* mobilization functions of a bacterial plasmid promote its transfer to plants. *Nature* **328**, 172–5

7. Chattoraj DK, Snyder KM, Abeles AL (1984) Plasmid P1 replication: Negative control by repeated DNA sequences. *Proc Natl Acad Sci USA* **82**, 2588–92

8. Cohn EJ, Edsall JT (1965) Proteins, Amino Acids and Peptides as Ions and Dipolar Ions. Hafner Publ Comp, New York

9. Davies GE, Stark GR (1970) Use of dimethyl suberimidate, a crosslinking reagent, in studying the subunit structure of oligomeric proteins. *Proc Natl Acad Sci USA* **66**, 651–6

10. de Boer HA, Comstock LJ, Vasser M (1983) The *tac* promoter: A functional hybrid derived from the *trp* and *lac* promoters. *Proc Natl Acad Sci USA* **80**, 21–5

11. de Graaf J, Crossa JH, Heffron F, Falkow S (1978) Replication of the nonconjugative plasmid RSF1010 in *Escherichia coli* K12. *J Bacteriol* **134**, 1117–22

12. Derbyshire KM, Willetts NS (1987) Mobilization of the nonconjugative plasmid RSF1010: A genetic analysis of its origin of transfer. *Mol Gen Genet* **206**, 154–60

13. Derbyshire KM, Hatfull G, Willetts N (1987) Mobilization of the nonconjugative plasmid RSF1010: A genetic and DNA sequence analysis of the mobilization region. *Mol Gen Genet* **206**, 161–8

14. Diaz R, Staudenbauer WL (1982) Replication of the broad-host-range plasmid RSF1010 in cell-free extracts of *Escherichia coli* and *Pseudomonas aeruginosa*. *Nucl Acids Res* **10**, 4687–702

15. Fiers W, Contreras R, Duerinck F, Haegeman G, Iserentant D, Merregaert J, Min Jou W, Molemans F, Raeymaekers A, Van de Berghe A, Volckaert G, Ysebaert M (1976) Complete nucleotide sequence of bacteriophage MS2 RNA: primary and secondary structure of the replicase gene. *Nature* **260**, 500–7

16. Filutowicz M, McEachern M, Greener A, Mukhopadhayay P, Uhlenhopp E, Durland R, Helinski D (1985) Role of the π initiator protein and direct nucleotide sequence repeats in the regulation of plasmid R6K replication. *In* Plasmids in Bacteria (Helinski DR, Cohen SN, Clewell, Jackson DA, Hollaender A, eds) Plenum, New York, pp. 125–40

17. Filutowicz M, Uhlenhopp E, Helinski DR (1985) Binding of purified wild-type and mutant π initiation proteins to a replication origin region of plasmid R6K. *J Mol Biol* **187**, 225–39

18. Garner MM, Revzin A (1981) A gel electrophoresis method for quantifying the binding of proteins to specific DNA regions: Application to components of the *Escherichia coli* lactose operon regulatory system. *Nucl Acids Res* **9**, 3047–60

19. Gefter ML, Hirota Y, Kornberg T, Wechsler JA, Barnoux C (1971) Analysis of DNA polymerases II and III in mutants of *Escherichia coli* thermosensitive for DNA synthesis. *Proc Natl Acad Sci USA* **68**, 3150–3

20. Griffith J, Hochschild A, Ptashne M (1986) DNA loops induced by cooperative binding of lambda repressor. *Nature* **322**, 750–2

21. Grinter NJ, Barth PT (1976) Characterization of Sm Su plasmids by restriction endonuclease cleavage and compatibility testing. *J Bacteriol* **128**, 394–400

22. Guerry P, van Embden J, Falkow S (1974) Molecular nature of two non-conjugative plasmids carrying drug resistance genes. *J Bacteriol* **117**, 619–30

23. Haring V (1986) Die Replikationsproteine des Plasmids RSF1010: überproduktion, Reinigung und Charakterisierung der plasmidkodierten Proteine (Doctoral Thesis) Technische Universität Berlin

24. Haring V, Scholz P, Scherzinger E, Frey J, Derbyshire K, Hatfull G, Willetts NS, Bagdasarian M (1985) Protein Rep C is involved in copy number control of the broad-host-range plasmid RSF1010. *Proc Natl Acad Sci USA* **82**, 6090–4

25. Heffron F, Rubens C, Falkow S (1975) Translocation of a plasmid DNA sequence which mediates ampicillin resistance: Molecular nature and specificity of insertion. *Proc Natl Acad Sci USA* **72**, 3623–7

26. Kim K, Meyer RJ (1985) Copy number of the broad host-range plasmid R1162 is determined by the amounts of essential plasmid-encoded proteins. *J Mol Biol* **185**, 755–67

27. Lanka E, Barth PT (1981) Plasmid RP4 specifies a deoxyribonucleic acid primase involved in its conjugal transfer and maintenance. *J Bacteriol* **148**, 769–81

28. Le Bowitz JH, Mc Macken R (1986) The *Escherichia coli* dna B replication protein is a DNA helicase. *J Biol Chem* **261**, 4738–48

29. Lin L-S, Meyer RJ (1986) Directly repeated, 20 bp sequence of plasmid R1162 DNA is required for replication, expression of incompatibility, and copy-number control. *Plasmid* **15** 35–47

30. Lin L-S, Kim Y-J, Meyer RJ (1987) The 20 bp, direct repeated DNA sequence of broad-host-range plasmid R1162 exerts incompatibility *in vivo* and inhibits R1162 DNA replication *in vitro*. *Mol Gen Genet* **208**, 390–7

31. Linney E, Hayashi M (1974) Intragenetic regulation of the synthesis of πX174 gene A proteins. *Nature* **249**, 345–8

32. Meyer R, Hinds M, Brasch M (1982) Properties of R1162, a broad-host-range, high-copy-number plasmid. *J Bacteriol* **150**, 552–62
33. Meyer RJ, Lin L-S, Kim K, Brasch MA (1985) Broad host-range plasmid R1162: Replication, incompatibility, and copy-number control. *In* Plasmids in Bacteria (Helinski DR, Cohen SN, Clewell DB, Jackson DA, Hollaender A, eds) Plenum, New York, pp. 173–88
34. Nordheim A, Hashimoto-Gotoh T, Timmis KN (1980) Location of two relaxation nick sites in R6K and single sites in pSC101 and RSF1010 close to the origins of vegetative replication: Implication for conjugal transfer of plasmid deoxyribonucleic acid. *J Bacteriol* **144**, 323–32
35. Roberts JD, McMacken R (1983) The bacteriophage lambda O replication protein: Isolation and characterization of the amplified initiator. *Nucl Acids Res* **11**, 7435–52
36. Scherzinger E, Bagdasarian MM, Scholz P, Lurz R, Rückert B, Bagdasarian M (1984) Replication of the broad-host-range plasmid RSF1010: Requirement for three plasmid-encoded proteins. *Proc Natl Acad Sci USA* **81**, 654–8
37. Schmidhauser TJ, Ditta G, Helinski DR (1988) Broad-host-range plasmid cloning vectors for Gram-negative bacteria. *In* Vectors. A Survey of Molecular Cloning Vectors and Their Uses (Rodriguez RL, Denhard DT, eds) Butterworth, pp. 287–382
38. Scholz P (1985) Strukturelle und funktionelle Analyse der Replikationsdeterminanten von RSF1010—ein Plasmid mit extrem breitem Wirtsspektrum (Doctoral Thesis) Freie Universität Berlin
39. Scholz P, Haring V, Scherzinger E, Lurz R, Bagdasarian MM, Schuster H, Bagdasarian M (1985) Replication of the broad-host-range plasmid RSF1010. *In* Plasmids in Bacteria (Helinski DR, Cohen SN, Clewell DB, Jackson DA, Hollaender A, eds) Plenum, New York, pp. 243–59
41. Siegel LM, Monty KJ (1966) Determination of molecular weights and frictional ratios of proteins in impure systems by use of gel filtration and density gradient centrifugation. Application to crude preparations of sulfite and hydroxylamine reductases. *Biochim Biophys Acta* **112**, 346–62
42. Tsygankov YD, Chistoserdov AY (1986) Genome organization of the plasmids of IncQ/p4 group and vector derivatives. *Genetika USSR* **11**, 2606–19
43. Willetts NS, Crowther C (1981) Mobilization of the nonconjugative IncQ plasmid RSF1010. *Genet Res* **37**, 311–6
44. Wittmann-Liebold B, Ashman K (1985) On-line detection of amino acid derivatives released by automatic Edman degradation of polypeptides. *In* Modern Methods in Protein Chemistry (Tschesche H, ed) Walter de Gruyter, Berlin, pp.303–27

CHAPTER 5

MOLECULAR GENETICS OF IncW PLASMIDS

Carrie R. I. Valentine and Clarence I. Kado

I. HISTORICAL BACKGROUND

The first IncW plasmid, designated S-a, was identified in 1968 by Watanabe and co-workers in Tokyo, Japan, in an epidemic strain of *Shigella* (152). These authors demonstrated that the determinants conferring Su, Sm, and Cm resistance could be transferred unsegregated in *Escherichia coli* by transduction with phage Plkc or by conjugation, but were segregated with respect to resistance markers and conjugal transferability after transduction in *Salmonella typhimurium* by phage P22 (which carried a small amount of DNA). In 1971, Hedges and Datta reported (61) that S-a conferred resistance also to Km and that this resistance factor, along with factors RA3 and RA4, constituted a new incompatibility group, which they proposed naming W after Watanabe. Factors RA3 and RA4 were isolated by Watanabe's group from *Aeromonas liquefaciens* (5) and conferred resistance to Su, Cm, and Sm. Hedges and Datta (61) showed that surface exclusion, or superinfection exclusion, was not a characteristic of IncW factors: i. e., a cell containing a W factor did not prevent another W factor from entering, but was rather 'dislodged' under selective pressure. Resistance factors at this time were categorized as being fi^+ (inhibits fertility of F factor) or fi^-, which has since been recognized as primarily a

Promiscuous Plasmids of Gram-Negative Bacteria   © 1989 Academic Press Limited
ISBN 0-12-688480-3   All rights of reproduction in any form reserved

characteristic of IncF plasmids (30); W factors were initially considered unique as fi^- factors that conferred Cm^r. [Early reports of fi^+ IncW plasmids (46,74) are thought to reflect the presence of other plasmids (32)]. Subsequently, many IncW factors were identified including R388 from *E. coli* in London, England, (61,34) and R_7K from *Proteus rettgeri* in Athens, Greece (27,78). Early in the literature describing these plasmids, the designation for S-a was shortened by many authors to Sa, which is the designation that we use. However, even up to current literature, some authors continue to use the original term, S-a. Similarly, R_7K has been written R7K.

Only three IncW plasmids, pSa, pR388,and pR7K, have been characterized physically by restriction enzyme mapping, heteroduplex mapping, or transposon mutagenesis. Of these, pSa and pR388 have been the most thoroughly studied (see below). Other IncW plasmids include a series of conjugative R-factors that conferred resistance to high levels of Tp in *E. coli* and *Klebsiella* sp. isolated from clinical laboratories in London hospitals (34). The transferable resistance factors conferred varying combinations of Su^r, Tc^r, Sm^r, Cm^r, and Ap^r in addition to Tp^r. The numbers assigned to them included R403–413 [IncW plasmid pR404 is not the same isolate as the plasmid aggregate designated R404 from *Salmonella enteritidis* (20,81), but isolates containing IncW pR404 contain two plasmids, 35–40 kb and 80 kb (156)] and R419–424, R388, and R389 [later designated 3879a (58 Md) to distinguish from cryptic plasmid, 3879b (104)]. Plasmid R409, which originally was reported to confer Tc^r as well as Su^r and Tp^r , has been shown to be identical to pR388 (66), which lacks Tc^r. Early in the subculture of this isolate (R409, Plasmid Reference Center, Stanford University), Tc^r was lost.

Another series of IncW plasmids, pHH720, pHH1191, pHH1303, pHH1307, pHH1188, pHH1302, and pHH1306 that conferred resistance to Tp as well as combinations of Gm, Su, Ap, Km and Hg, was isolated from *E. coli*, *Enterobacter*, and *Klebsiella* at Hammersmith Hospital in 1980 in London, England (31). Plasmid pHH720 was considered to be identical to pR388. All except plasmids pHH1303 and pHH1307 from *Enterobacter* and pHH1302 and pHH1306 from *E. coli* were demonstrated to be conjugative. Since the molecular mass of pHH1302 (Gm^r, Tp^r, Su^r, Ap^r, Hg^r) was 27 Md, (31,104) and therefore greater than pR388 (Tp^r and Su^r), which was 21 Md, evidence for transposition of Gm^r, Ap^r and Hg^r was sought (31). The nonconjugative character of 64 Md pHH1307, (104) and pHH1302, which both conferred Gm^r, allowed for the identification of Tn*733* and Tn*734* (presumably the same transposon based on *Eco*RI digests) by introducing another conjugative IncW plasmid (R389a) into *recA* strains of *E. coli* and selecting for transfer of Gm^r to a recipient strain (104,35). The size of this transposon was identified as 8.7 kb and produced a Gm acetyltransferase, AAC3.

Two other IncW plasmids, pOH30221 of 38 Md and pOH3017, were isolated in Obihiro, Japan, and carried determinants for the utilization of *cis*-aconitate (citrate utilization) and tricarballylate (124). pOH30221 was nonconjugative. The prototype IncW plasmid, pSa is cit-negative. Citrate utilization appeared in *E. coli* derived from animals and was traceable to these IncW and other IncHI plasmids. A further IncW plasmid, R27, was isolated from a clinical isolate of *E. coli* in London (112,35); this is apparently different from the pR27 isolated from *Salmonella typhimurium* (127,82), which has become the prototype for IncHI (143). Several plasmids, pKMR207-1, pKMR208-1, pKMR209, pKMR210, and pKMR212, that have been isolated in Krasnodar, USSR, (37) are incompatible with the plasmid pR388, although it was incorrectly considered to be an IncFI plasmid if they used the same pR388 that has been widely studied. IncW plasmids have also been identified in *Myxococcus virescens* by phage specificity (96).

IncW plasmids include the smallest (20–25 Md) conjugative plasmids (150,137) and yet little has been done to characterize their transfer functions until recently (39). They have been exploited as cloning vectors and used widely to identify mobile genetic elements because of their efficient conjugal transfer. The following sections detail much of what is known of their molecular biology.

II. PHYSICAL AND GENETIC CHARACTERIZATION OF THE IncW PLASMIDS.

IncW plasmids range in molecular mass from 21 Md to 64 Md (104). Those plasmids whose size has been determined are shown in Table I. The three smallest of these, pSa, pR388, and pR7K have a GC content of 62% (42) and have also been characterized by restriction enzyme mapping. The genetic and physical maps of these three are described in detail and shown in Figs. 1–3.

A. Physical and Genetic Map of pSa

Figure 1 shows a restriction enzyme map of pSa, with genetic functions so far identified based on information from many authors (145,72,136,149, 150,55,156). The size of pSa earlier reported as 29.6 kb (136) omitted of the 11 kb *Sst*II fragment (lower left sector) from the restriction enzyme map (72). Several subsequent reports contain this error (137,139,140), but this does not affect most considerations since few restriction sites appear in this fragment (1 *Bst*EII and 2 *Pvu*I sites). Therefore, some ambiguity does appear in the earlier literature about the size of cloning vector pSa322 (139,140).

1. Clustering of Restriction Enzyme Sites around Resistance Genes The restriction enzyme sites are clustered in one sector of the pSa map that contains the four antibiotic resistance determinants, Cm^r, Su^r Sm/Sp^r and Km^r/Gm^r/Tb^r, a phenomenon also observed for other plasmids. Possible explanations have been discussed by Ward and Grinsted (150). Two major hypotheses have been

TABLE I. *Physical Size of IncW Plasmids.*

Plasmid	Size (Md)	(kb)	Determined by	Ref.
pSa	23–25		Alkaline sucrose density	42
			gradient centrifugation	113
		39	Restriction fragment lengths	72
pR388	21		Agarose gel electrophoresis	104
			Alkaline sucrose density	
			centrifugation	42
		33	Restriction fragment lengths	145
pR7K	20		Alkaline sucrose density	
			centrifugation	42
	22		Heteroduplex mapping	55
		35	Restriction fragment lengths	150
pHH1302	27		Agarose gel electrophoresis	35
pR389a	58		Agarose gel electrophoresis	104
pHH1307	64		Agarose gel electrophoresis	104

offered: 1) transfer between bacteria of different genera selects for the loss of restriction sites, so the most recently acquired sequences (antibiotic resistance genes) have had less time to lose their sites than the rest of the plasmid (144), and 2) restriction enzymes may be involved in *in vivo* site-specific recombination. This suggests that the resistance genes were acquired because of the convenience of nearby restriction sites. It has been demonstrated (21) that plasmid and eukaryotic DNA will reassort at *Eco*RI sites *in vivo* in cells containing *Eco*RI. Probably each of these hypotheses account for some of the clustering observed.

2. Location of the Streptomycin and Kanamycin Resistance Genes From DNA sequence data (141) an open reading frame has been identified for the Sm[r] gene and the carboxy terminal of the Km[r] gene, which together compose an operon. Further evidence and other information are contained under Section IV.C.D. The sequence of this open reading frame has been compared (141) to the Sm/Sp adenyltransferase gene (*aad*A) of plasmid R538-1 (69), which is almost identical to the sequence of the corresponding gene of Tn7 (48). The R538-1 gene has 263 (assuming N-terminus previously assigned, ref.48) amino acids compared to 262 for Tn7; two other base changes do not change the amino acid sequence. The streptomycin resistance gene of pSa has 88% nucleotide homology with R538-1 and contains six additional amino acids at the N-terminus before homology begins. The alignment of the amino acid sequences of pSa and pR538-1 is exact for the 263 amino acids of R538-1, with dispersed base substitutions (81% identity 212 amino acids), most of which

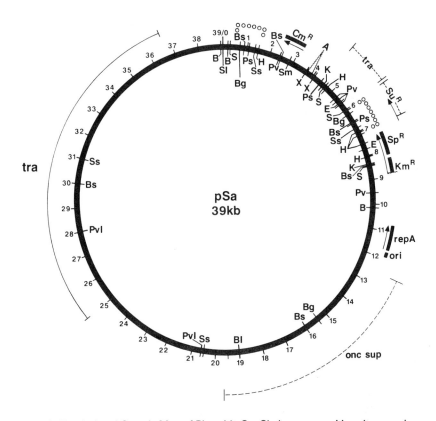

Figure 1. Physical and Genetic Map of Plasmid pSa. Six-base recognition sites are shown with the O map position taken at the unique *Sal*I site: A-*Aha*III; B1-*Bal*I; B-*Bam*HI; Bg-*Bst*II; Bs-*Bst*EII; E-*Eco*RI; H-*Hind*III; K-*Kpn*I; Ps-*Pst*I; Pv-*Pvu*II; Pvl-*Pvu*I; S1-*Sal*I; Sm-*Sma*I; S-*Sst*I (*Sac*I); Ss-*Sst*II (*Sac*II); X-*Xho*I. Open circles indicate direct repeats. Solid bars indicate locations of genetic traits and arrows, the direction of transcription; speckled boxes are not precisely defined: Cm^r, chloramphenicol resistance; Su^r, sulfanilamide resistance; Sp^R, spectinomycin resistance; Km^R, kanamycin resistance; tra, conjugative transfer; repA, replication protein A; ori, origin of replication; onc sup, oncogenic suppression. *Pvu*I sites are known only for the *Bg*1II-*Sal*I fragment 15.5 kb to 39 kb. *Bg*1I and *Hind*II sites have also been mapped on the 9.3 *Bg*1II kb fragment (150), but the accuracy of the mapping is not sufficient to place these sites relative to all the other sites now known. Two *Aha*III sites are near the *Xho*I sites at 4.0, but are not mapped relative to these sites. Two *Ava*II sites are confirmed from the sequence data of the kanamycin gene (139); the enzymes *Hin*fI, *Dde*I, and *Hae*II have been used to map the origin of replication (137). Since there must be other unknown sites for these four-base recognition enzymes on pSa, we have not included them on this map. See text for basis of genetic assignments. The map positions in kb of the known restriction enzyme sites are mostly from ref.143; *Pvu*I, *Bst*EII, and *Aha*III are from ref.154; *Bst*EII sites are confirmed from sequence data (Murray & Shaw, pers. comm.; 884).

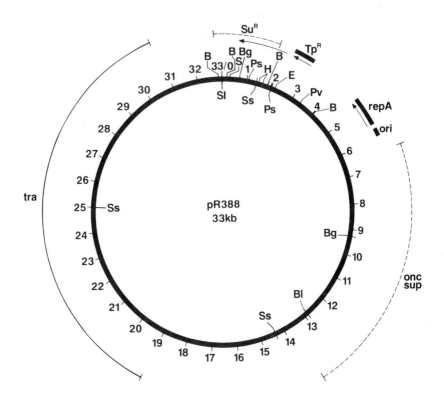

Figure 2. Physical and Genetic Map of Plasmid pR388. Enzyme sites and symbols are as described in Figure 1. The O map position taken at the *Sal*I site. The map is largely identical to pSa except for the lack of DNA between the two direct repeats of pSa yielding only one copy of this region in pR388 that confers sulfanilamide resistance, and for the substitution of trimethoprim resistance where streptomycin and kanamycin resistance appear in pSa. See text for assignment of genetic functions. Presumably the *Bst*EII sites of pSa not deleted in pR388 are present also, but have not been reported.

occur near the N-terminus. The open reading frame for the enzyme from pCN1 (identical to Tn7) has been confirmed by sequencing the first ten amino acids from the purified protein (23). Therefore, we consider that the assigned reading frame of pSa for Sm/Sp resistance is confirmed. An open reading frame for the C-terminus of the Km^r gene has been identified at the 5' end of the Sm^r gene and shows limited homology to a known aminoglycoside phosphorylating gene (141). Although good evidence exists for this assignment, we do not yet consider that it has been confirmed.

3. Location of the Chloramphenicol Resistance Genes The chloramphenicol resistance determinant has been cloned as a 1 kb fragment (36,−pHH1) and

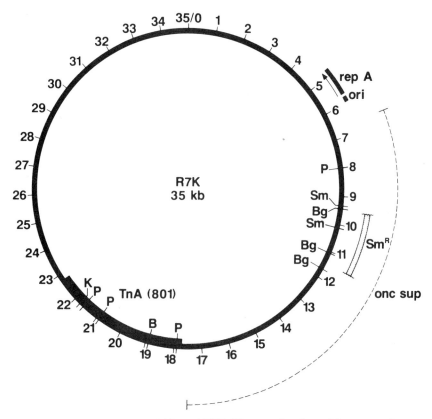

Figure 3. Physical and Functional Map of R7K. The map is oriented for comparison to the map of pR388; the region 0 to 8 kb corresponds to the R388 map. The first *Pst*I site is set at 8 kb; set text for basis of alignment. Restriction enzyme sites as for Figure 1. The double region corresponds to the position of TnA(801), which confers ampicillin resistance (Ap^R). The shaded region is the approximate location of the Sm^r gene. The dashed line represents the maximum region that hybridizes to the virulence suppression region of pSa. See text for basis of assignments.

has recently been sequenced (Murray and Shaw, pers. comm.). The structural gene is centred around the *Pvu*II and *Sma*I sites: the N-terminus is 180 nucleotides to right of *Sma*I, and the C-terminus 30 nucleotides to left of *Pvu*II. *4. Location of the Sulfanilamide Resistance Gene* The sulfanilamide resistance gene corresponds to a 1 kb region that is duplicated in direct, homologous repeats (145,84). Only one of these repeats confers sulfanilamide resistance and both are being sequenced by CIRV in collaboration with Bruce A Roe (University of Oklahoma, Norman). Several authors have observed that the deletion of sequences to the left of the *Bgl*II site inactivates Su^r (150,136,83,145). Recent work on pR388 indicates that a promoter for its Su^r

structural gene exists between the Bam HI and Eco RI sites near map position 2 kb, with the structural gene to the left of the Hind III sites (132). Since the promoter region diverges between these two plasmids it is likely that the Su^r gene is part the Km^r/Sp^r operon.

5. *Direct Homologous Repeats* The presence of extended homologous repeats (1 kb) explains the observation of many authors that Cm^r is spontaneously lost (30,109,61,55,43,136,72). The loss of Cm^r is associated with a 5.4 kb deletion corresponding to the region between the two repeats, resulting in one repeat (145). The deletion is frequent, at 1% after 8 logs of growth, and is dependent on the *recA* function of *E. coli* (72). Thus it is assumed that the deletion is the result of a single crossover event between the two homologous repeats. Sequencing data almost complete (CRIV) indicates that the repeats are probably identical. The left copy probably lacks a promoter. The function of the Su^s left repeat is not known. It may have been acquired by pSa from recombination with the Su^r gene of a plasmid conferring both Cm^r and Su^r. The region between the two repeats (perhaps it can replicate in species other than *E. coli*) may be the remnants of such a plasmid. Selection for Cm^r will cause retention of the second copy, but there would have been no selective pressure to maintain Su^r because of the right copy. Although this sequence of events is easily imagined, most duplications in bacteria are tandem repeats (129). Separated repeats do include insertion elements (105) and mutants of the *arg* and *trp* operons (8,73). No evidence has been obtained for the transposition of the Cm^r determinant, as would be expected if these repeats were insertion elements. Normally, duplications are removed in bacteria by homologous recombination, as does happen in this case without selective pressure for retention of the Cm^r gene.

6. *Origin of Replication* The origin of replication, which confers the wide-host-range of this plasmid, has been cloned by rescue of a chloramphenicol acetyltransferase cassette (137,138). It consists of an initiation origin and a *rep*A protein that cover 1.2 kb and has been mapped relative to the Bam HI site at 10 kb. The details of this function are described in section III, Origin of DNA Replication.

7. *Conjugal Transfer* The conjugal transfer locus of pSa has been defined by transposon mutagenesis in recent work (39). The Tra region consists of three closely spaced blocks covering aproximately 11 kb. It contains ten complementation groups, six of which contribute to pilus synthesis. The origin of conjugal transfer is at the end of the region, at 25 kb. These results confirm earlier reports of a Tn*1* insertion (63) that affected pilus formation (15). Previously, a section near 5 kb on the map was identified as reducing conjugal transfer efficiency (136); however, Dymock and Warner (39) did not detect quantitative differences in conjugation efficiency with a *Kpn*I deletion including the same region. If there is a function at this second locus which affects transfer, it is not required for conjugation.

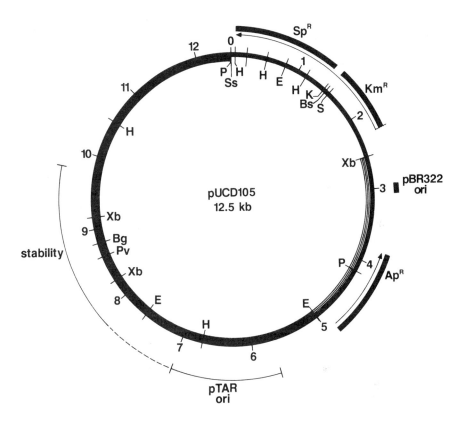

Figure 4. Physical and genetic map of pUCD105. Single line shows sequences derived from pSa; double thickness line indicates sequences from pTAR and the solid and striped line indicates sequences from pBR322. Restriction emnzyme site abreviations as for Figure 1 with the addition of Xb - *Xba*I. The *Xba*I site near the pBR322 origin was created from a *Pvu*II site with linkers (156).

8. Suppression of Oncogenesis of Agrobacterium tumefaciens The locus effecting oncogenic suppression of the tumor-inducing (Ti) plasmid of *A. tumefaciens* (Section VII), has been located on a 7.6 kb partial *Sau*3A fragment that is positioned in the lower right sector (156). Initially, it was reported that the deletion derivative of pSa, pSa151 did not suppress oncogenesis of *A. tumefaciens*; thus it was concluded that the suppression genes were located somewhere on the left half of pSa (136). However, it was subsequently found that one of the cloning vectors derived from pSa734 (a pSa151–pBR322 recombinant) did not permit complementation of Ti virulence mutants when the wild-type gene was cloned in this vector (25,156). Other vectors derived from pSa [pSa4 and pUCDl; pUCD2; pUCD100 (156)] were able to provide the necessary virulence function to a nonfunctional mutant Ti plasmid. It was shown that under nonselective conditions pSa151, pSa734, pUCD2, and pSa4 are gradually lost from *Agrobacterium* strains (2%/division), although the parent plasmid, pSa is stable (25). In spite of this loss, they are retained sufficiently to demonstrate a difference in complementation ability *in planta*; however, pSa734 (the pSa151 derivative) gave variable results in complementation ability (156). To conclusively demonstrate that the 7.6 kb *Sau*3A fragment contained all the necessary oncogenic suppression gene(s), this fragment was inserted into a vector designed for stability in *Agrobacterium*, pUCD105 (Figure 4), and complete suppression of virulence was achieved (156). Furthermore, the 23 kb *Sal*I–*Bgl*II fragment of pSa inserted in pBR325 (pSa325, Figure 5) with a 9 kb insert containing a replication origin from an *Agrobacterium* plasmid (pTAR) for stable maintenance (pSa325.1, pSa325.2—two orientations of pTAR origin) did not attenuate virulence to any extent. The *Bgl*II site at map position 15 apparently interrupts the region conferring oncogenic suppression since DNA from either side does not affect virulence, but the whole region does. This result shows that the previous assignment of oncogenic suppression (136) to the upper left sector was not correct. Recent work has narrowed the range assigned to oncogenic suppression.

B. Physical and Genetic Map of pR388

1. Antibiotic Resistance Determinants (Tp^r, Su^r) Plasmid pR388 has been shown to be homologous to pSa by heteroduplex mapping (55); it also has a very similar restriction enzyme map (150,145). However, pR388 lacks the Cm^r region between the two direct repeats of pSa (retaining one copy of these repeats as well as Su^r) while Tp^r is substituted for the Sm^r and Km^r genes of pSa (Figure 2). The region encoding dihydrofolate reductase, which confers Tp^r has been sequenced and located precisely on the map (135,158). A region 300 bp upstream from the structural gene sequence is required in *cis* for expression of Tp^R and contains typical promoter sequences suggesting that the functional promoter for this gene is somewhat distant from the structural gene (135).

The position of the Su^r gene has been discussed in conjunction with pSa; it has been identified by transposon mutagenesis, but has not yet been sequenced and compared to the pSa sequence. Depending on the origin of R388, the sequence here may be a composite of the two repeats of pSa. By analogy with the data from pSa, it is assumed that sequences to the left of the *Bgl*II site are required for Su^r. The identity of restriction sites compared to pSa resumes close to the N-terminal of the Tp^r gene and several hundred bases to its C-terminal.
2. Origin of Replication, Conjugal Transfer and Oncogenic Suppression The

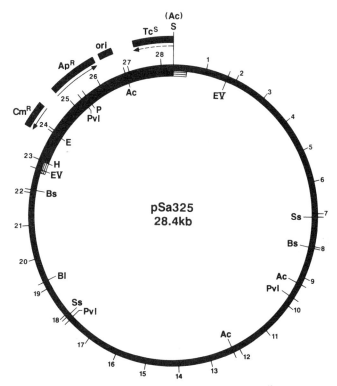

Figure 5. Physical and genetic map of pSa325 (D. Zaitlin, unpublished). The origin is taken at the unique *Sal* I site. The single thickness line represents sequences form pSa, the double thick line indicates sequences from pBR325, and the solid line and striped line sequences from pBR327. Abbreviations as in Figure 1 with the addition of Ac for *Acc* I and Ev for *Eco* RV.

replication origin has been assigned based on analogy to pSa, although the pR388 origin has been shown to reside between the nearby *Bam* HI and *Bgl* II sites (149). Two Tn A (Tn*801*) insertions have been shown to reduce the efficiency of conjugal transfer 100-fold and are indicated at those positions by 'tra' (149). The accuracy of mapping these positions is not precise, since the location of the *Sst* II site in this region was not known at the time. However, it would appear that the Tn*1* insertion that inactivated W-pilus formation of pSa is between these two locations. On this assumption, the site of pilus formation of pSa is place on the pR388 map by analogy. The outer limits of the transfer region have also been placed on the map by analogy to pSa. A derivative of pR388 containing a Tn*10* insertion, pUB5554 (150), has been shown to cause oncogenic suppression of *A. tumefaciens*, just as pSa does (156). By analogy with the restriction map of pSa, the oncogenic suppression region is indicated on the map of pR388.

C. Physical and Genetic Map of pR7K

1. Restriction Enzyme Map Relative to pR388 Although R7K has been shown by heteroduplex mapping to have extensive homology to pR388 and pSa (55), none of the restriction enzyme sites that cut both pR7K and either of these other plasmids clearly correspond. Consequently, there is an ambiguity in the alignment of the map of pR7K relative to these two other maps, resolved here by accepting the assignment of Gorai *et al.* (55) of the larger heteroduplex loop to the naturally-occurring TnA (Tn*801, 802*) insertion of pR7K. This assignment was based on the size of this loop and that of TnA (5 kb, ref.64) and preliminary restriction enzyme data not reported. Subsequent restriction maps of pR7K have been reported (150), which indicated that the other heteroduplex loop probably corresponds to Sm^r. The restriction sites in this region that define an area of non-identity to pR388 total 3.7 kb, and therefore was considered the smaller heteroduplex loop; however, a short continuation on either side of nonhomologous sequence would make this loop as large or larger than the TnA loop. Based on the assumption that the TnA loop is the larger loop, the restriction enzyme map of Ward and Grinsted (150) and the heteroduplex mapping of Gorai *et al.* (55) have been used to prepare the functional and physical map of pR7K (Figure 3). The maps aligned by Ward and Grinsted (150) actually reverse the orientation of pR7K relative to pR388 compared to the map of Gorai *et al.* (55). This is not immediately obvious since the location of the Su^r determinant was not known for the earlier work. However, the earlier work is correct and the map of Ward and Grinsted can be corrected by reversing the order of Su^r and Tp^r on pR388 and extending the *rep* region to the left instead of the right. This correction has been confirmed by hybridization data (156) in which the 7.6 kb fragment containing the gene for oncogenic suppression from pSa (see Figure 1) hybridized strongly to the 10 kb *Pst*I fragment of pR7K, which contains the region between the two loops and the Sm^r loop. Faint hybridization was found to the 22 kb *Pst*I fragment. This result is clearly consistent with the map of Gorai *et al.* (55), but not with that of Ward and Grinsted (150). Both of these groups assumed the larger loop was that of TnA; if this is not correct, then the origin of replication would be about 25 kb on the pR7K map, with the oncogenic suppression region extending past the TnA insertion toward the Sm^r region. The *Bam*HI-*Bgl*II fragment of pR388 that contains the origin of replication hybridizes exclusively to the 26 kb *Bam*HI-*Bgl*II fragment of pR7K; however, this does not distinguish between the two possible locations for the replication origin.

2. Antibiotic Resistance Genes (Ap^r, Sm^r) The assignment of TnA to pR7K is based on identity of restriction enzyme maps to the transposon Tn*801* and Tn*802* and the ampicillin resistance conferred by R7K (55,150). The Sm^r locus is defined by the observation that deletion of the two small *Bgl*II fragments inactivates Sm^r.

3. Replication Origin and Oncogenic Suppression Genes The replication origin and oncogenic suppression genes have been assigned based on the homology to pSa demonstrated by heteroduplex mapping and by hybridization of the 7.6 kb oncogenic suppression region described above. The replication origin was set by comparison with the map of pR388—no defined character exists at the origin. The first *Pst*I site was arbitrarily set at 8 kb, which is based on an estimate from the heteroduplex maps of the distance between the Sm^r loop of R7K and the Cm^r loop of pSa. Again, this restriction enzyme map reveals that sites are clustered at the regions of antibiotic resistance genes.

III. ORIGIN OF DNA REPLICATION

The W-plasmids represent the smallest, naturally-occurring group of transmissible plasmids that still are able to carry a large variety of antibiotic resistance determinants (55,150,156). The conserved region of the plasmids mainly comprise the transfer functions and a region involved in DNA replication. Genetical analysis have localized the replication origins within a 5.5 kb region for pSa and R388 (150) and a 4.0 kb region for pSa (136). The origin of replication was subsequently isolated and mapped within a 0.48 kb region of pSa (137). Although the molecular circuitry of initiation of replication of W-plasmids is not well understood, electron microscopic analysis of the isolated origin of pSa suggests that the replication may proceed in both directions, apparently in a bidirectional mode of replication (137). At least one plasmid-encoded diffusible gene product is required for efficient replication of pSa. This product is a 35 kDa polypeptide encoded by the *repA* gene situation next to the origin (137). A protein of 35 kDa, designated *pi* and encoded by the IncX plasmid R6K, is involved in the formation of nascent RNA during the DNA intiation process (70). Interestingly, R6K will partially complement *repA* mutants of pSa (138). Nucleotide sequence comparisons of the origins of pSa with that of R6K revealed a near homologous 13 bp sequence present twice in the origin of pSa and 14 times in the origin of R6K. The 13 bp sequence may serve as the binding sites for the initiator protein (138). Also a protein of 33 kDa encoded by a RepA1 gene of R100 (IncFII) is required to initiate plasmid DNA synthesis (117), and a protein of 29 kDa encoded by a putative gene associated with the origin of replication of the mini-F (IncFI) may have similar function (98). pSa replication is dependent on chromosomal genes *dnaB*, *dnaF* and *dnaE* (110). A partial requirement for *dnaA* also exists.

IV. ANTIBIOTIC RESISTANCE GENES

The physical locations of the antibiotic resistance genes of IncW plasmids are discussed under Section II.A.2–4. This section describes the characteristics of these genes and their products, if known.

A. Trimethoprim Resistance

The dihydrofolate reductase gene of pR388, the most thoroughly characterized of the antibiotic resistance genes of Inc W plasmids, is highly homologous to a similar gene from the plasmid pR67 (16). pR67 is an Inc 6 plasmid isolated from *Citrobacter* (112). Both genes code for 78 amino acids of which 61 are identical (135,16). The two enzymes are tetramers (Table II), a property unique among dihydrofolate reductases, since all other characterized dihydrofolate reductase enzymes from a variety of sources are either monomers of approximately 20 k Da (130,135 for literature summary) or, for bacteriophage T4 and Tn7, dimers (114).

The properties of purified dihydrofolate reductases from R-factors have been compared to the chromosomal enzyme (Table III) and have been categorized into two groups (Types III and IV have been identified more recently, ref 76a,153a), pR388 and pR67 enzymes falling into Type II (112). Type I enzymes, represented by pR483 (126) and Tn7 (47), exhibit I_{50} values several thousand-fold higher than the chromosomal enzyme and are expressed in several-fold higher activity levels in the cell. Type II enzymes are even more resistant, with I_{50} values several hundred-fold higher than Type I enzymes for trimethoprim, methotrexate, and aminopterin and have activity levels in cells equal to or less than the chromosomal enzyme. Their great sensitivity to methotrexate and aminopterin is unique to this class of dihydrofolate reductases (112,4).

No homology has been detected between the pR67 gene and the *E. coli* gene, which is considerably longer (159 amino acids to 78 amino acids) (130). Several authors have suggested that the origin of the plasmid enzyme may be bacteriophage, initially because the molecular weights of the whole enzymes are closer (112,4, 130). However, the bacteriophage T4 enzyme has been sequenced and been shown to be homologous to the *E. coli* chromosomal enzyme (114); furthermore, the T4 enzyme has a subunit size of 23 000 and forms a dimer (114). Also, antibody to type II enzymes does not cross react with type I or T4 enzyme (47). The 9 kDa subunit of the reductase from Tn402 was antigenically related to the pR67 subunit, but no evidence has been found for transposition of the pR388 enzyme (35). It has been suggested that the type II enzymes are derived from a mutant oxidoreductase (47), which is consistent with the lower turnover number of the enzyme (158).

Some authors have noted that the pR388 enzyme seemed to be larger than that of the pR67 protein on gels (see sizes, Table II (158,47) and suggested that read-through at the carboxy-terminus might add 1.6 k Da (158). Deliberate addition of nucleotide sequences that replaced the last two amino acids of pR67 (which are different from the pR388 sequence) with an additional sequence

TABLE II. *Molecular Weight of Dihydrofolate Reductases From pR388 and pR67.*

Plasmid	Form of Protein	Size, (kDa)	Method	Ref
pR388 *Inc*W	Whole enzyme	35	gel filtration	3
		36	gel filtration	158
	subunit	8.4	acrylamide gel	158
		10.5	acrylamide gel	47
	in vitro product	10	acrylamide gel	135
	subunit	8.266	DNA sequence	135
				158
	subunit	8.4	acrylamide gel	158
		10.5	acrylamide gel	47
	in vitro product	10	acrylamide gel	135
	subunit	8.266	DNA sequence	135
				158
		8.020	amino acid composition	158
pR67 *Inc*6	Whole enzyme	36	gel filtration and acrylamide gel	128
	subunit	8.5	acrylamide gel	128
		9	antibody to reductase inacrylamide gels	47
		8.444	DNA sequence	16
		8.444	amino acid sequence	130

of twenty amino acids (16) did not inactivate Tp^r. This success, combined with the small size of the gene (234 bp), has led to the use of the pR388 reductase gene as a protein-fusion cloning vector (46). Vectors have been developed that contain the structural sequence from pR388 with unique cloning sites (*Bsp*RI or *Pvu*II for different reading frames) at the C-terminus, which terminate translation immediately 3' of the cloning site. Insertions that add 6 to 80 amino acids do not destroy Tp^R. The vectors have been shown to be useful for cloning antigenic determinants of proteins. This fusion construction is more advantageous than β-galactosidase vectors because of the small size of the reductase protein, which leads to a higher molar concentration of protein in the bacterial cell. An efficient purification scheme for the fusion protein is available (146). The enzyme is stable under denaturing conditions and at high temperature (157) and therefore serves well as a purification vector.

The structural sequence for pR388 has also been used as a selective marker for a cloning vector for *Saccharomyces cerevisiae* attached to the yeast promoter, TRP5 (94). A fragment from the 3' end of TRP5 also increased expression of the protein. Since the enzyme is resistant to aminopterin and methotrexate, it was found to be as good a selective marker for yeast as the available nutritional markers. The copy number of the plasmid vector

TABLE III. *Biological and Enzymatic Constants for Dihydrofolate Reductases from pR388, pR67 and E. coli*

Source of enzyme	MIC^{a}Tp	I$_{50}$^{b}Tp	I50MTX	K$_i$^{c}Tp	K$_m$^{d}dihydrofolate	Ref
pR388	3000 µg/ml (10mM)	0.4 mM, pH 7.5	—	—	—	3
pR67	>2,000 µg/ml (6.6 mM)	0.18 mM, pH 6.0		0.15mM, pH 6.0	8.3 µM, pH 6.0	4
		5.7 mM pH?				47
		7.0 mM, pH 7.0	1.1 mM, pH 7.0		4.1 µM, pH 7.0	112
E.coli K12J5	0.15 µg/ml (0.5 µM)	7 nM, pH 7.0	4 nM, pH 7.0		1.24 µM, pH 7.0	112
114	0.2 µg/ml (0.69 µM)	18 nM, pH 7.5	—	—	—	3
		10 nM, pH 6.0	—	6.0 nM, pH 6.0	20 µM, pH 6.0	3

^{a} minimum inhibitory concentration
^{b} 50% inhibitory concentration
^{c} inhibitor constant
^{d} Michaelis-Menten constant

increased with drug selection and with a weaker promoter. It was therefore seen to have uses as a selective marker for transformation, a marker to determine promoter strength and a vehicle to detect copy number mutants. The ability to increase the copy number of the plasmid under selection with a weak promoter lends itself also to use for the over-production of proteins. Although the enzyme from pR67 is more resistant to Tp than the enzyme from pR388, this characteristic of plasmid amplification as a result of weak expression is a beneficial result of choosing the Tp^r gene of pR388 as the selective marker.

The pR67 reductase has also been introduced into cloning vectors for selection. A vector for transformation of mouse fibroblasts (106) contained SV40 promoter and cap sites; splicing and polyadenylation sites were from the globin gene. Methotrexate-resistant transformants were successfully selected. Vectors for transformation of *Drosophila* cells (12) and plant cells (67) have also successfully used the pR67 enzyme.

B. Sulfanilamide Resistance

Until recently, plasmid-coded Sm^r genes were categorized as type I or II (100) depending on whether they produced a resistant dihydropteroate synthase (153,126) or produced altered cell permeability to the drug. Although one report has demonstrated altered permeability (1), most assignments to type II were based on failure to detect an altered enzyme (111,100,133). However this could simply be due to a more labile, resistant dihydropteroate synthase (57), and this has been confirmed for plasmids pR388 and pR22259 (134) after cloning their Su^r determinants into amplifiable vectors with stronger promoters. This increased expression of the gene sufficiently to permit rescue of a temperature-sensitive mutant of dihydropteroate synthase of the host (*E. coli*) and to detect resistant enzyme activity *in vitro* (134). Currently, all carefully studied Su^r plasmids have been shown to confer an altered dihydropteroate synthase (134).

Consequently, plasmid Su^r genes are now categorized according to the heat lability of their dihydropteroate synthetase, type I being the labile enzymes and, type II, heat stable. Several previously studied R-factors have been shown by restriction enzyme mapping and hybridization to have the same Su^r gene as pR388 including R1, R100 and R6 (134). Recent isolates (pGS01, pGS02) have also been shown to contain homologous resistance determinants (134,147). Out of 63 Su^r clinical isolates collected in Sri Lanka from 1980 to 1982, all but one hybridized to probes from either Type I or II genes (5 to both). Type I and II (pGS04, pGS05), appeared at equivalent frequencies.

Although the original classification, that type I synthases are found on large, conjugative plasmids associated with multiple drug resistances and with Sm^r genes that produce adenylylating enzymes (type 2, Sm^r) and that type II synthases are found on small nonconjugative plasmids either alone or with Sm^r

genes that phosphorylate the drug (type I, Sm^r) is still generally true, the type II genes have recently been found also on large, conjugative plasmids (100,132,147). The type II DNA probe has been observed to hybridize to chromosomal DNA (147). This was interpreted as a possible transposon, but homology to the host enzyme was not ruled out by the reported experiments.

C. Streptomycin and Spectinomycin Resistance

Sm^r and Sp^r are both inactivated by cloning fragments into the *Eco*RI site of vectors derived from this portion of pSa (83,136). A protein product of about 30 kDa has been identified that correlates with Sm^r in minicells (139) in agreement with the DNA sequence, which predicts a protein of molecular weight 29 948. This gene is highly homologous to that of Tn7, whose reading frame has been confirmed by a limited protein sequence (23) (Section II.A.2.). The enzyme of Tn7 is an *a*minoglycoside *ad*enylating enzyme, ADD(3″)(9), with similar specificity (Sm^r and Sp^r); adenylation occurs at the 3″ hydroxyl of the amino–hexose III ring of streptomycin or the 9-hydroxyl on the actinamine ring of spectinomycin (48). Comparison of the sequences of related genes from different plasmids indicates that one copy of the repeats found at the ends of Tn7 is found at the C-terminus, but not the N-terminus of pSa (18) suggesting that this gene of pSa is derived from Tn7, but no longer transposes. The same structure is found at the C-terminus only of resistance genes of pDG0100, pR538 and pR388. Further homology exists between pSa, pDG0100 and R538 that extends at least until the *Hind*III site at 7.1 kb on the pSa map. The terminal repeat element of Tn7 has been lengthened from 54 to 59 bp as a result of these comparisons. The pSa version is 56 bp long and has diverged more than the other plasmids from the consensus sequence.

It is thought that the Sm^r/Sp^r gene is the second gene of an operon that codes first for Km^r, because insertions into the *Kpn*I site, which inactivate Km^r, usually inactivate Sm^r (139,25). The expression of Sm^r resistance in a clone that does express resistance is dependent on the orientation of the inserted fragment. Filter-binding experiments designed to detect the complexing of RNA polymerase to DNA fragments, indicated that no RNA polymerase binding site was present between the *Eco*RI and *Kpn*I sites near this gene, but such binding did occur between the *Kpn*I and *Bam*HI sites further upstream, consistent with the location of the promoter for Km^r resistance which would also serve the Sm^r gene (141). The sequence of this region has demonstrated 53 bp between the Sm^r and a potential reading frame for the Km^r gene (141). RNA from this region could form a stem loop structure with a free energy value of −68 kcal/mol. No typical promoter sequences appear in this region. Thus, it seems that the Sm^r/Sp^r gene does not have its own promoter.

D. Kanamycin Resistance

Small deletions inactivate resistance to Km, Gm and Tm simultaneously, as well as insertions into the *Kpn*I site in this region (136). A 2 kb cloned fragment also confers all three resistances. Similar specificity has been noted for acetylating enzymes (11,17), although they may act poorly on Tm. The apparent C-terminus of this gene has been sequenced (141) and compared to the sequence of an aminoglycoside phosphorylating enzyme, *APH* of Tn*5*, which is homologous to Tn*903* and *Streptomyces fradiae* at the C-terminus region (141). Only limited homology was found in the region that has been sequenced for the pSa gene; it is known that Tn*5* confers higher levels of resistance to neomycin than does pSa and, therefore, they are not identical in their specificities. The type of enzyme conferred by the Km^r gene of pSa has not been determined.

E. Chloramphenicol Resistance

Plasmid pSa specificies a type II chloramphenicol acetyl transferase (49). IncW plasmids RA3 and RA4 also confer type II *cat* genes (121). Type II is an infrequently encountered plasmid *cat* gene and is distinguished from both type I and III by electrophoretic mobility and by immune precipitation (51). The *cat* gene from pSa has recently been sequenced (Murray and Shaw, pers. comm.) and contains 240 amino acids. Substantial homology was found to the Type I gene coded by Tn*9* (2), although an earlier report indicated no hybridization to the Tn*9* derivative, pBR325 (72). Another earlier report also suggested that since Tn*9* causes deletions, the chloramphenicol determinant of pSa may be causing its own deletion by the same mechanism (136). However, characterization of these deletions, described above under Section II.A.5. (145), indicates that they may be mediated through direct homologous repeats some distance away from the Cm^r gene. No evidence has been found for transposition of Cm^r (136,72). However recently pSa has been shown to form cointegrates with a nonhomologous plasmid of *Zymomonas mobilis* (131).

F. Ampicillin Resistance

It has been described above under Section II.C.2. that pR7K contains TnA (801) based on restriction sites (149) and hybridization (65). β-lactamases will not be reviewed here.

V. GALACTOSE OPERON REGULATION

Genetic and enzymatic evidence suggests that plasmid pSa (and Rts1 of IncT) carries a galactose epimerase gene that recombines readily with the chromosome of *E. coli* (45). The galactose operon of *E. coli* consists of three genes in the order epimerase, transferase, and kinase (71). A variety of mutants that reduced epimerase production were complemented by pSa, but not by mutants reducing production of the other two gene products. The complementation (measured by ability to ferment galactose) was variable, occurring in only 70% of transconjugants and this galactose fermentation was unstable, reversing back and forth in subsequent generations. However, 100% of transconjugants conferred a galactose resistant phenotype (epimerase mutants are killed by galactose), which was taken as an indication of lesser amounts of epimerase present. Inducible epimerase activity was clearly demonstrated in the strains containing pSa. The variability, then, seemed to reflect varying amounts of enzyme, which was ascribed to a pSa epimerase. Recombination was suggested by relief of a polar mutation in the epimerase gene (reducing transferase production) by the plasmid (demonstrated to be present in normal size). In addition, normal but not $recA1$ cells produced Gal$^-$ colonies at a rate of 2–5% when containing pSa. Segregants of the polar mutant carrying pSa that lost antibiotic resistance markers and plasmid DNA still retained the typical complementation characteristics of the strains containing the whole plasmid. Apparently, the epimerase gene on the plasmid was expressed at low levels and therefore its presence always relieved galactose sensitivity, but the Gal$^+$ phenotype required higher amounts that were provided when the homologous plasmid gene integrated into the chromosomal site under the control of the chromosomal promoter and became subject to induction (45). Relief of the polar mutation was interpreted as confirming that two functional genes were adjacent on the same messenger RNA.

Although integration of a pSa gene into the gal operon is one possibility, it is remarkable that an integration based on $recA$ recombination of homologous sequences is so efficient that 70% of transconjugants express the Gal$^+$ phenotype. The deletion of the region between the repeated sequences of pSa, which provide 1 kb of homologous sequences, occurs at only a 1% frequency. No direct, confirming experiments were done, such as demonstrating the presence of epimerase activity in cells that were deleted for the chromosomal epimerase gene or hybridization of pSa DNA to chromosomal DNA. Presumably the epimerase mutants are point mutations that produce a polypeptide that is poorly functional; indeed the polar mutant does exhibit some epimerase activity upon induction. The possibility that pSa produces a product that binds to the altered epimerase of $E. coli$, stabilizing it, and restoring its activity (or suppressing mutations found only in the epimerase gene) has not been ruled out. The fact that a polar mutation was complemented by the plasmid could also be explained by interpreting this mutation as an operator mutant that binds repressor more tightly, thus responding poorly to induction. Plasmid pSa may produce a product that binds to the repressor (or operator), thus reducing the tightness of binding and increasing inducibility. Thus, it has not been unequivocally demonstrated that the effect of pSa is more than an effect on expression of the *gal* operon or the activity of the epimerase.

The genetic location of these pSa functions which increase epimerase activity and alter *gal* operon expression have not been determined. It would be interesting to know if they coincide with the function that suppresses oncogenesis in *Agrobacterium* or suppresses synthesis of the protein that is required in *Agrobacterium* for attachment (Section VII).

VI. CONJUGATIVE TRANSFER

Plasmids of the W incompatibility group are self-transmissible and, as shown in Table IV, are conjugatively transferred between different members of the Enterobacteriaeceae (152,61,33,30), Pseudomonadaceae (75,76,60,139), Rhizobiaceae (139,83) as well as other Gram-negative bacteria (139).

The conjugative character of pR388 has been widely exploited for the mobilization of transposons (36,99,120,9,10,92,29) including bacteriophage Mu (26) or for the formation of cointegrate plasmids (59,7,155,103). pSa has also been used to mobilize transposons in *Pseudomonas aeruginosa* (77). IncW plasmids are limited in their ability to mobilize other plasmids by conjugation: IncFI plasmids are more effective in mobilizing ColE1 than pR388 (151), and pSa did not mobilize cloned *Thiobacillus ferrooxidans* plasmids in *E. coli* (115).

Although there is no direct evidence that W-pili are required for conjugation, a qualitative correlation between low piliation and poor donor ability in *Salmonella* abony SQ401–1(pSa) has been demonstrated (15). Unlike the pili of other R-factors, pili determined by pSa, pR388 and pR7K appear with pointed tips (14,15). The conjugative transfer of pR388 and pSa is inhibited by RP1 and R6K (107). Interestingly, in spite of the fact that RP1 and R6K are in different incompatibility groups, they may possess origin of transfer functions in common with those of pSa and pR388. Indeed the IncW fertility inhibition regions have been mapped to two broad regions of RP1, one of which contains the Tra1 function while the other has no transfer function (154).

The *tra* region of pSa has recently been characterized by transposon mutagenesis (39) (II.A.7).

VII. ONCOGENIC SUPPRESSION OF *AGROBACTERIUM TUMEFACIENS*

Previous work using pSa to mobilize the large Ti (for tumor-inducing) plasmid *A. tumefaciens* revealed an interesting phenomenon whereby unlike other R-plasmids, pSa somehow suppressed the ability of *A. tumefaciens* to cause tumors in plants (88,86,87). The plasmid could be transferred into and stable

TABLE IV. *Host range of* IncW *group plasmids*

Species	Plasmid	Reference
Acinetobacter calcoaceticus	pSa, pR388	68
Aeromonas liquefaciens	RA3, RA4	61
Aeromonas salmonicida	pSa	5,109
Agrobacterium tumefaciens	pSa	88
	R388	156,22
	R7K	156
Agrobacterium rhizogenes	pSa	148
Alcaligenes eutrophus	pSa	139,140
Enterobacter	pHH1191, pHH1193, pHH1303, pHH1307	???
Erwinia amylovora	pSa	139,140
Erwinia carotovora subsp. *carotovora*	pSa	122
Erwinia herbicola	pSa	J. Shaw, unpublished
Erwinia rubrifaciens	pSa	139
Erwinia stewartii	pSa	139
Escherichia coli J53 and other strains e.g. CR34–3, K12	pSa, R388	33,107,31
Klebsiella spp.	pSa	34
Klebsiella K9	pHH720 (pR388)	31
Klebsiella pneumoniae	pSa	139,140
Legionella pneumophila	pSa	38
Methylophilus methylotrophus	pSa	100
Myxococcus virescens	pBL1005, pBL1009, pBL1010, pBL1011, pBL1013, pBL1014, pBL1016, pBL1017, pBL1021, pBL1024	96
Myxococcus xanthus	pSa	80
Proteus rettgeri	pSa, R7K	27
Proteus mirabilis	pSa	33
Providentia stuartii	pSa	140
Pseudomonas aeruginosa	pSa, R388, R7K	15, 142 ???
Pseudomonas fluorescens	pSa	139,140
Pseudomonas glumae	pSa	120
Pseudomonas putida	pSa	139
Pseudomonas solanacearum	R388	95
Pseudomonas syringae pv. *glycinea*	pSa	122
Pseudomonas stutzeri	pSa	139,140
Rhizobium leguminosarum	pSa	139,140
Rhizobium trifolii	pSa	139,140
Salmonella enteritidis	pSa	19

TABLE IV. *continued*

Species	Plasmid	Reference
Salmonella typhimurium	pSa, R388, R7K	15,14
	pSa	119,152
Salmonella ordonez	pSa	118
Serratia marcescens	pSa	139,140
Shigella sp.	pSa	152
Shigella flexneri	pSa, R388, R7K	15,14
Vibrio cholerae	pSa	57
Xanthomonas campestris pv. *campestris*	pSa	122
Xanthomonas campestris		
pv.*malvacearum*	pSa	50
Zymomonas mobilis	pSa	

maintained by both octopine- and nopaline-utilizing *A. tumefaciens* strains, and it had no effect on Ti plasmid maintenance or some other functions such as octopine utilization or opine-mediated conjugal transfer function (43). This effect has been also observed with certain other members of the W-group plasmids such as pUB5554 (a Tn*10* insert into pR388, ref.150) (156), and pR388 (6,87), but is not universal to all W-group members (156). Hybridization analysis of the indicated plasmids showed that they are very closely related to pSa (156,150,55). The effect on oncogenicity is specific to these plasmids, and is not *A. tumefaciens* strain specific (88,156,43,22). pSa also affects the rhizogenicity of *A. rhizogenes* (148). When pSa is cured from *A. tumefaciens* transconjugants or when their Ti plasmids are genetically transferred to an appropriate recipient, the resultant strains lacking pSa regain oncogenicity (43). Restriction endonuclease analysis of plasmid DNA isolated from transconjugants harboring pSa showed no difference in Ti plasmid cleavage patterns when compared to plasmid DNA isolated from the oncogenic parent strain, indicating that pSa does not induce detectable permanent genetic alteration of the Ti plasmid (43).

A genetic map of pSa was constructed to determine which part of the plasmid molecular was responsible for inhibiting the oncogenic phenotype in *A. tumefaciens* (136). The region for oncogenesis suppression (OS) mapped in a section required for conjugative transfer of pSa. Mapping studies of the 39 kb molecule were further refined by co-workers from our laboratory (156,72), and the OS region was localized to a 7.6 kb sector of pSa, which encoded six polypeptides of 35, 27, 26, 20.5 and 14.3 kDa in *E. coli* minicells and in an *in vitro* coupled-transcription/ translation system (156). Preliminary studies have narrowed this sector to 4.5 kb. Further analyses are necessary to determine which of the polypeptides are encoded in this sector and required for OS activity. The characterization of the *tra* region (39) shows that this function is not part of the conjugative transfer region.

The nature of OS by pSa has been explored by several workers (156,85,86, 87,102,89,22). New *et al.* (102) confirmed that pSa causes loss of virulence of *A. tumefaciens* on pinto bean leaves. Using the pinto bean leaf assay, they also reported that a lipopolysaccharide (LPS) preparation from pSa-containing strains was unable to inhibit tumor initiation ability of the virulent wild-type strain, unlike the LPS preparation from the wild-type strain. Although no qualitative differences in polyacrylamide electrophoretic patterns were observed between the two LPS preparations, it was suggested that the presence of pSa in *A. tumefaciens* results in the production of a sufficiently modified LPS that it no longer exhibits site binding activity (102). Conversely, Lobanok *et al.* (85,86,87) and Chernin *et al.* (22), who also confirmed that pSa suppresses oncogenicity, reported that pSa and pR388 decreased the ability of *A. tumefaciens* to produce the plant growth hormone indole 3-acetic acid (IAA) and suggested that this phenomenon is responsible for the oncogenic deficient phenotype. Wild-type *A. tumefaciens* produced about three times more IAA than did the same strain containing pSa. Chemical complementation with IAA resulted in the restoration of oncogenicity of a strain containing pR388 but no restoration was observed with the same strain containing pSa (22). Avdienko *et al.* (6) were unable to detect defects in the LPS of *A. tumefaciens* cells harboring pSa or pR388, and contended that LPS changes noted by New *et al.* (102) are plasmid- and strain-specific since the same 1D1 strain carrying pR388 does not show differences in LPS, which pSa displayed in 1D1. Matthysse (89) reported that strain 1D1, A6-Ce-12, and C58 carrying pSa attached to carrot cells with kinetics that were indistinguishable from those of the parent strains when the binding was measured in Murashige and Skoog tissue culture medium. However, when the bacteria were grown in medium containing no auxin, and the plant cells were depleted of auxin, then only the wild-type strains bound to the plant cells. Strains containing pSa do not bind to carrot cells in the absence of auxin. It was concluded that the effect of auxin is on the bacteria rather than on the plant cells because the bacteria required auxin to bind to heat-killed carrot cells. No differences were found in the size of LPS between bacteria with and without pSa and in the presence or absence of auxin. However, it was also reported that tetracycline inhibited the binding of bacterial cells containing pSa in auxin-containing medium, and that bacteria containing pSa growing in the absence of auxin lacked surface polypeptides, which were present in auxin-treated pSa-containing bacteria or wild-type bacteria.

One of these proteins (91) corresponds to a protein identified as missing from mutants of *A. tumefaciens* that fail to bind (90). Thus, these results give the appearance that auxin produced either internally by *Agrobacterium* or externally, regulates the expression of a protein that is required for productive attachment of *A. tumefaciens* to plant cells. Whether internal production of auxin responds to the presence of a potential wound site of a plant is not known.

Based on the available information, it is clear that the molecular basis of OS by pSa has not been unequivocally demonstrated. Fine mapping studies of the OS region of pSa will likely pinpoint the exact gene or sets of genes responsible for this activity in order to explain the nature of OS activity (Johnson and Kado, unpublished).

VIII. IncW-SPECIFIC BACTERIOPHAGES

Several bacteriophages have been isolated that are specific to cells containing IncW plasmids or to a limited number of incompatibility groups. PRD1 and PR4 are double-stranded DNA viruses containing lipid that infect cells carrying IncW, IncN, and IncP plasmids Bradley (13). However, PRD1 will not infect cells containing R7K even though it does infect cells containing pSa or pR388 (108). Originally, it was reported that these viruses absorbed to the cell wall rather than to the sex pili, but it has been observed that PR4 does attach to the tips of W-pili (13). Phage K7, also a double-stranded DNA phage (97), is specific for IncW plasmids [Sa,R7K, and plasmids from *Myxococcus virescens* identified as IncW based on plasmid specificity (96)]. Newly-released DNA from this phage contains single-stranded ends.

IX. RESTRICTION-MODIFICATION SYSTEM

Evidence for a restriction-modification enzyme system on pSa has been obtained with bacteriophage K7, specific for IncW plasmids (97). Phage grown in cells containing either of two other IncW plasmids, pBL1016 and pBL1024 (96, and see above VIII), produced equivalent plaque-forming units when plated on cells containing either plasmid, but showed 10^4 fewer plaques when plated on cells containing pSa. Phage grown in pSa-containing cells plated with the high frequency on these cells. No further characterization of this system has been done.

X. CLONING VECTORS

IncW plasmids have been reported to exhibit stable maintenance in many genera of Gram-negative bacteria (Table IV); consequently, the origin of replication from either pSa or pR388 has been incorporated into a wide range of broad-host-range cloning vectors. As the vectors developed initially, together with several cosmid derivatives are presented in Chapter 9, only the more recent vectors, and particularly those derivatives of pSa which have been designed for a particular purpose, will be described here.

A. Tracer vectors

Plasmid vectors containing a reporter gene or gene-set that allow one to follow the fate of an organism on a substrate or in an environment without disrupting the system are known as tracer vectors. Vectors containing the bioluminescent (*lux*) gene set fit under this definition. The *lux* operon of *Vibrio fischeri* that produces light in a ventricular bucal organ of the Japanese popcorn fish was incorporated in the broad-host-range pSa based vector pUCD4 (*Xba*I substituted for *Pvu*II of pUCD2, ref.25). Initially the operon was freed of its own complicated promoter and regulatory system (41) and the 'front end' of the operon reconstructed to operate under the control of the promoter of the tetracycline resistance gene. The resulting construction pUCD607 (122) contained a constitutive *lux* gene system. Bacteria containing pUCD607 bioluminesce and therefore can be followed either visually or with light measuring devices including film. An example of the use of this tracer vector is in studies of the infection of a cauliflower plant inoculated with *Xanthomonas campestris* pv. *campestris* containing pUCD607 (122). In this case, the pathogenic bacteria can be seen invading a leaf through its veins far in advance of disease symptoms. The infection process can be measured without destroying the leaf or taking the leaf off the plant.

Tracer vectors are useful in monitoring the movement of bacteria in the environment. Mutants or strains used in the biological control of pests and plant diseases could be monitored using a tracer vector such as pUCD607. This is particularly important when genetically engineered bacteria are released into the environment.

B. Promoter monitoring vectors

The vectors pUCD206B contains a promoterless gene for Cm acetyltransferase activity (CAT), and resistance gene to Nm/Km (24). It contains two origins of replication, derived from pBR327 and from pSa. Included with the pBR327 origin is its *bom* site for transfer using a helper plasmid. The polylinker site adjacent to the CAT cassette permits the cloning of promoter active fragments generated by *Sau*3A digestions.

Another representative of pSa-derived promoter monitoring vectors is pUCD615 (Figure 6) (116). This vector was derived from pUCD5 (127) and contains two selectable markers (Km^r, Ap^r), two origins of replication (pBR327 and pSa), a lambda *cos* site and a promoterless *lux* cassette. A polycloning site is adjacent to the *lux* cassette to permit the cosmid cloning of promoter active fragments. The strength of the promoters can be measured by the amount of light produced. Clones containing promoter active fragments can be easily screened by placing a plate of bacterial colonies containing the clones over X-ray film in a dark room. The brightness of the colony will reflect the degree of promoter activity.

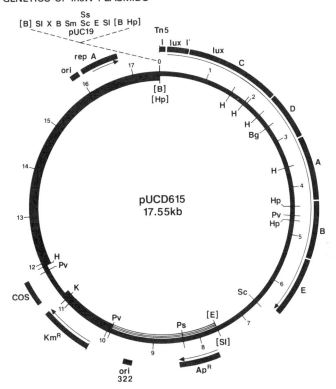

Figure 6. Physical and Genetic Map of pUCD615. The O map position is taken at the insertion of the pUC19 polylinker. Abbreviations for restriction enzymes are as for Figure 1, with the addition of Hp-*Hpa*I and the substitution of Sc for *Sac*I(*Sst*I). Restriction sites marked with brackets indicate the location of sites that are missing in pUCD615 as a consequence of the construction. The solid double line represents sequences from pSa, the single line indicates sequences from the *lux* operon of *Vibrio fischeri* (40), another single line indicates sequences from the cohesive termini of bacteriophage lambda (*cos*), and the solid and striped section, sequences from pBR322. Arrows indicate direction of transcription. The pSa sequences originate from pSa151, the recircularized 13 kb *Sst*II(*Sac*II) fragment of pSa; this *Sst*II site occurs at 12 kb. The vector was assembled (116) by combining the 9.9 kb *Eco*RI-*Bam*HI fragment of pUCD5B (52) with a 7.6 kb fragment containing the *lux* genes in the plasmid pJE347 (40) and part of the pUC19 polylinker.

C. IS and Tn entraping vectors

Many bacteria contain cryptic insertion sequences (IS) and transposable (Tn) elements. Because IS elements usually contain no detectable genes unrelated to insertion function, and because cryptic Tn elements contain undefined genes, which may or may not be selectable, positive selection of these elements is therefore not possible. Generally, the detection and isolation of suspected IS and Tn elements has been tedious. The development of the broad-host-range vector pUCD800 (54), utilizing the origin of DNA replication of pSa and the isolation of the structural gene *sacB* from *Bacillus subtilis* (53) have made it possible to positively select ('entrap') these elements in Gram-negative bacteria. The vector is based on the fact that the *sacB* gene product, levansucrase, which catalyzes transfructorylation from sucrose to various acceptors, is lethal in many Gram-negative bacteria when grown in the presence of sucrose (0.25% to 5%) on agar medium. Therefore, spontaneous insertions or deletions in *sacB* can be detected by *sacB* inactivation and the growth of the bacterial cell on medium containing sucrose.

pUCD800 was derived from pUCD5, which contains the origins of DNA replication of pSa and pBR322, resistance conferring genes to kanamycin, ampicillin and tetracycline, and a lambda *cos* site (25), and from pLS306 [a derivative of pJH101 (44)], which contains the origin of replication of pBR322, a chloramphenicol resistance gene, and the *sac*B gene. pUCD800 can be transferred by transformation or by conjugation promoted by triparental matings with pRK2013. Transformants or transconjugants appearing on agar medium containing sucrose (since sensitivity to sucrose varies among different bacteria, various concentrations should be tested) can be individually examined by rapid miniscreening (e. g. ref.77) for pUCD800 containing an insert. The vector containing the insert can be maintained in *E. coli* and used for characterization.

XI. CONCLUSIONS

Plasmids of the IncW group comprise some of the smallest conjugative plasmids naturally present. Interestingly, the domain containing the origin of DNA replication is also small and yet comprises genetic determinants that confer broad-host-range characteristics on plasmids of this group. This characteristic was exploited in the construction of a series of vehicles useful in the isolation of genes and movable genetic elements, and in sensing and quantifying gene expression. Because of the utility of the IncW origin, it is likely that vectors with useful attributes will be constructed in the future. Additional information will be needed on maintaining the stability of chimeric plasmids, and this could be developed in the study of the partitioning functions of IncW plasmids.

The antibiotic genes on IncW plasmids seemed to be grouped where restriction enzyme sites are clustered, a feature common to R-plasmids, which remains to be explained. Some of the IncW plasmid genes have been useful in studies with eukaryotic systems. They include dihydrofolate reductase, which confers resistance to trimethoprim in bacteria and methotrexate in mammalian cells, and chloramphenicol acetyltransferase, which confers resistance to chloramphenicol in bacteria and in plant cells. As our knowledge of the antibiotic genes grow, it is likely that these genes will continue to be useful in the construction of various molecular tools.

Besides the antibiotic resistance phenotype, IncW plasmids have the unique feature of conferring the suppression of oncogenicity of *A. tumefaciens*, a pathogen well known for its ability to transfer part of its genetic material (the T-DNA of the Ti plasmid) into the chromosomes of plants. The mechanism of oncogenic suppression has not been elucidated but is clearly a function involving virulence of this organism. Specific for pSa is its effect on the *gal* operon of *E. coli*. The mechanism by which the operon is affected is not understood and hence in depth studies are required.

Although there are many IncW plasmids, only pSa and pR388 have received appreciable characterization. Clearly, additional research is needed for the complete physical and functional characteristics of these interesting macromolecules. The proven utility of the IncW plasmids warrants such additional studies.

ACKNOWLEDGEMENTS

The studies summarized in this chapter were supported by NIH grant CA-11526 from the National Cancer Institute, DHHS. We acknowledge the generous and patient clarification of details of pSa vector maps by R. C. Tait and communication of data before publication by D. Zaitlin, I. Murray and W.V. Shaw and by A.G. Matthysse.

XII. REFERENCES

1. Akiba R, Yokota T (1962) Studies on mechanism of transfer of drug resistance in bacteria. 18. Incorporation of ^{35}S-sulfathiazole into cells of the multiple-resistant strain and artificial sulfonamide-resistance strain of *E. coli*. *Med Biol* 63, 155–9.
2. Alton NK, Vapnek D (1979) Nucleotide sequence analysis of the chloramphenicol resistance transposon Tn9. *Nature* 282, 864–9
3. Amyes SGB, Smith JT (1974) R-factor trimethorprim mechanism: an insusceptible target site. *Biochem Biophys Res Comm* 58, 412–8
4. Amyes SGB, Smith JT (1976) The purification and properties of the trimethoprim-resistance dihydrofolate reductase mediated by the R-factor, R388. *Eur J Biochem* 61, 597–603
5. Aoki T, Egusa S, Ogata Y, Watanabe T (1971) Detection of resistance factors in fish pathogen *Aeromonas liquefaciens*. *J Gen Microbiol* 65, 343–9
6. Avdienko ID, Sakharova MN, Ovadis MI, Zoz NN, Chernin LS (1985) A study of the host range and mechanism of antitumorigenic activity of R-plamids from the IncW group of *Agrobacterium tumefaciens* strains of dif-

ferent origin. *Genetika* **21**, 1438–48

7. Balligand G, Laroche Y, Cornelis G (1985) Genetic analysis of virulence plasmid from a serogroup 9 *Yersinia enterocolitica* strain: autoagglutination. *Infect Immun* **48**, 782–6

8. Beeftinck F, Cunin R, Glansdorff N (1974) Arginine duplications in recombination proficient strains of *Escherichia coli* K12. *Mol Gen Genet* **132**, 241–53

9. Bennett PM, Grinsted J, Richmond MH (1977) Transposition of Tn A does not generate deletions. *Mol Gen Genet* **154**, 205–11

10. Bennett PM, Richmond MH (1976) Translocation of a discrete piece of deoxyribonucleic carrying an *amp* gene between replicons in *Escherichia coli*. *J Bacteriol* **126**, 1–6

11. Benveniste R, Davies J (1971) Enzymatic acetylation of aminoglycoside antibiotics by *Escherichia coli* carrying an R-factor. Biochemistry **10**, 1787–96

12. Bourouis M, Jarry B (1983) Vectors containing a prokaryotic dihydrofolate reductase gene transform *Drosophila* cells to methotrexate resistance. *EMBO J* **2**, 1099–104

13. Bradley DE (1976) Absorption of the R-specific bacteriophage PR4 to pili determined by a drug resistance plasmid of the W compatibility group. *J Gen Microbiol* **95**, 181–5

14. Bradley DE (1978) W-pili: characteristics and interaction with lipid phages specific for N, P-1, and W group plasmids. *In* Pili (Bradley DE, Raizen E, Fives-Taylor P, Ou J, eds) Int.Conf.Pili, Washington, D.C. pp. 319–38

15. Bradley DE, Cohen DR (1976) Basic characterization of W-pili. *J Gen Microbiol* **97**, 91–103.

16. Brisson N, Hohn T (1984) Nucleotide sequence of the dihydrofolate-reductase gene borne by the plasmid R67 and conferring methotrexate resistance. *Gene* **28**, 271–5

17. Brzezinska M, Beneviste R, Davies J, Daniels PJL, Weinstein J (1972) Gentamycin resistance in strains of *Pseudomonas aeruginosa* mediated by enzymatic *N*-acetylation of the deoxystreptamine moiety. *Biochemistry* **11**, 761–6

18. Cameron FH, Obbink DJG, Ackerman VP, Hall RM (1986) Nucleotide sequence of the AAD(2′′) aminoglycoside adenylyltransferase determinant *aadB*. Evolutionary relationship of this region with those surrounding *aadA* in R538–1 and *dhfr*II in R388. *Nucl Acids Res* **14**, 8625–35

19. Causey SC, Brown LR (1978) Transconjugant analysis: limitations on the use of sequence-specific endonuclease for plasmid identification. *J Bacteriol* **135**, 1070–9

20. Cebrat S, Kurowska E, Rogolinski J (1979) Genetic properties of the *Salmonella enteritidis* R404 plasmid aggregate. IV. Reconstruction of R404 plasmid aggregate and separation of twelve genetically distinct derivative forms.

Acta Microbiolica Polonica **28**, 29–37

21. Chang S, Cohen SN (1977) *In vivo* site-specific genetic recombination promoted by the *Eco*RI restriction endonuclease. *Proc Natl Acad Sci USA* **74**, 4811–5

22. Chernin LS, Lobanok EV, Fomicheva VV, Kartel NA (1984) Crown gall-suppressive *Inc*W R-plasmids cause a decrease in auxin production in *Agrobacterium tumefaciens*. *Mol Gen Genet* **195**, 195–9

23. Chinault AC, Blakesley VA, Roessler F, Willis DG, Smith CA, Cook RG, Fenwick Jr RG (1986) Characterization of transferable plasmids from *Shigella flexneri* 2a that confer resistance to trimethoprim, streptomycin, and sulfonamides. *Plasmid* **15**, 119–31

24. Close TJ, Tait RC, Kado CI (1985) Regulation of Ti plasmid virulence genes by a chromosomal locus of *Agrobacterium tumefaciens*. *J Bacteriol* **164**, 774–81

25. Close TJ, Zaitlin D, Kado CI (1984) Design and development of amplifiable broad-host-range cloning vectors: analysis of the *vir* region of *Agrobacterium tumefaciens* plasmid pTiC58. *Plasmid* **12**, 111–8

26. Coelho A, Maynard-Smith S, Symonds N (1982) Abnormal cointegrate structures mediated by gene B mutants of phage Mu: their implications with regard to gene function. *Mol Gen Genet* **185**, 356–62

27. Coetzee JN, Datta N, Hedges RW (1972) R-factors from *Proteus rettgeri*. *J Gen Microbiol* **72**, 543–52

28. de la Cruz F, Avila P (1987) Study of the transfer functions of the IncW plasmid R388. Abstract, EMBO Workshop on Promiscuous Plasmids of Gram-negative Bacteria. July, 1987, Birmingham, England.

29. de la Cruz F, Grinsted J (1982) Genetic and molecular characterization of Tn*21*, a multiple resistance transposon from R100.1. *J Bacteriol* **151**, 222–8

30. Datta N (1975) Epidemiology and classification of plasmids. *In* Microbiology 1974 (Schlessinger D, ed) American Society for Microbiology, Washington, D.C., USA pp. 9–15

31. Datta N, Dacey S, Hughes V, Knight S, Richards H, Williams G, Casewell M, Shannon KP (1980a) Distribution of genes for trimethoprim and gentamicin resistance in bacteria and their plasmids in a general hospital. *J Gen Microbiol* **118**, 495–508

32. Datta N, Hedges RW (1971) Compatibility groups among *fi*⁻ R-factors. *Nature* **234**, 220–2

33. Datta N, Hedges RW (1972a) Host ranges of R-factors. *J Gen Microbiol* **70**, 453–60

34. Datta N, Hedges RW (1972b) Trimethoprim resistance conferred by W plasmids in Enterobacteriaceae. *J Gen Microbiol* **72**, 349–55

35. Datta N, Nugent M, Richards H (1980b) Transposons encoding trimethoprim or gentamicin resistance in medically important bacteria. *Cold*

Spring Harbor Symp Quant Biol **45**, pt. 1, 45–51

36. Diaz-Aroca E, de la Cruz F, Ortiz JM (1984) Characterization of the new insertion sequence IS*91* from an alpha-hemolysin plasmid of *Escherichia coli*. *Mol Gen Genet* **193**, 493–9

37. Dorokhina OV, Korotyaev AI (1984) Study on incompatibility properties of pKMR plasmids determining antibiotic resistance and capacity for production of colonization antigen. *Antibiotiki* **29**, 91–5

38. Dreyfus LA, Iglewski BH (1985) Conjugation- mediated genetic exchange in *Legionella pneumophila*. *J Bacteriol* **161**, 80–4

39. Dymock D, Warner PJ (1987) Studies on the transfer of the Inc W plasmid pSa. Abstract, EMBO Workshop on Promiscuous Plasmids of Gram-negative Bacteria. July, 1987. Birmingham, England.

40. Engebrecht J, Simon M, Silverman M (1983) Bacterial luminescence: Isola tion and genetic analysis of functions from *Vibrio fischeri*. *Cell* **32**, 773–81

41. Engebrecht I, Simon M, Silverman M (1985) Measuring gene expression with light. *Science* **227**, 1345–7

42. Falkow S, Guerry P, Hedges RW, Datta N (1974) Polynucleotide sequence relationships among plasmids of the I compatibility complex. *J Gen Microbiol* **85**, 65–76

43. Farrand SK, Kado CI, Ireland CR (1981) Suppression of tumorigenicity by the Inc W R-plasmid pSa in *Agrobacterium tumefaciens*. *Mol Gen Genet* **181**, 44–51

44. Ferrari FA, Ferrari E Hoch JA (1982) Chromosomal location of a *Bacillus subtilis* DNA fragment uniquely transcribed by sigma-28-containing RNA polymerase. *J Bacteriol* **152**, 780–5

45. Fietta A, Grandi G, Malcovati M, Valentini G, Sgaramella V, Siccardi AG (1981) R-factor-mediated suppression of the galactose-sensitive phenotype of *Escherichia coli* K-12 *galE* mutants. *Plasmid* **6**, 78–85

46. Fleming MP, Datta N, Grunberg RN (1972) Trimethoprim resistance determined by R-factors. *Brit Med J* **1**, 726–8

47. Fling ME, Elwell LP (1980) Protein expression in *Escherichia coli* minicells containing recombinant plasmids specifying trimthoprim-resistance dihydrofolate reductases. *J Bacteriol* **141**, 779–85

48. Fling ME, Kopf J, Richards C (1985) Nucleotide sequence of the transposon Tn*7* gene encoding an aminoglycoside-modifying enzyme, 3′′ (9)-*O*-nucleotidyl-transferase. *Nucl Acids Res* **13**, 7095–106

49. Foster TJ, Shaw WV (1973) Chloramphenicol acetyltransferases specified by *fi⁻* R-factors. *Antimicrob Ag Chemother* **3**, 99–104

50. Gabriel DW, Burges A, Lazo GR (1986) Gene-for-gene interactions of five cloned avirulence genes from *Xanthomonas campestris* pv. *malvacearum* with specific resistance genes in cotton. *Proc Natl Acad Sci USA* **83**, 6415–9

51. Gaffney DF, Foster TJ, Shaw WV (1978) Chloramphenicol acetyltransferases determined by R-plasmids from Gram-negative bacteria.

J Gen Microbiol **109**, 351–8

52. Gallie DR, Novak S, Kado CI (1985) Novel high- and low-copy number stable cosmids for use in *Agrobacterium* and *Rhizobium*. *Plasmid* **14**, 171–5

53. Gay P, LeCoq D, Steinmetz M, Ferrari E, Hoch JA (1983) Cloning structural gene *sacB*, which codes for exoenzyme levansucrase of *Bacillus subtilis*: expression of the gene in *Escherichia coli*. *J Bacteriol* **153**, 1424–31

54. Gay P, LeCoq D, Steinmetz M, Berkelman T, and Kado CI (1985) Positive selection procedure for entrapment of insertion sequence elements in Gram-negative bacteria. *J Bacteriol* **164**, 918–21

55. Gorai AP, Heffron F, Falkow S, Hedges RW, and Datta N (1979) Electron microscope heteroduplex studies of sequence relationships among plasmids of the W incompatibility group. *Plasmid* **2**, 485–92

56. Grinsted J, Bennett PM, Higginson S, Richmond MH (1978) Regional preference of insertion of Tn*501* and Tn*802* into RP1 and its derivatives. *Mol Gen Genet* **166**, 313–20

57. Hamilton-Miller JMT (1979) Mechanisms and distribution of bacterial resistance to diaminopyrimidines and sulphonamides. *J Antimicrob Chemother* (Suppl.B) **5**, 61–73

58. Hamood AN, Sublett RD, Parker CD (1986) Plasmid-mediated changes in virulence of *Vibrio cholerae*. *Infect Immun* **52**, 476–83

59. Harayama S, Oguchi T, Iino T (1984) Does Tn*10* transpose via the cointegrate molecule? *Mol Gen Genet* **194**, 444–50

60. Hedges RW (1972) Resistance to spectinomycin determined by R-factors of various compatibility groups. *J Gen Microbiol* **72**, 407–9

61. Hedges RW, Datta N (1971) *fi⁻* R-factors giving chloramphenicol resistance. *Nature New Biol* **234**, 220–1

62. Hedges RW, Datta N, Fleming MP (1972) R-factors conferring resistance to trimethoprim but not sulphonamides. *J Gen Microbiol* **73**, 573–5

63. Hedges RW, Jacob AE (1974) Transposition of ampicilin resistance from RP4 to other replicons. *Mol Gen Genet* **132**, 31–40

64. Heffron F, Rubens C, Falkow S (1975) Translocation of a plasmid DNA sequence which mediates ampicillin resistance; molecular nature and specificity of insertion. *Proc Natl Acad Sci USA* **72**, 3623–7

65. Heffron F, Sublett R, Hedges RW, Jacob AE and Falkow S (1975) The origin of the TEM β-lactamase gene found on plasmids. *J Bacteriol* **122**, 250–6

66. Heidorn JV, Valentine CRI (1986) Restriction enzyme map of cryptic plasmid accompanying pR711b and pR409 and pR388. *J Basic Microbiol* **10**, 621–5

67. Herrera-Estrella L, De Block M, Messens E, Hernalsteens JP, Van Montagu M, Schell J (1983) Chimeric genes as dominant selectable markers in plant cells. *EMBO J* **2**, 987–95

68. Hinchcliffe E, Vivian A (1980) Restriction mediated by pAV2 affects the

transfer of plasmids in *Acinetobacter calcoaceticus*. *J Gen Microbiol* **121**, 419–23

69. Hollingshead S, Vapnek, D (1985) Nucleotide sequence analysis of a gene encoding a streptomycin/spectinomycin adenyltransferase. *Plasmid* **13**, 17–30

70. Inuzuka M, Helinski DR (1978) Requirement of a plasmid-encoded protein for replication *in vitro* of plasmid R6K. *Proc Natl Acad Sci USA* **75**, 5381–5

71. Ippen K, Shapiro JA, Beckwith JR (1971) Transposition of the *lac* region to the *gal* region of the *Escherichia coli* chromosome: Isolation of *lac* transducing bacteriophages. *J Bacteriol* **108**, 5–9

72. Ireland CR (1983) Detailed restriction enzyme map of crown-gall suppressive IncW plasmid pSa, showing ends of deletion causing chloramphenicol sensitivity. *J Bacteriol* **155**, 722–7

73. Jackson EN, Yanofsky C (1973) Duplication- translocations of trypophan operon genes in *Escherichia coli*. *J Bacteriol* **116**, 33–40

74. Jacob AE, Shapiro JA, Yamamoto L, Smith DI, Cohen SN Berg D (1977) Plasmids studied in *Escherichia coli* and other enteric bacteria. *In* DNA Insertion Elements, Plasmids and Episomes. (Bukhari AI, Shapiro TA, Adhya SL, eds), Cold Spring Harbor Laboratory, pp. 635–6

75. Jacoby GA (1975) Properties of R-plasmids in *Pseudomonas aeruginosa*. *In* Microbiology–1974 (Schlessinger D, ed.) American Society for Microbiology, Washington, DC. pp. 36–42

76. Jacoby GA (1975) R-plasmids determining gentamicin or tobramycin resistance in *Pseudomonas aeruginosa*. *In* Drug-inactivating enzymes and antibiotic resistance. (Mitsuhashi S, *et al*, eds) Avicenum, Czechoslovak Medical Press, Prague, Springer-Verlag, Berlin. pp. 287-

76a. Joyner SS, Fling ME, Stone D, Baccanari DP (1984) Characterization of an R-plasmid dihydrofolate reductase with a monomeric structure. *J Biol Chem* **259**, 5851–6

77. Kado CI, Liu S-T (1981) Rapid procedure for detection and isolation of large and small plasmids. *J Bacteriol* **145**, 1365–73

78. Kontomichalou P (1971) R-factors controlling resistance to the penicillins. Thesis, University of Athens.

79. Krishnapillai V, Royle P, Lehrer J (1981) Insertions of the transposon Tn*1* into the *Pseudomonas aeruginosa*. *Genet* **97**, 1495–511

80. von Kruger WMA, Parish JH (1981) β-Lactamase activity and resistance to penicillins in *Myxococcus xanthus*. *Arch Microbiol* **130**, 150–4

81. Lachowicz Z (1971) Drug resistance of *Salmonella enteritidis* bacilli. *Arch Immunol Therap Experiment* **19**, 851–9

82. Lawn AM, Meynell E., Meynell GG, Datta N (1967) Sex pili and the classification of sex factors in the Enterobacteriaceae. *Nature* **216**, 343–6

83. Leemans J, Langenakens J, De Greve H, Deblaere R, Van Montagu M,

Schell J (1982) Broad-host-range cloning vectors derived from the W-plasmid, pSa. *Gene* **19**, 361–4

84. Lehman MJ (1986) DNA sequencing of the direct repeats of multiple drug-resistance plasmid, pSa. Master's Thesis, Oral Roberts University.

85. Lobanok EV, Fomicheva VV, Chernin LSA (1982) Correlation between tumorigenicity and phytohormonal activity in *A. tumefaciens*. Proc Metabolic Plasmids, Tallin, (Heinaru A, ed) Tartu State University Estonian SSR, USSR. pp. 146–8

86. Lobanok EV, Fomicheva VV, Kartel NA (1983) R-plasmid effect on the oncogenic properties of *Agrobacterium tumefaciens*. *Dokl Akad Nauk USSR* **27**, 462–4

87. Lobanok EV, Fomicheva VV, Kartel NA, Chernin LS (1983b) Effect of plasmid pSa on the oncogenic activity and production of beta-indoleacetic acid by *Agrobacterium tumefaciens* containing pTi 15955. *Dokl Akad Nauk USSR* **269**, 967–70

88. Loper JE, Kado CI (1979) Host-range conferred by the virulence-specifying plasmid of *Agrobacterium tumefaciens*. *J Bacteriol* **139**, 591–6

89. Matthysse AG (1987) Effect of the presence of the plasmid pSa and of auxin on the attachment of *Agrobacterium tumefaciens* to plant host cells. *In* Molecular Genetics of Plant-Microbe Interactions. (Verma DPS, Brisson N, eds) Martinus Nijhoff Publishers, Dordrecht/ Boston/ Lancaster. pp. 11–13

90. Matthysse AG (1987) Characterization of non-attaching mutants of *Agrobacterium tumefaciens*. *J Bacteriol* 169, 313–23.

91. Matthysse AG (1987) The effect of the plasmid pSa on the attachment of *Agrobacterium tumefaciens* to carrot cells. *App Env Microb* 53, 2574–82

92. McCombie WR, Hansen JB, Zylstra GJ, Maurer B, Olsen RH (1983) *Pseudomonas* streptomycin resistance transposon associated with R-plasmid mobilization. *J Bacteriol* **155**, 40–8

93. Meyers JA, Sanchez D, Elwell LP, Falkow S (1976) Simple agarose gel electrophoretic method for identification and characterization of plasmid deoxyribonucleic acid. *J Bacteriol* **127**, 1529–37

94. Miyajima A, Miyajima I, Arai KI, Arai N (1984) Expression of plasmid R388-encoded type II dihydrofolate reductase as a dominant selective marker in *Saccharomyces cerevisiae*. *Mol Cell Biol* **4**, 407–14

95. Morales VM, Sequeira L (1985) Suicide vector for transposon mutagenesis in *Pseudomonas solanacearum*. *J Bacteriol* **163**, 1263–4

96. Morris DW, Ogden-Swift SR, Virrankoski-Castrodeza V, Ainley K, Parish JH (1978) Transduction of *Myxococcus virescens* by coliphage P1CM: generation of plasmids containing both phage and *Myxococcus* genes. *J Gen Microbiol* **107**, 73–83

97. Morris DW, Virrankoski-Castrodeza V, Ainley K, and Parish JH (1980) Bacteriophage K7, a double-stranded DNA phage that infects strains of

Escherichia coli harbouring drug resistance factors of incompatibility group W. *Arch Microbiol* **126**, 271–5

98. Murotsu T, Matsubara K, Sugisaki H, Takanami M (1981) Nine unique repeating sequences in a region essential for replication and incompatibility of the mini-F plasmid. *Gene* 15, 257–71.

99. Muster CJ, Shapiro JA, MacHattie LA (1983) Recombination involving transposable elements: role of target molecule replication in Tn*1*-Ap-mediated replicon fusion. *Proc Natl Acad Sci USA* **80**, 2314–7

100. Nagate T, Inoue M, Inoue K, Mitsuhashi S (1978) Plasmid-mediated sulfanilamide resistance. *Microbiol Immunol* **22**, 367–75

101. Nesvera J, Hochmannova J, Holubova I, Cejka K (1987) Transfer of IncW plasmids into methylotrophic bacteria by conjugation and mobilization. *App Microbiol Biotech* **26**, 147–8

102. New PB, Scott JJ, Ireland CR, Farrand SK, Lippincott BB, Lippincott JA (1983) Plasmid pSa causes loss of LPS-mediated adherence in *Agrobacterium*m. *J Gen Microbiol* **129**, 3657–60

103. Nugent ME, Hedges RW (1979) Recombinant plasmids formed *in vivo* carrying and expressing two incompatibility regions. *J Gen Microbiol* **114**, 467–70

104. Nugent ME, Datta N (1980) Transposable gentamicin resistance in IncW plasmids from Hammersmith Hospital. *J Gen Microbiol* **121**, 259–62

105. Nyman K, Nakamura K, Ohtsubo H, Ohtsubo E (1981) Distribution of the insertion sequence IS*1* in Gram-negative bacteria. *Nature* **289**, 609–12

106. O'Hare K, Benoist C, Breathnach R (1981) Transformation of mouse fibroblasts to methotrexate resistance by a recombinant plasmid expressing a prokaryotic dihydrofolate reductase. *Proc Natl Acad Sci USA* **78**, 1527–31

107. Olsen RH, Shipley PL (1975) RP1 properties and fertility inhibition among P, N, W and X incompatibility group plasmids. *J Bacteriol* **123**, 28–35

108. Olsen RH, Siak J-S, Gray RH (1974) Characteristics of PRD1, a plasmid-dependent broad host range DNA bacteriophage. *J Virol* **14**, 689–99

109. Olsen RH, Wright CD (1976) Interaction of *Pseudomonas* and Enterobacteriaceae plasmids in *Aeromonas salmonicida*. *J Bacteriol* **128**, 228–34

110. Ortiz-Melon JM, Andres I (1982) Analysis of chromosomal and extrachromosomal functions involved in the replication of bacterial plasmids. *Rev Esp Fisol* **38**,Supl. 259–70.

111. Pato ML, Brown GM (1963) Mechanisms of resistance of *Escherichia coli* to sulfonamides. *Arch Biochem Biophys* **103**, 443–8

112. Pattishall KH, Acar J, Burchal JJ, Goldstein FW, Harvey RJ (1977) Two distinct types of trimethoprim-resistant dihydrofolate reductase specified by R-plasmids of different compatibility groups. *J Biol Chem* **252**, 2319–23

113. Portnoy DA, Moseley SL, Falkow S (1981) Characterization of plasmids

and plasmid-associated determinants of *Yersinia enterocolitica* pathogenesis. *Infec Immun* **31**, 775–82

114. Purohit S, Mathews CK (1984) Nucleotide sequence reveals overlap between T4 phage genes encoding dihydrofolate reductase and thymidylate synthase. *J Biol Chem* **259**, 6261–6

115. Rawlings DE, Woods DR (1985) Mobilization of *Thiobacillus ferroxidans* plasmids among *Escherichia coli* strains. *Appl Environ Microbiol* **49**, 1323–5

116. Rogowsky P, Chimera JA, Close TJ, Shaw JJ, and Kado CI (1987) Regulation of the *vir* genes of *Agrobacterium tumefaciens* plasmid pTiC58. *J Bacteriol* **169**, 5101–12

117. Rosen J, Ryer T, Inokuchi H, Ohtsubo H, Ohtsubo E (1980) Genes and sites involved in replication and incompatibility of an R100 plasmid derivative based on nucleotide sequence analysis. *Mol Gen Genet* **179**, 527–37

118. Roussel A, Carlier C, Gerbaud G, Chabbert YA, Croissant D, Blangy D (1979) Reversible translocation of antibiotic resistance determinants in *Salmonella ordonez*. *Mol Gen Genet* **169**, 13–25

119. Sanderson KE, Stocker BAD (1981) Gene *rfaH*, which affects lipopolysaccharide core structure in *Salmonella typhimurium*, is required also for expression of F-factor functions. *J Bacteriol* **146**, 535–41

120. Schmitt R, Bernhard E, Mattes R (1979) Characterization of Tn*1721*, a new transposon containing tetracycline resistance genes capable of amplification. *Mol Gen Genet* **172**, 53–65

121. Shaw WV (1983) Chloramphenicol acetyltransferase: enzymology and molecular biology. *CRC Crit Rev Biochem* **14**, 1–46

122. Shaw JJ, Kado CI (1986) Development of a *Vibrio* bioluminescence gene-set to monitor phytopathogenic bacteria during the ongoing disease process in a non-disruptive manner. *Bio/Technology* **4**, 560–4

123. Shaw JJ, Kado CI (1987) Direct analysis of the invasiveness of *Xanthomonas campestris* mutants generated by Tn*4431*, a transposon containing a promoterless luciferase cassette for monitoring gene expression. *In* Molecular Genetics of Plant-Microbe Interactions. (Verma DPS, Brisson N, eds) Martinus Nijhoff publishers, Dordrecht/Boston/Lancaster. pp. 57–60

124. Shinagawa M, Makino S, Hirato T, Ishiguro N, Sato G (1982) Comparison of DNA sequences required for the function of citrate utilization among different citrate utilization plasmids. *J Bacteriol* **151**, 1046–50.

125. Skold O (1976) R-Factor-mediated resistance to sulfonamides by a plasmid-borne, drug-resistant dihydropterase synthase. *Antimicrob Ag Chemother* **9**, 49–54

126. Skold O, Widh A (1974) A new dihydrofolate reductase with low trimethoprim sensitivity by an R-factor mediating high resistance to trimethoprim. *J Biol Chem* **249**, 4324–5

127. Smith HR, Grindley NDF, Humphreys GO, Anderson ES (1973) Interac-

tions of group H resistance factors with the F factor. *J Bacteriol* **115**, 623–8
128. Smith SL, Stone D, Novak P, Baccanari DP, Burchall JJ (1979) R-plasmid dihydrofolate reducatase with subunit structure. *J Biol Chem* **254**, 6222–5
129. Starlinger P (1977) DNA rearrangements in procaryotes. *Ann Rev Genet* **11**, 103–26
130. Stone D, Smith S (1979) The amino acid sequence of the trimethoprim-resistant dihydrofolate reductase specified in *Escherichia coli* by R-plasmid R67. *J Biol Chem* **21**, 10857–61
131. Strezelecki AT, Goodman AE, Rogers PL (1987) Behaviour of the Inc W plasmid in *Zymomonas mobilis*. *Plasmid* **18**, 46–53
132. Swedberg G (1987) Organization of two sulfonamide resistance genes on plasmids of Gram-negative bacteria. *Antimicrob Ag Chemother* **31**, 306–11
133. Swedberg G, Skold O (1980) Characterization of different plasmid-borne dihydropteroate synthases mediating bacterial resistance to sulfonamides. *J Bacteriol* **142**, 1–7
134. Swedberg G, Skold O (1983) Plasmid-borne sulfonamide resistance determinants studied by restriction enzyme analysis. *J Bacteriol* **153**, 1228–37
135. Swift G, McCarthy BJ, Heffron F (1981) DNA sequence of a plasmid-encoded dihydrofolate reductase. *Mol Gen Genet* **181**, 441–7
136. Tait RC, Lundquist RC, Kado CI (1982a) Genetic map of the crown gall suppressive Inc W plasmid pSa. *Mol Gen Genet* **186**, 10–5
137. Tait RC, Close TJ, Rodriguez RL, Kado CI (19892b) Isolation of the origin of replication of the Inc W-group plasmid pSa. *Gene* **20**, 39–49
138. Tait RC, Kado CI, Rodriguez RL (1983) A comparison of the origin of replication of pSa with R6K. *Mol Gen Genet* **192**, 32–8
139. Tait RC, Close TJ, Lundquist RC, Hagiya M, Rodriguez RL, Kado CI (1983) Construction and characterization of a versatile broad host range DNA cloning system for Gram-negative bacteria. *Bio/Technology* **1**, 269–75.
140. Tait RC, Close TJ, Hagiya M, Lundquist RC, Kado CI (1983) Construction of cloning vectors from the Inc W plasmid pSa and their use in analysis of crown gall tumor formation. *In* Genetic Engineering in Eukaryotes. (Lurquin PF, Kleinhofs A, eds) Plenum Publishing Corporation. pp. 111–23
141. Tait RC, Rempel H, Rodriguez RL, Kado CI (1985) The amino-glycoside-resistance operon of the plasmid pSa: nucleotide sequence of the streptomycin-spectinomycin resistance gene. *Gene* **36**, 97–104.
142. Tardif G, Grant RB (1983) Transfer of plasmids from *Escherichia coli* to *Pseudomonas aeruginosa*: characterization of a *Pseudomonas aeruginosa* mutant with enhanced recipient ability for enterobacterial plasmids. *Antimicrob Ag Chemother* **24**, 201–8
143. Taylor DE, Brose EC, Kwan S, Yan W (1985) Mapping of transfer regions within incompatibility group HI plasmid R27. *J Bacteriol* **162**, 1221–6

144. Thomas CM, Meyer R, Helinski DR (1980) Regions of broad-host-range plasmid RK2 which are essential for replication and maintenance. *J Bacteriol* **141**, 213–22

145. Valentine CRI (1985) One-kilobase direct repeats of plasmid pSa. *Plasmid* **14**, 167–70

146. Vermersch PS, Klass MR, Bennett GN (1986) Use of bacterial DHFR-II fusion proteins to elicit specific antibodies. *Gene* **14**, 289–97

147. Vinayagamoorthy T, Theivendrarajah K (1986) Incidence of PGS01 and pGS04 type sulphonamide resistance genes among clinical isolates collected from hospitals in Sri Lanka. *Singapore Med J* **4**, 312–6

148. Vlasak J, Ondrej M (1985) The effect of Sa and miniSa plasmids on the induction of plant crown galls and hairy roots by *Agrobacterium tumefaciens* and *Agrobacterium rhizogenes*. *Folia Microbiol* **30**, 148–53

149. Ward JM, Grinsted J (1978) Mapping of functions in the R-plasmid R388 by examination of deletion mutants generated *in vitro*. *Gene* **3**, 87–95

150. Ward JM, Grinsted J (1982) Physical and genetic analysis of the IncW plasmids R388, Sa, and R7K. *Plasmid* **7**, 239–50

151. Warren GJ, Saul MW, Sherratt DJ (1979) ColE1 plasmid mobility: essential and conditional functions. *Mol Gen Genet* **170**, 103–7

152. Watanabe T, Fruse C, Sakaizumi S (1968) Transduction of various R-factors by phage P1 in *Escherichia coli* and by phage P22 in *Salmonella typhimurium*. *J Bacteriol* **96**, 1791–6

153. Wise Jr EM, Abou-Donia MM (1975) Sulfonamide resistance mechanism in *Escherichia coli*: R-plasmids can determine sulfonamide-resistant dihydropteroate synthases. *Proc Natl Acad Sci USA* **72**, 2621–5

153a. Young H-K, Amyes SGB (1986) A new mechanism of plasmid trimethoprim resistance. Characterization of an inducible dihydrofolate reductase. *J Biol Chem* **261**, 2503–5

154. Yusoff K, Stanisich VS (1984) Location of a function on RP1 that fertility inhibits IncW plasmids. *Plasmid* **11**, 178–81

155. Yamamoto T, Motegi A, Takei T, Okayama H, Sawai T (1984) Plasmid R46 provides a function that promotes *recA*-independent deletion, fusion and resolution of replicon. *Mol Gen Genet* **193**, 255–62

156. Zaitlin D (1984) Genetic characterization of *Agrobacterium tumefaciens* virulence suppression encoded on the IncW R-plasmid pSa. Ph.D. Dissertation, University of California, Davis.

157. Zolg JW, Hanggi U, Zachau HG (1978) Isolation of a small DNA fragment carrying the gene for a dihydrofolate reductase from a trimethoprim resistance factor. *Mol Gen Genet* **164**, 15–29.

158. Zolg JW, Hanggi UJ (1981) Characterization of a R-plasmid-associated, trimethoprim-resistant dihydrofolate reductase and determination of the nucleotide sequence of the reductase gene. *Nucl Acids Res* **9**, 697–710

CHAPTER 6

IncN GROUP PLASMIDS AND THEIR GENETIC SYSTEMS

V.N. Iyer

I. INTRODUCTION

Inc N group plasmids all of which are conjugative have been isolated from natural populations of *Aeromonas* spp., *Enterobacter* spp., *Escherichia coli*, *Klebsiella aerogenes*, *Proteus* spp., *Providencia* spp., *Salmonella* spp., *Shigella* spp., and *Vibrio cholerae* (2, 27, 3). This attests to their promiscuity. Phylogenetically, these bacterial species are in the Gamma-3 subdivision of the purple bacteria, a group of nonphotosynthetic bacteria with diverse metabolisms and ecological niches (67). A systematic examination of the extent of this promiscuity has not been reported. It is likely that plasmid promiscuity in nature is the outcome of a balance of factors such as the efficiency with which a plasmid is transmitted intercellularly, the stability with which it is maintained and the efficiency with which it may be reacquired if lost from a cell or population of cells. This is different from intracellular stability in a bacterial host under the conditions of a laboratory culture or industrial fermentor. Inc N group plasmids have also been reported to be transferred in the laboratory and maintained with selection in *Caulobacter crescentus*, *Myxococcus*, *Pseudomonas* and *Rhizobium* species but in these instances the plasmid was as a rule lost from a fraction of the population when selection for its maintenance was released (22, 40, 42, Selvaraj and Iyer, unpublished observations). This review is not intended to be exhaustive. Its aim is to highlight those features of Inc N group plasmids that confer a degree of distinctiveness to the group and to suggest that the study of the genetic systems associated

Promiscuous Plasmids of Gram-Negative Bacteria
ISBN 0-12-688480-3
© 1989 Academic Press Limited
All rights of reproduction in any form reserved

with them is likely to be rewarding.

The attributes of these plasmids that have genetic specificity are:

(a) the essential and minimal replicon;

(b) accessory systems that are only conditionally necessary and which may affect the viability of the host or the plasmid and

(c), the conjugative transfer system. Plasmid-associated recombination systems and a system that increases error-prone DNA repair in the host are also often and perhaps always associated with this group of plasmids and will be reviewed briefly even though these systems are also associated with other plasmid groups. Plasmid-associated phenotypes that are mentioned but not considered further are those that determine resistance to antibiotics, inhibition of the fertility of Inc P group plasmids (*fip*) and the production of an endonuclease (*nuc*) (see Figure 1 and references quoted in its appendix). The review will conclude with a summarized assessment of the uses of these systems, in particular the *tra* system.

II. THE MINIMAL REPLICON

Structural and functional analysis has begun to be undertaken on four members of Inc N group plasmids with different natural origins. Their overall organization is summarized and shown in Figure 1 in a manner that facilitates comparisons. The segment shown as *rep* in this figure has usually been defined as a region that is shared by all deletion derivatives of a plasmid, a region the loss or inactivation of which cannot be tolerated or a region that can rescue, by *in vitro* fusion and transformation, nonreplicating DNA fragments containing dominant markers. In Figure 1, the differences in the size of the *rep* region are probably only a reflection of the current stages of analysis for each plasmid. It is likely that *rep* functions of Inc N group plasmids will be contained in a single region of about 2 kb. In pCU1, this is contained in a single *Pvu* II fragment (34). When this fragment was linked to the *polA*-dependent plasmid, pBR329, the resultant plasmid was *polA*-independent (pCU1 and its derivatives are *polA*-independent). This *Pvu* II fragment could also rescue nonreplicating tetracycline resistance-determining DNA fragments from heterologous plasmids. The region has been sequenced and has some distinctive features which are illustrated in Figure 2 (the complete sequence is in 34). Within the region there are two families of nearly identical deoxyribonucleotide sequences. The first of these is a thirteen-member family of tandem direct repeats each member consisting of thirty-seven base-pairs of G-C rich sequences flanked by A-T rich sequences. All of these thirteen members have the hexadeoxynucleotide sequence TGTGGG(A). The second and smaller family of direct repeats consists of thirty base-pairs of a different sequence repeated once and separated from one another by a sequence derived from the central part of the thirty base-pairs. In each member of this family, the heptanucleotide

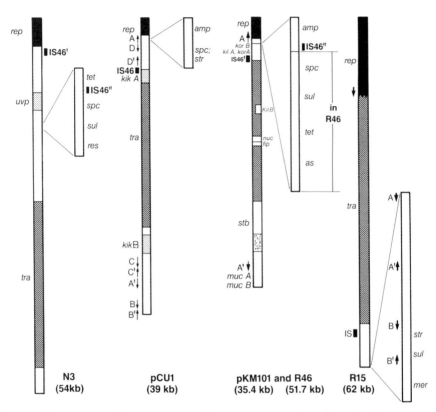

Figure 1. Comparison of the structural and functional organization of four IncN group plasmids of different origin. The circular maps are aligned and displayed as linear diagrams to facilitate comparisons. The solid pairs of arrows (e. g. A and A') indicate nucleotide sequences with extensive and sufficient similarity to form intrastrand duplex structures that can be visualized (32). IS46 is an Insertion Sequence element. Abbreviations: *rep*, replication; *uvp*, ultraviolet protection; *tra*, conjugative transfer; *amp, as, mer, spc, sul, str*, resistance to ampicillin, arsenate, mercuric chloride, spectinomycin, sulfonamide and streptomycin respectively; *kik*, killing of *K. pneumoniae*; *kil*, killing of *E. coli* K12; *kor*, overriding of the effect of *kil*; *nuc*, nuclease; *fip*, fertility inhibition of P-group plasmids; *stb*, stability of plasmid; *muc*, mutability by UV or by chemicals. The open rectangles with genetic regions that emerge from the main map are part of the map of each plasmid and placed to indicate their relative position.

sequence AGGTGGG is repeated. The pentanucleotide sequence ACAGG is also found in this replicon region (Figure 2). Similar sequences of five to seven base-pairs and direct sequence repeats of different length and number are a structural feature of regions containing the origins of replication of a number of different replicons of relatively narrow and broad host range (24). An additional structural feature that can be detected in the pCU1 replicon is a nine-teen base-pair sequence which is partially within one of the members of the thirty base-pair family and partially outside it and for which there is an inverted repeat sequence straddling each member of the thirteen-member family; this is shown at the bottom of Figure 2 and could potentially form intrastrand hairpin loops. Sequence similarities are not detected between either of these families of repeated sequences and those that have been observed at the origins of vegetative replication of other plasmids (F, R6K, RK2, pSC101, pSa, RSF1010). The *rep* region of the IncN group plasmid pKM101 is also being sequenced and there are indications of similarities to that of pCU1 (Hall, R., pers. commun.).

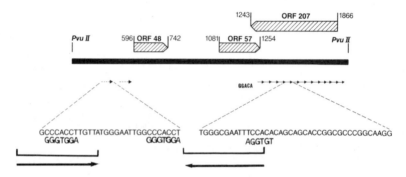

Figure 2. A single DNA fragment of pCUI of about 2 kb bounded by *Pvu* II sites and containing all *cis* and *trans*-acting regions required for the stable maintenance of the plasmid in *E. coli* (the *rep* region). See Figure 1 for the position of this fragment relative to the rest of the plasmid. The sequence reads from right to left on the conventional genetic map (32) ORF48, ORF57 and ORF207 are open reading frames of the indicated amino acid residues. The arrows indicate two different families of direct repeats of thirteen and two members respectively. The thirteen member family is tandem. The members of the smaller family (arrows with broken lines) are separated by a partial sequence of the member. The sequences of members of a family are very similar but not identical. For details see (32). The square brackets indicate a nineteen base-pair sequence that could form hairpin-like loops. The sequence ACAGG and the sequences shown as double strands are meant to highlight sequences that have often been observed at vegetative replication origins (see 24).

It is emphasized that this small region of pCU1 must contain not only the functional origin or origins of vegetative replication but also all plasmid deter-minants of plasmid replication, copy number, incompatibility and stability in *E. coli* K12. Furthermore and at least for sixty generations without selec-tion, it is maintained stably in *Klebsiella pneumonie*, *Proteus mirabilis*, *Alcaligenes faecalis*, *Serratia liquefaciens* and *Salmonella typhimurium* (Rajen-drakrishnan, B., personal commun.). Figure 2 also shows three open reading frames (ORFs) that may encode polypeptides of 48, 57 and 2107 amino acids respectively with the largest ORF including the large family of direct repeats. The two small ORFs have appropriately placed *E. coli* consensus promoter sequences. The unusual larger ORF does not. These features of the *rep* region of an IncN group plasmid are sufficiently different from those of the *rep* and *oriV* regions of other plasmids to warrant detailed functional analyses. In both pCU1 and in pKM101, the *rep* region, and also the *mucAB* genes, are contained in a region that is bounded by long inverted repeats that have been observed to form transposon-like homoduplex structures *in vitro* (32). Transposition of this structure has not been detected.

III. NONESSENTIAL SYSTEMS THAT AFFECT VIABILITY OF THE HOST OR THE PLASMID

In pKM101, two genes called *kilA* and *kilB* have been identified and mapped (see Figure 1) whose expression leads to lethality to the *E. coli* K12 host (65). In strains carrying pKM101 this potential lethality is prevented by two additional genes *korA* and *korB*. Expression of both *kor* genes is required to prevent the lethality caused by each of the two *kil* genes. It is of interest that insertion mutations in either *kilA* or *kilB* affect other phenotypes. Tn*5* insertions in *kilA* (65) or close to it (35) reverse a pKM101-mediated phenotype called Slo (giving rise to tiny colonies on Glucose Minimal Agar). Tn*5* insertions in *kilB* are conjugatively transfer-deficient suggesting that *kilB* is either required for transfer or is part of the same transcriptional unit as this *traD* gene (Figures 1, 3). There is evidence suggesting that genes with effects similar to *kil* and *kor* also exist in pUC1. It has been observed that some regions on pCU1 can be cloned into the pl5A replicon in *E. coli* K12 only in the presence of other regions. Also, when Tn*5* insertions about 3 kb anticlockwise from the *rep* regions of this plasmid were mapped, they were found to be accompanied by deletions extending in the direction of the *rep* region but not including it (33). One interpretation of the latter observation is that the primary insertion event inactivated a *kor* gene and that deletions promoted by a plasmid recombination system remove a cognate *kil* gene closer to but not including the *rep* region. In pKM101 additional *kil* genes are thought to be present at a similar position (Perry and Elledge; quoted in 65). It is possible that in both of these plasmids additional *kil* and *kor* genes remain to be identified. These observations invite comparisons with regions on the P-group plasmid RK2 that have similar effects on cell viability and/or plasmid replication. However, the lethality of the *kilA* and *kilB* genes of RK2 is not prevented by the presence of the IncN group plasmid N3. Thus if *kor* genes exist in N3, which is likely, they are not sufficiently similar to the *kor* genes of RK2 to afford protection against *kilA* and *kilB* of RK2.

When *E. coli* strains carrying any of several N group plasmids are mated with some strains of *K. pneumoniae*, only 1 to 10% of the recipients survive

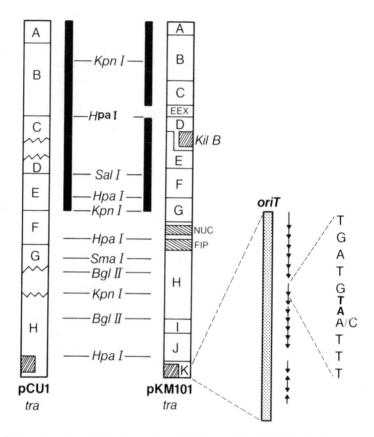

Figure 3. A comparison of the *tra* region of pCU1 and of pKM101. The lettered open boxes indicate different complementation groups with the dark bar indicating the regions determining N pilus synthesis and assembly. Eex, NUC and FIP are regions that interrupt *tra* of pKM101 (entry exclusion, nuclease and fertility inhibition of P-group plasmids). The hatched box contains the origin of conjugative transfer with the arrows indicating direct and inverted repeats. The sequence of one of a thirteen member family of repeats is also shown.

(45). This phenotype has been designated as the Kik (*ki*lling of *K*lebsiella) phenotype to distinguish it from the operationally different Kil phenotype seen in *E. coli* K12 (previous paragraph). It is quite possible that the molecular basis of the Kil and Kik phenomena will be related. However, in the present state of our knowledge, it is useful to use distinctive terminologies for them. The Kik phenotype is detected by mating plasmid-free *Klebsiella* with *E. coli* carrying an N group plasmid and determining the viability of *Klebsiella*. This phenotype is not determined by a substance that diffuses extracellularly from *E. coli* to the *Klebsiella* and transfer of DNA to the *Klebsiella* is required (45,

23). Transposon and deletion mutagenesis and DNA cloning experiments have defined three regions on pCU1 that are required optimally for this phenotype (55; Figure 1): First, the conjugative *tra* region composed of several complementation groups and mutations in any of which abolished both transfer and the Kik phenotype. Secondly, two regions called *kikA* and *kikB* mapping at or near either end of *tra*. The *kikA* region is at present defined by Tn5 insertions which abolish the Kik phenotype without affecting significantly the Tra phenotype. Insertion mutations of the presumptive *kikB* region have not yet been obtained. However, when this region was not included in clones that carried *kikA* and *tra*, the Kik phenotype was found to be reduced and variable.

The question of whether products of one or more of the *tra* genes are needed directly in *Klebsiella* for this phenotype or only indirectly as a means of efficiently transporting the two *kik* regions into *Klebsiella* was addressed by constructing *in vitro* a derivative of pCU1 from which the *oriT* region (see Section III) was deleted. This derivative does not show the Kik or Tra phenotypes. However, and presumably by providing all *trans*-acting Tra functions to an *oriT* (N)-containing plasmid, it is a helper plasmid of *oriT* (N)-containing plasmids in mating reactions. It was used as a helper for the transfer of a compatible recombinant plasmid composed of the plasmid replicon pACYC184 into which was inserted a segment of pUC1 containing *kikA*, *kikB* and the *tra* region. Tra⁻ Tn5 insertions were introduced separately into each of the *tra* genes of this recombinant plasmid and tested with and without the helper plasmid in mating experiments with *Klebsiella*. Only with the helper did all such mutant plasmids promote the killing of *Klebsiella* (Gill, M.Sc. Dissertation, Carleton University 1985' Rotheim, personal commun.). For this phenotype, *kikA* and *kikB* may therefore be the only regions required to be transmitted stably into *Klebsiella*. However, a possible involvement of the *oriT* region and/or of the mating process itself for this phenotype is an open question. Since full expression occurred when *kikA* and *kikB* were maintained by the heterologous replicon pACYC184, we can conclude that the phenotype can be manifested independent of the *rep* region of pUC1.

As indicated earlier, the Kik phenotype has been observed with all N group plasmids examined. It is therefore likely that regions analogous to *kikA*, *kikB* or both will be present on other N group plasmids. In pKM101, a region needed for this phenotype maps at a position similar to that of *kikB* (Winans and Walker, pers. commun.). Mutations in the known *kil* genes do not however affect the Kik phenotype. The mechanism of the Kik phenotype is not yet understood but it is known that mutations in all other known genetic regions of the plasmid and affecting other phenotypes (except Tra) do not affect the Kik phenotype. The immunity to this phenotype acquired by the surviving *K. pneumoniae* is particularly intriguing because the surviving cells and their progeny continue to be immune even upon losing the plasmid. (Gill, S., M.Sc. Dissertation, Carleton University, 1986). Regions on the plasmid required for this mutator activity have been identified and are currently being investigated.

IV. CONJUGATIVE PLASMID TRANSFER SYSTEM

Productive mating reactions involving a strain carrying an N group plasmid and a plasmid-free strain occur more efficiently on the surfaces of solid media than in liquid culture (20). This has been attributed to surface tension and the resulting intracellular physical pressure (51). N^+ strains produce rigid pili with a characteristic morphology (7) that is shared by pili produced by the promiscuous P-and W group plasmid (8) and also by plasmids of the group X (9) and one of the two pili produced by group I_2 (10). These pili are short (relative to the pili specified by other plasmid groups), rigid, 10 nm thick in electron micrographs and with one end characteristically pointed. Although morphologically similar, the pili specified by these different plasmid groups have not been found to be serologically related. In electron micrograph preparations, N group pili have been difficult to observe as structures attached to the cell surface but they have been easy to visualize as detached structures of variable length. Such detachment may occur by extrusion from the cell or as a consequence of the manipulative procedures. The filamentous bacteriophage IKe (29) has been found to be attached to the pointed tips of these detached N group pili (7) and to the surface of N^+ cells (12) and it has been suggested that the latter observation may be a consequence of the retraction of the N pilus (7). The tips of these detached N group pili have also been observed to be attached to other serologically-unrelated phages: phage PRD1, a lipid-containing phage with an icosohedral head and tail (and closely-related phages) and phage X, a filamentous curled phage. Although all these three phages attached to the pointed tips of N pili, their attachment specificities are reported to be different. Thus, IKe and the related phage I_2 have an attachment specificity that includes N and P-1 group pili (10), PRD1 has a specificity that includes the tips of N,P-1 and W group pili (6) and X has an attachment specificity that includes the tips of at least N, P-10 and X group pili (11). Furthermore, there is now good circumstantial evidence for an essential role for N pili in conjugation specified by the N-*tra* system (see below). It is therefore likely that the tips of these pili also play an important role in conjugation. Studies on the structure, biochemistry and physiology of these pili invite attention.

The *tra* regions of pKM101 and of pCU1 have been analysed by restriction-site mapping, insertion mutagenesis, cloning and genetic complementation (63,54). In both cases, a subset of the *tra* region is sufficient for N pilus formation and sensitivity to the N pilus-specific phages even when this subset is separated from the rest of the *tra* region and cloned into a heterologous nonconjugative plasmid. In pKM101, this N pilus region of *tra* is interrupted by a region, *eex*, mutations in which affect entry exclusion (but not conjugation efficiency). A second such interruption contains a gene, *fip*, required for the inhibition of fertility of coresident P-group plasmids and a separate gene *nuc*, for the production of a nuclease. A comparison of the *tra* region(s) of pKM101 and of pCU1 is in Figure 3. In both of these plasmids, the location of a region on the plasmid that is required to be present *in cis* for conjugative transfer has been defined. This region is referred to as the *oriT* region to indicate that it contains the origin of transfer. Recently, in both R46 (pKM101) and in pCU1 the limits of *ori*T have been defined and the region of about 600 bp sequenced (15; Patterson, pers. commun.). Features of this sequence that are of interest are the presence of thirteeen direct 11 bp repeats, three different pairs of 10 bp inverted repeats and an A-T rich segment (Figure 3). Direct repeat sequences have not been observed in the *ori*T regions of any other transfer system examined so far. Deletion of the 11 bp direct repeats resulted in a substantial reduction but not a complete elimination of *oriT* activity. There is also circumstantial evidence that the region of the inverted repeats contains the *nic* site, a site essential for *oriT* activity. Computer-assisted comparisons of the *oriT* sequences of R46 to the *oriT* of pSC101 and a sequence overlapping the pairs of inverted repeats of *oriT* of R46. The *oriT* sequences of F, RK2, ColE1, RSF1010, CloDF13 and pSC101 have revealed only limited homology between a region adjacent to the *nic* site of *oriT* regions of pCU1 and of R46 are nearly identical with a difference of only a few base pairs.

In naturally-occurring plasmids, the minimal N type of plasmid replicon is as a rule associated with an N *tra* system. However, there is some circumstantial evidence for the occurrence in nature of plasmids consisting of the N-*tra* system and IncF type replicons (1). In any event, the N-*tra* system can be dissociated from its native plasmid and reassociated with other plasmid replicons with the full retention of *tra* functions (15, 53, 63).

V. PLASMID RECOMBINATION SYSTEMS

Figure 1 indicates that the IncN group plasmids that have been examined all tend to contain either transposon-like structures, known Insertion Sequence elements or both. Such structures are often associated with either duplicative or conservative site-specific recombination events (30). It has been shown that R46 does encode a site-specific recombination system analogous to those of the transposons of the Tn*A* family (19). The existence of such systems condition the structural stability of a plasmid and may therefore indirectly influence plasmid maintenance. There is at present no evidence for a more direct role of plasmid recombination systems in the maintenance of IncN group plasmids.

VI. A SYSTEM THAT INCREASES ERROR-PRONE REPAIR

All or most plasmids of the N group enhance the susceptibility of *E. coli* or *S. typhimurium* to chemical and ultraviolet mutagenesis (58, 60). However, this ability is more widespread among the plasmids of Gram-negative bacteria (but it has not been reported in the other promiscuous plasmid groups of these bacteria). pKM101 enhances chemical and ultraviolet-induced mutagenesis and has played an important role in the development of sensitive and simple tests for the detection of chemical mutagens that are potential carcinogens (39). The mechanism of this enhancement of mutagenesis is not well understood but it has been the subject of insightful analysis. Inducible DNA repair in *E. coli* has been reviewed recently (36,60) and is not considered in detail here. Briefly, a current concept is that in *E. coli* and in *S. typhimurium*, mutagenesis induced by ultraviolet irradiation (UV) and certain chemical agents requires the participation of the bacterial genes *umuD* and *umuC* whose products along with those of other bacterial genes such as *recA* and *lexA* constitute the SOS regulatory network of DNA repair. pKM101 carries analogs of these two genes called respectively *mucA* and *mucB* and the plasmid suppresses the loss of induced mutability of *umuDC* mutants. Both the *umuDC* and *mucAB* operons are under the transcriptional control of the LexA repressor. Upon activation by an inducing signal, the activated RecA protein interacts with the LexA repressor. This interaction inactivates the LexA repressor allowing the induction of proteins such as UmuD and UmuC or MucA and MucB that are otherwise transcriptionally repressed by LexA. To account for the observation that activated RecA is continually required for this process, it has been suggested that the products of *mucA* (and *umuD*) may also be altered post-translationally by direct interaction with the activated RecA protein (37). Although the plasmid *mucA* and *mucB* genes together can suppress the phenotypes of the *umuDC* mutants, MucA alone cannot substitute for UmuD in a *umuD umuC*$^+$ host. Neither can MucB alone substitute for UmuC in *umuD*$^+$ *umuC* hosts. This implies that suppression requires a specific interaction between the two Muc proteins. It has been suggested (59) that the presence of *mucAB* genes on a promiscuous plasmid like pKM101 (R46) may be related to the observation that a number of Gram-negative bacteria in nature such as *P. mirabilis* are not mutable by UV and some chemicals unless they carry plasmids like pKM101 (26).

VII. POTENTIAL APPLICATIONS

The observation that the N *tra* system is efficient and has a broad host-range prompted the *in vitro* construction of host-dependent plasmid suicide vectors in which the *tra* region of pCU1 (Figure 1, 3) was cloned into the small plasmid pACYC184 with the p15A replicon that is not maintained in many Gram-negative bacteria (47). Alternatively a vector that is smaller and which carries only *oriT* of pCU1 has been constructed (Love, personal commun.). In this case, the *trans*-acting *tra* functions are provided by a helper that has its own *oriT* deleted (Figure 4). By introducing various transposons into these two

TABLE I. *Assessment of the usefulness of suicide plasmid vectors based on the N transfer system of pCU1 and the instability of the p15ᵃ replicon (pGS-and related vectors)*[a]

Bacterial species	Vector	Laboratory (and published references to the study, if any)	Summarized[*] uses and/or observed limitations
Agrobacterium tumefaciens	pGS9	Y.H. Cen, Guandong Insitute of Microbilogy, Guangzhou, China	Auxotrophs; non-tumorogenic mutants
Agrobacterium tumefaciens Biotype 2, strain H73	pGS9	J.A. Thompson, Lab. for Molec. and Cell Biology, P.O. Box 30947, Bramfontein 2017, South Africa	Auxotrophs
Azospirillum brasilensis (29710)	pGS9	W. Klingmuller, Lehrstuhl für Genetik Universität Bayreuth, Universitätstraße 30, Fed. Rep. Germany	Auxotrophs, auxin biosynthesis, nitrogen fixation mutants
Azospirillum lipoferum	pGS9 pGS81	R. Baly, Université Claude-Bernard Lyon 1 43 Boulevard Dev. 11 Nov. 1918, 69622 Ville urbanne, Lyon, France	Auxotrophs
Azospirillum sp.	pGS9	M. Bazzicalupo, Dipartimento de Biología Animal e Genetica, Via Romana 17, 50125 Firenze, Italy	Auxotrophs
Azotobacter chroococcum	pGS18	R.L. Robson, AFRC Unit of Nitrogen Fixation, Univ. of Sussex, Brighton BN1 9RQ, U.K. [44]	Nitrogen fixation
Azotobacter vinelandii	pGS9	H.K. Das, School of Environmental Sciences, Jawaharlal Nehru University, New Delhi, India	Nitrogen fixation
	pGS9	C. Kennedy, AFRC Unit of Nitrogen Fixation, Univ. of Sussex, Brighton BN1 9RZ, U.K	Nitrogen fixation
	pGS9	J. Casadesus, Dept. de Genetica, Facultad de Biología, Univ. de Sevilla, Apartado 1095, Spain	Auxotrophs, nitrogen fixation
Bradyrhizobium sp. (*Arachis*)	pGS9	K.J. Wilson, Dept. of Genetics, Harvard Medical School, Cambridge, Mass. U.S.A. [66]	Root nodulation; nitrogen fixation
Bordetella pertussis	pGS9	J.G. Coote, Microbiology Dept., Univ. of Glasgow, Glasgow G61 1QH, Scotland, U.K.	Auxotrophs
Pseudomonas atlantica (marine pseudomonad)	pGS9 pGS18	D.H. Bartlett, Agouron Inst., 505 Coast Boulevard South, La Jolla, California 92037, U.S.A.	Both plasmids maintained stably and not suicidal
Pseudomonas fluorescens (3580; Pf-5)	pGS9	J. Loper, U.S. Dept. of Agriculture, Hortic. Crops Research Lab., 3420 N.W. Orchard Av, Corvallis, Oregon 97330, U.S.A.	Transfer not detected ($<5 \times 10^{-10}$)

TABLE I. continued

Bacterial species	Vector	Laboratory (and published references to the study, if any)	Summarized* uses and/or observed limitations
Pseudomonas fluorescens (that is wheat-root-colonising and fungal pathogen-suppressive	pGS9	L. Thomashow, Root Disease and Biological Control Research Unit, 367 Johnson Hall, Washington State University, Pullman, WA 99164-6430, U.S.A. [56]	Mutants affecting the production of a phenazine antibiotic and pathogen-suppressivity
Pseudomonas paucimobilis	pGS9 pGS81	R. Baly, Université Claude-Bernard Lyon 1 43 Boulevard Dev. 11 Nov. 1918, 69622 Ville urbanne, Lyon, France	Auxotrophs
Pseudomonas putida (ATCC23973)	pGS9	J.K. Fredrickson, Terrestrial Sciences Section, Environmental Sciences, Battelle Pacific N.W. Labs., P.O. Box 999, Richland, Washington 99352, U.S.A.	Tn5-labelling and subsequent population studies on the bacteria in field soils
Pseudomonas syringae pv. *phaseolica*	pGS9	D.A. Cooksey, Dept. of Plant Pathology, Univ. of California, Riverside, 92521 U.S.A. [41]	Phaseolotoxin-negative
Pseudomonas syringae pv. *syringae* (several strains)	pGS9 pGS18 pGS27	A.K. Chaterjee, Dept. of Plant Pathology, Kansas State University, Manhattan, Kansas, U.S.A.	High level of intrinsic resistance to ampicillin and chloramphenicol by strains preclude the usefulness of pSG18 and pSG27; pSG9 found suitable for 10 of 13 strains tested
Pseudomonas syringae pv. *tabaci* (strain BR2 cured of plasmid pBPW1)	pGS9	P.D. Shaw, Dept. of Plant Pathology, College of Agriculture, 1102 South Goodwin Av., Urbana, Ill., U.S.A. 6180	Plant pathogenicity
Pseudomonas syringae pv. *tabaci* (strain BR2)	pGS18	P.D. Shaw, Dept. of Plant Pathology, College of Agriculture, 1102 South Goodwin Av., Urbana, Ill., U.S.A. 6180	Ampicillin-resistance lost upon release of selection
Pseudomonas syringae pv. tomato	pGS9	D.A. Cooksey, Dept. of Plant Pathology, Univ. of California, Riverside, 92521 USA [41]	Non-toxigenic; Tn5-labeling of indigenous plasmids
Pseudomonas syringae pv. tomato	pGS9	D.A. Cuppels, Agriculture Canada Research Center, London, Ontario N6A 5B7 Canada[16]	Auxotrophs

TABLE I. *continued*

Bacterial species	Vector	Laboratory (and published references to the study, if any)	Summarized* uses and/or observed limitations
Pseudomonas sp.	pGS9	P. Dion, Département de Phytologie, Université Laval, Quebec, Canada	Auxotrophs, octopine and nopaline catabolism, labeling indigenous plasmids
Rhizobium fredei (japonicum)	pGS9	L.D. Kuykendall, U.S. Dept. of Agriculture, Beltsville, Maryland, U.S.A.	Auxotrophs, nitrogen fixation
Rhizobium fredei (japonicum)	pGS6, pGS9	D.P.S. Verma, Dept. of Biology, McGill Univ., Montreal, Quebec, Canada [41, 46]	Construction of pGS8: MuD1 (Kan Lac)
Rhizobium leguminosarum (several strains)	pGS9	S. Ma, Dept. of Biology, Carleton University, Ottawa, Ontario K1S 5B6	Auxotrophs; nitrogen fixation
Rhizobium loti	pGS9	P. Hengen, Dept. of Biology, Carleton Univ., Ottawa, Ontario K1S 5B6 (M.Sc. dissertation, Univ. of Vermont, U.S.A.	Auxotrophs; root-nodulation
Rhizobium melitoti (Rmd 2011)	pGS9 pGS27	S. Kumar, Biotechnology Centre, Indian Agricultural Research Institute, New Delhi 110012, India	Auxotrophs; nitrogen fixation
Rhizobium melitoti (JJ1-C10)	pGS9 pGS27	G. Selvaraj, Plant Biotechnology Institute, Saskatoon, Sask., Canada [49]	Auxotrophs, root nodulation nitrogen fixation
Rhizobium melitoti (S14)	pGS9	H. Guilmette, Allelix Inc., 6850 Goreway Drive, Mississauga, Ontario L4V 1P1, Canada	Auxotrophs
Vibrio cholerae	pGS9	M.J. Voll, Dept. of Microbiology, Univ. of Maryland, College Park, Maryland 20743, U.S.A.	Vector maintained stable and not suicidal
Xanthomonas campestris	pGS9	M.J. Daniels, Dept. of Genetics, John Innes Institute, Colney Lane, Norwich NR4 7OB, U.K.	Tn5 transposition inefficient

[a] It will be noted that this Table of shared experiences contains relatively few references to observations that have as yet been published. I am grateful to all the individuals who have communicated their observations to me prior to such publication and whose addresses are shown.

The attention of the reader is also drawn to the fact that following the early observation of Beringer [55] and of Denarie [17] a number of different suicide vectors systems have been developed for bacteria other than *E. coli* and its close relatives [see chapter by R. Simon]. This affords a choice for the investigator and for the strain and species being studied.

types of suicide vectors, efficient delivery systems have been developed for insertion mutagenesis and subsequent manipulations in a variety of Gram-negative bacteria. The species in which these vectors have been tested and their usefulness are listed with comments in Table I. In general the usefulness of one or more of these suicide vectors depends on

(a) the efficiency with which the N-*tra* system functions in matings between the *E. coli* donor and the test recipient;

(b) the suicide effect, i. e. the degree of instability of the p15A-driven replicon;

(c) the relative frequency of single insertions of the transposon in contrast to other more complex events and

(d) the randomness with which the transposon is targeted.

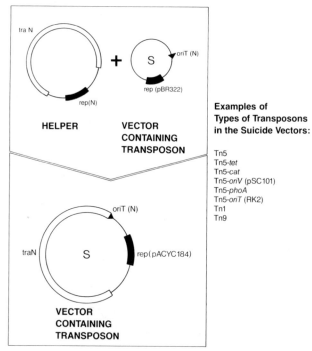

Figure 4. A schematic diagram of two kinds of suicide vectors that have been developed utilizing the efficient N transfer system and replicons (S) that are suicidal in a wide range of Gram-negative bacteria (See Table I). The dark rectangles in the S vectors indicate the replicon that is suicidal; *tra* N is the entire conjugative region of pUC1. A variety of transposons can be inserted and potentially used with these vectors and only some of those that we have constructed or used are shown. Tn5', *transposon* Tn5 Tn5-*tet* and Tn5-*cat*, Transposon Tn5 with dominant markers for tetracycline-resistance and chloramphenicol-resistance respectively; *ori*V, origin of vegetative replication; *pho*, *E. coli* alkaline phosphatase, *ori*T, origin of conjugative transfer.

Table I which is based on the collective experience of several laboratories suggests that pGS9 bearing Tn5 meets desirable criteria for several Gram-negative species. Since several derivatives of Tn5 carrying different markers are now available, the usefulness of this type of vector can be extended. It may also be noted that Tn5 itself confers resistance to streptomycin and bleomycin in some species of bacteria (25, 38, 43, 48) thus allowing the use of these phenotypes in a situation where this can be advantageous (see Chapter 8 for a more extensive discussion of transposon mutagenesis). The derivative pGS9 carrying the ori T region of RK2 has also been used successfully in our laboratory to mobilize the symbiotic plasmids of various Rhizobium species into Agrobacterium tumefaciens. As with any suicide vector used for insertion mutagenesis, strain differences are seen to exist in the efficiency and frequency with which simple insertions with the intended transposon arise relative to other unintended or more complex events. It is therefore desirable to subject mutants after they are obtained to a further genetic or structural analysis. The fact that the rep region of pCU1 is selfcontained in a relatively small region of 2 kb suggest an obvious use of this system for the development of cloning vectors. Structural and functional studies with this objective are currently in progress.

ACKNOWLEDGEMENTS

I am grateful to the Medical Research Council and the Natural Sciences and Engineering Research Council of Canada for operating research and equipment grants that supports the research in our laboratory. I take this opportunity to acknowledge the infectious enthusiasm of students and colleagues who have shared my interest in this topic and contributed energetically to its understanding. For a version of Figure 3 I thank Paul Hengen.

VIII. REFERENCES

1. Anderson ES, Threlfall EJ, Carr JM, Frost JA (1974) Transferable drug resistance in Salmonellae in South and Central America. Proc Soc Gen Microbiol 1, 66
2. Aoki T, Kitao T, Arai T (1977) R-plasmids in fish pathogens. In Plasmids, Medical and Theoretical Aspects, (Mitsuhashi S, Rosival L and Krcmery V, eds), Springer-Verlag, Berlin pp 39–45
3. Arai T, Yao-Sugawa K, Aoki, T (1980) The stability of R-plasmids in Vibrio cholerae. Keio J Medicine 29, 125–32
4. Bender CL, Stone HE, Sims JJ, Cooksey DA (1987) Reduced pathogen fitness of Pseudomonas syringae pv. tomato Tn5 mutants defective in coronatin production. Phytopath Mol Plant Path 30 (in press).
5. Beringer JE, Benyon JL, Buchanan-Wollaston AV, Johnston AWB (1978)

Transfer of the drug-resistance transposon Tn5 to *Rhizobium. Nature* **276**, 633–4

6. Bradley DF (1978) W-pili: characteristics and interaction with lipid phages specific for N, P-1 and W group plasmids. *In* Pili, (Bradley DE, Raizen E, Fives-Taylor P, Ou J eds) Intl Conferences on Pili, Washington DC.

7. Bradley DE (1979) Morphology of pili determined by the N incompatibility group plasmid N3 and interaction with bacteriophages PR4 and IKe. *Plasmid* **2**, 632–6

8. Bradley DE (1980) Morphological and serological relationships of conjugative pili. *Plasmid* **4**, 155–69

9. Bradley DE (1980) Determination of pili by conjugative bacteria carrying drug resistance plasmids of the incompatibility groups B, C, H, J, K, M, V, and X. *J Bacteriol* **141**, 828–37

10. Bradley DE, Coetzee JN, Hedges RW (1983) Inc I$_2$ plasmids specify sensitivity of filamentous bacteriophage IKe. *J Bacteriol* **154**, 505–7

11. Bradley DE, Coetzee JN, Bothma T, Hedges RW (1981) Phage X: a plasmid-dependent, broad host-range, filamentous bacterial virus. *J Gen Microbiol* **126**, 389–96

12. Brodt P, Leggett F, Iyer R (1974) Absence of a pilus receptor for the filamentous phage IKe. *Nature* **249**, 856–8

13. Brown AMC, Willetts NS (1981) A physical and genetic map of the Inc N plasmid R46. *Plasmid* **5**, 188–201

14. Brown AC, Coupland GM, Willetts NS (1984) Characterization of IS*46*, an insertion sequence found on two Inc N plasmids. *J Bacteriol* **159**, 472–81

15. Coupland AC, Coupland GM, Willetts NS (1987) The origin of transfer (*ori T*) region of the conjugative plasmid R46: characterization by deletion analysis and DNA sequencing. *Mol Gen Genet* **208**, 219–25

16. Cuppels DA (1986) Generation and characterization of Tn5 insertion mutations in *Pseudomonas syringae* pv tomato. *Appl Env Microbiol* **51**, 323–7

17. Denarie J, Rosenberg C, Bergeron B, Boucher C, Michel M, Barate BM (1977) Potential for RP4::Mu plasmid for *in vivo* genetic engineering of Gram-negative bacteria. *In* DNA Insertion Elements, Plasmids and Episomes, (Bukhari AI, Shapiro JA, Adhya SL eds), Cold Spring Harbor Laboratory, New York.

18. Datta N, Hedges RW (1971) Compatibility groups among *f*⁻ R-factors. *Nature* **234**, 222–3

19. Datta N, Hughes MV (1983) Plasmids of the same Inc groups in Enterobacteriaceae before and after the medical use of antibiotics. *Nature* **306**, 616–7

20. Dennison S, Baumberg S (1975) Conjugational behaviour of N plasmids in *Escherichia coli* K-12. *Mol Gen Genet* **138**, 323–31

21. Dodd HM, Bennett PM (1983) R46 encodes a site-specific recombina-

tion system interchangeable with the resolution function of Tn A. *Plasmid* **9**, 247–61

22. Ely B (1979) Transfer of drug resistance factors to the dimorphic bacterium *Caulobacter crescentus. Genetics* **91**, 371–80

23. Gill S, Iyer VN (1982) Nalidixic acid inhibits the conjugal transfer of conjugative N incompatibility group plasmids. *Can J Microbiol* **28**, 256–8

24. Filutowicz M, McEachern M, Greener A, Mukhopadhyay, P., Uhlenhopp E, Durland R, and Helinski D (1985) Role of the initiation protein and direct nucleotide sequence repeats in the regulations of plasmid R6K replication. *In* Plasmids in Bacteria, (Helinski DR, Cohen SN, Clewell DB, Jackson DA, Hollander A, eds), Plenum Press, NY, pp 125–30

25. Geniolloud O, Garrido MC, Moreno F (1984) The transposon Tn5 carries a bleomycin resistance determinant. *Gene* **32**, 225–32

26. Hofemeister J, Kohler H, Fillipor VD (1979) Repair and plasmid R46 mutation requires inducible functions in *Proteus mirabilis. Mol Gen Genet* **176**, 265–73

27. Jacob AE, Shapiro JA, Yamamoto L, Smith DI, Cohen SN, Berg D (1977) Plasmids studied in *Escherichia coli* and other enteric bacteria. *In* DNA insertion elements, plasmids and episomes, (Bukhari AI, Shapiro JA, Adhya SL eds) Cold Spring Habor Lab, NY, pp 607–56

28. Khatoon H, Iyer RV (1971) Stable coexistence of *fi⁻* R-factors in *Escherichia coli. Can J Microbiol* **17**, 669–75

29. Khatoon H, Iyer RV, Iyer VN (1972) A new filamentous bacteriophage with sex factor specificity. *Virology* **48**, 145–55

30. Kleckner N (1981) Transposable elements in prokaryotes. *Ann Rev Genet* **15**, 341–404

31. Konarska-Kozlowska M, Iyer VN (1981) The genetic and physical basis of variability in *Escherichia coli* strains carrying a reference IncN group plasmid. *Can J Microbiol* **27**, 616–26

32. Konarska-Kozlowska M, Thatte V, Iyer VN (1983) Inverted repeats in the DNA of plasmid pCU1. *J Bacteriol* **153**, 1502–12

33. Konarska-Kozlowska M, Iyer VN (1984) A minimal region of an IncN group plasmid required for its maintenance in *Escherichia coli. In* Plasmids in Bacteria, (Helinski DR, Cohen SN, Clewell DB, Jackson DA, Hollaender A eds) Plenum Press, New York, pp 864

34. Kozlowski M, Thatte V, Lau PCK, Visentin LP, Iyer VN (1987) The minimal replicon of the promiscuous plasmid PCU1. *Gene* **58**, 217–28

35. Langer PJ, Shanbruch WG, Walker GC (1981) Functional organization of plasmid pKM101. J Bacteriol **145**, 1310–6

36. Little JW, Mount DC (1982) The SOS regulatory system of *Escherichia coli* K12. *Cell* **29**, 11–22

37. Marsh L, Walker GC (1987) New phenotypes associated with *mucAB*: alteration of MucA sequence homologous to *lexA* cleavage sites. *J Bacteriol*

169, 1818-23

38. Mazodier P, Giraud E, Glasser F (1982) Tn*5* dependent streptomycin resistance in *Methylobacterium organophilum*. *FEMS Microbiol Lett* **13**, 27-30

39. McCann J, Spingarn NE, Kobori J, Ames BN (1975) Detection of carcinogens as mutagens: bacterial tester strains with R-factor plasmids. *Proc Natl Acad Sci USA* **72**, 979-83

40. Molina AM, Babudri N, Tamara M, Venturina S, Monti-Bragadin C (1982) Enterobacteriaceae plasmids enhancing chemical mutagenesis and their distribution among incompatibility groups. *FEMS Microbiol Lett* **5**, 33-7

41. Olson ER, Sadowsky MJ, Verma DPS (1985) Identification of genes involved in the *Rhizobium*-legume symbiosis by Mu-Dl (Kan, Lac) generated transcription fusions. *BioTechnology* **3**, 143-9

42. Parish JH (1975) Transfer of drug resistance of *Myxococcus* from bacteria carrying drug-resistance factors. *J Gen Microbiol* **87**, 198-210

43. Putnoky P, Kiss GB, Oh I, Kondorosi A (1983) Tn*5* carries a streptomycin resistance determinant downstream from the kanamycin resistance gene. *Mol Gen Genet* **191**, 228-84

44. Robson RL (1986) Nitrogen fixation in strains of *Azotobacter chroococcum* bearing deletions of a cluster of genes coding for nitrogenase. *Arch Microbiol* **146**, 74-9

45. Rodriguez M, Iyer VN (1981) Killing of *Klebsiella pneumoniae* mediated by conjugation with bacteria carrying antibiotic-resistance plasmid of the group N. *Plasmid* **6**, 141-7

46. Rostas K, Sista PR, Stanley J, Verma DPS (1984) Transposon mutagenesis of *Rhizobium japonicum*. *Mol Gen Genet* **197**, 230-5

47. Selvaraj G, Iyer VN (1984) Suicide plasmid vehicles for insertion mutagenesis in *Rhizobium meliloti* and related bacteria. *J Bacteriol* **156**, 1292-1300

48. Selvaraj G, Iyer VN (1984) Transposon Tn*5* specifies streptomycin resistance in *Rhizobium* spp. *J Bacteriol* **158**, 580-9

49. Selvaraj G, Hooper I, Shantharam S, Iyer VN, Barran L, Wheatcroft R, Watson RJ (1987) Derivation and molecular characterization of symbiotically-deficient mutants of *Rhizobium meliloti*. *Can J Microbiol* **33**, 739-47

50. Singh M, Klingmuller W (1986) Transposon mutagenesis in *Azospirillum brasilense*: isolation of auxotrophic and Nif⁻ mutants and molecular cloning of the mutagenized *nif* DNA. *Mol Gen Genet* **202**, 136-42

51. Singleton P (1983) N-mating: influence of surface tension. *FEMS Microbiol Lett* **19**, 179-82

52. Tardif G, Grant RB (1980) Characterization of the host range of N-plasmids. *In* Plasmids and Transposons, (Stuttard C, Rozee KR eds), Academic Press, New York, pp 351-9

53. Thatte V, Iyer VN (1983) Cloning of a plasmid region specifying the N system of bacterial conjugation. *Gene* **21**, 227–36

54. Thatte V, Bradley DE, Iyer VN (1985) N conjugative transfer system of plasmid pCU1. *J Bacteriol* **163**, 1229–36

55. Thatte V, Gill S, Iyer VN (1985) N conjugative transfer system of plasmid pCU1. *J Bacteriol* **163**, 1296–99

56. Thomashow LS, Weller DM, Cook RJ (1986) Molecular analysis of phenazine antibiotic synthesis by *Pseudomonas fluorescens* strains 2–79. *Third Intl Symp on the molecular genetics of plant-microbe interactions.* McGill Univ, Montreal, pp 22

57. Tschape H (1977) Genetic characterization of transferable plasmids presumed to be unloaded. *In* Plasmids, Medical and Theoretical Aspects, (Mitsuhashi S, Rosival L, Krcmery V eds) Springer-Verlag, Berlin, pp 109–15

58. Upton C, Pinney RJ (1983) Expression of eight unrelated Muc$^+$ plasmids in eleven DNA repair-deficient *E. coli* strains. *Mutation Res* **112**, 261–73

59. Walker GC (1984) Mutagenesis and inducible DNA responses to deoxyribonucleic acid damage in *Escherichia coli*. *Microbiol Rev* **48**, 60–93

60. Walker GC (1985) Inducible DNA repair systems. *Ann Rev Biochem* **54**, 425–57

61. Watanabe T, Nishida H, Ogata C, Arai T, Sato S (1964) Episome-mediated transfer of drug resistance in Enterobacteriaceae. VII. Two types of naturally occurring R-factors. *J Bacteriol* **88**, 716–26

62. Winans SC, Walker GC (1983) Genetic localization and characterization of a pKM101-coded endonuclease. *J Bacteriol* **154**, 1117–25

63. Winans SC, Walker GC (1985) Conjugal transfer system of the IncN plasmid pKM101. *J Bacteriol* **161**, 402–10

64. Winans SC, Walker GC (1985) Fertility inhibition of RP1 by IncN plasmid pKM101. *J Bacteriol* **161**, 425–7

65. Winans SC, Walker GC (1985) Identification of pKM101-encoded loci specifying potentially lethal gene products. *J Bacteriol* **161**, 417–24

66. Wilson KJ, Anjaiah V, Nambiar PTC, Ausubel FM (1987) Isolation and characterization of symbiotic mutants of *Bradyrhizobium* sp. (Arachis) strain NC92: mutants with host-specific defects in nodulation and nitrogen fixation. *J Bacteriol* **169**, 2177–86

67. Woese CR, Weisburg WG, Hahn CM, Paster BJ, Zablev LB, Lewis BJ, Macke TJ, Ludwig W, Stachenbrandt E (1985) The phylogeny of purple bacteria: the Gamma subdivision. *System Appl Microbiol* **6**, 25–33

CHAPTER 7

USE OF IncP PLASMIDS IN CHROMOSOMAL GENETICS OF GRAM-NEGATIVE BACTERIA

Dieter Haas and Cornelia Reimmann

I. INTRODUCTION

The genetic analysis of Gram-negative bacteria has greatly benefitted from the use of the broad-host-range IncP plasmids RP1, RP4, RK2, and R68 and their derivatives. These plasmids have become very valuable for chromosome mobilization and as cloning vehicles in a wide range of Gram-negative bacteria. Thus, in principle, it is no longer necessary to search for plasmids that would be suitable as sex factors or cloning vectors in each individual bacterial genus or species of interest, but instead the existing array of IncP (and IncQ) plasmids can be applied. In practice, however, there may be some limitations in that the antibiotic resistance determinants of IncP plasmids might not all be expressed in a particular bacterial isolate or the transposition functions (which are important for chromosome mobilization and R' plasmid formation) may not be active in all hosts in which IncP plasmids replicate.

There is a vast body of empirical information on the use of IncP plasmids as genetic tools in Gram-negative bacteria. Less information is available on molecular mechanisms, e. g. the processes of chromosome mobilization and R' formation, and the data that have been obtained for *Escherichia coli*

Promiscuous Plasmids of Gram-Negative Bacteria   © 1989 Academic Press Limited
ISBN 0-12-688480-3   All rights of reproduction in any form reserved

or *Pseudomonas* cannot automatically be extrapolated to any Gram-negative bacterium. In this review we try to bring together some concepts of how IncP plasmids interact with other replicons and how these interactions can be exploited in the genetic analysis of chromosomal genes.

The main emphasis will be on the IncP plasmids RP1, RP4, RK2 and R68 which, as discussed in Chapter 3, are virtually indistinguishable. The transposable elements of these plasmids have received different designations but it is not known whether sequence differences exist. The transposon carrying the TEM-2 β-lactamase gene has been called Tn*1* (on RP4; 52), Tn*401* or Tn*801* (on RP1; 19,9), or more generally Tn*A* (on R68 and other plasmids; 120), and the insertion sequence near the kanamycin resistance determinant has been designated IS*8* (on RP4; 29) or IS*21* (on R68; 120). In this review we will use Tn*1* and IS*21* as generic designations for these two transposable elements. The role of the newly discovered tellurite resistance transposon Tn*521* of RP4 (15) cannot yet be assessed.

II. THE F-PLASMID AS A PARADIGM; USE OF Tn5-*OriT* INSERTIONS

The F-plasmid of *E. coli* is the best known conjugative plasmid of Gram-negative bacteria and hence most ideas about conjugation and mobilization are derived from studies of this system. Conjugal transfer of F is initiated at the origin of transfer (*ori T*). A single strand is then transferred unidirectionally to the recipient, with the 5' end leading. Double-stranded plasmid molecules are generated in both the donor and the recipient by conjugal DNA synthesis (see Chapter 2). Stable integration of F into the *E. coli* chromosome results in Hfr donor strains which transfer the *ori T*-proximal part of F and adjacent chromosomal genes to recipient cells at high frequencies (42). In this situation, both *ori T* and the *tra* genes are present on the mobilized replicon in *cis*. Polar chromosome transfer also occurs when *ori T* alone is inserted into the chromosome and the *tra* genes are located in *trans* on an autonomous plasmid, which *per se* may have little or no chromosome mobilizing ability (= Cma) (14). Thus, the presence in *cis* of *ori T* on the mobilized replicon is assumed to be essential for mobilization (119).

By analogy with the F system, the *ori T* of RP4/RK2 may be inserted into a replicon to be mobilized; the *tra* functions can be supplied in *cis* or in *trans* by an IncP plasmid. The construction of Tn5-*ori T*$_{RP4/RK2}$ (= Tn5-Mob) (123,108) permits random insertion of *ori T*$_{RP4/RK2}$. Polarized chromosome transfer from *ori T*$_{RP4/RK2}$ mediated by an autonomous IncP plasmid has been demonstrated in *E. coli*, *Rhizobium meliloti* (108,123), and *Bradyrhizobium japonicum* (D. Kuykendall, personal communication).

The autonomous F-plasmid mobilizes the *E. coli* chromosome at low frequencies. Some of the recombinants formed in F$^+$ x F$^-$ matings are due to the presence of stable Hfr donors in the F^+ population (24). Stable integration of F comes about by homologous recombination between insertion sequences (IS2, IS3) which are present on the F-plasmid and on the chromosome at various locations. In addition, the transposable element Tn1000 of F can promote the formation of stable Hfr strains (51). Most recombinants arising in F^+ x F^- crosses appear to result from unstable insertion of F into the donor chromosome; the same transposable elements (IS2, IS3, Tn1000) probably participate in this process (17). By analogy again, we can state that IncP plasmids mobilize other replicons and, in particular, the chromosome via the formation of stable or transient cointegrates (see IV and V).

F' plasmids, which carry a segment of chromosomal DNA, transfer the chromosome unidirectionally from the region of homology (91). A region of 'portable homology' (65) can also be created by the insertion of a transposon (e. g., Tn10) into the F-plasmid and into the chromosome, and homologous recombination between the two transposon copies permits polar chromosome transfer as in the situation above. Again, the same principles apply to IncP R-plasmids (7,63,68). Table I summarizes the mechanisms of mobilization in Gram-negative bacteria.

III. INTERACTIONS OF IncP PLASMIDS WITH OTHER REPLICONS: ROLE OF TRANSPOSABLE ELEMENTS

Transposon Tn1 (5.0 kb) on RP4 is a member of the Tn3 family (53). These transposons mostly use a replicative pathway of transposition (37). In a first step, the Tn1/3 transposase (the tnpA gene product) catalyzes the formation of a cointegrate between the donor and the target replicon. In the cointegrates the two replicons are joined together by transposon copies in the same orientation (Figure 1). The cointegrates are resolved by the Tn1/3 site-specific recombination system which involves resolvase (the tnpR gene product) and an internal resolution site (IRS, located between the tnpA and tnpR genes) (53). Alternatively, the host recombination system may resolve the cointegrates by homologous recombination (Figure 1). Inactivation of the resolvase function permits the physical isolation of cointegrate plasmids (53). For instance, the tnpR gene of Tn1 on RP1 has been inactivated by insertion of a trimethoprim (Tp) resistance determinant into the unique BamHI site, which is located within tnpR (see Chapter 3 Figure 1). The resulting recombinant plasmid pME134 forms stable cointegrates with other plasmids at frequencies of ca. 10^{-4} in recombination-deficient E. coli strains (93). Cointegrates between pME134 and the chromosome of P. aeruginosa are formed at low frequencies (ca. 10^{-7}); the integrated plasmid is bordered by the Tn1Tp determinants in direct orientation as expected and promotes unidirectional chromosome transfer (93). The low frequency of pME134 integration is due to the fact that Tn1/3 transposes to chromosomal sites much less frequently

TABLE I. *Modes of mobilization*

Location of *tra* genes	Requirements	
	Mobilizing plasmid	Mobilizable replicon (chromosome, plasmid)
In *trans*	*tra*$^+$ e.g. RP1	*oriT*$^+$ [a]; e.g.insertion of Tn5-*oriT*
	i) *tra*$^+$ plus transposable element(s); e.g. RP1 (with Tn*1*), pULB113 (with mini-Mu), or R68.45 (with (IS*21*)$_2$)	None
In *cis*, i.e. formation of a cointegrate between mobilizing and mobilizable replicon[b]	ii) *tra*$^+$; e.g. RP1	Transposable element(s) analogous to (i)
	iii) *tra*$^+$ plus homology with mobilizable replicon; e.g. R′ carrying chromosomal insert	Homology with mobilizing plasmid; Rec$^+$ host

[a] Additional mobilization (*mob*) functions may be required; e.g.the *mob* genes of ColE1 are needed for F-dependent mobilization of a replicon carrying *oriT*$_{ColE1}$ (14).

[b] The stability of such a cointegrate is not considered here.

than it does to plasmid targets (first observed in *E. coli*; 67). When the resolution of the plasmid-chromosome cointegrate is not prevented by *tnpR* and host *recA* mutations, the plasmid will be excised, leaving behind a single Tn*1* copy in the chromosome (of *E. coli* or *P. aeruginosa*) (26,100,50,47).

The insertion element IS*21* (2.1 kb) of R68 displays low transpositional activity (120,93,94). IS*21* has the unusual ability to form tandem repeats. In this configuration, first found on R68.45 (Figure 1), the transpositional activity is elevated at least a thousand-fold (98,69,120,94). Plasmids carrying an IS*21* duplication form cointegrates with other plasmids in *E. coli* at 10^{-3}–10^{-5} (detected by mating-out assays; 99,94), and it seems that R68.45 mobilizes the bacterial chromosome by an analogous mechanism (120; cf. section V). Cointegrates between R68.45 and another replicon carry a single IS*21* copy at each junction (Figure 1; 99). The fact that plasmids with (IS*21*)$_2$ readily give cointegrate structures does not imply that IS*21* itself uses a replicative ('cointegrate') pathway of transposition, which is typical of the Tn*1/3* elements. Current evidence strongly suggests that plasmids with (IS*21*)$_2$ show a conservative, IS*21*-mediated mode of transposition (94). The IS*21*-encoded transposase appears to recognize predominantly the junction of (IS*21*)$_2$. R68.45 and related plasmids seem to be opened between the two IS*21* elements and inserted as a whole into target replicons. R68.45 can therefore be viewed as

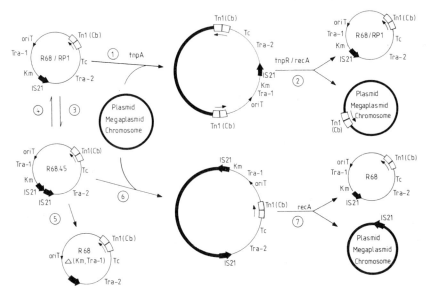

Figure 1. Transposition pathways of Tn*1* and IS*21*. 1, Replicative, *tnpA*-dependent transposition of Tn*1*; 2, cointegrate resolution by resolvase (*tnpR*) or homologous recombination of host (*recA*); 3,duplication of IS*21* (see text); 4, loss of one IS*21* copy, probably via homologous recombination; 5, IS*21*-promoted deletion of one IS*21* copy and an adjacent region: either Km^R and Tra1 (46) or Tra2 (not shown; 23); 6, conservative, IS*21*-mediated transposition; 7, cointegrate resolution by homologous recombination.

a large transposon whose ends come together forming a circular molecule (94).

In the tandem repeat of R68.45, the two IS*21* copies are separated by 3 bp (our unpublished data). The left 11 bp terminal inverted repeat of the right-hand IS*21* element (near Tra-2) carries a typical '−35' sequence (TTGACA) of the *E. coli* consensus promoter (N. Willetts, personal communication). The right-hand 11 bp repeat of the left element (near the kanamycin resistance determinant) includes a potential imperfect '−10' region (TATAAG), at an adequate distance (18 bp) from the '−35' region. Although transcription initiation at this potential promoter has not yet been demonstrated, it is clear that the junction region contains a promoter reading into the left IS*21* element (103). In *E. coli* the transposition functions of the left element are a hundred times more active than those of the right element (94). It therefore seems likely that the putative junction promoter is responsible for the elevated transpositional activity of the left element. Since R68.45 mobilizes the chromosome of many different bacteria other than *E. coli* (Table II), the host range of the transpositional activity of the IS*21* tandem duplication must be relatively broad. In some bacteria, e. g. *Proteus mirabilis* and *Pseudomonas cepacia* 249 (22,73), R68.45

does not display Cma. Is this a consequence of poor IS21 transposase expression in these bacteria?

There is circumstantial evidence that IS21 tandem duplications are not always identical at the sequence level, since independent isolates of IncP plasmids carrying IS21 duplications may have different Cma's. For instance, pMO61 is indistinguishable from R68.45 in restriction analysis, but is much better at mobilizing the *P. putida* chromosome than is R68.45 (16). An R68.45-like derivative, pBLM2, has higher Cma than has R68.45 in *Rhodobacter capsulatus* (formerly *Rhodopseudomonas capsulata*) (75,125). While it seems clear that IS21 does not code for a resolvase (120,99), little is known about the IS21 transposition functions at the biochemical level. The IS21 sequence has the coding capacity for two or more proteins, one of which would be the transposase. Another protein might be involved in stimulation of transposition; an indication for such a function comes from the finding that IS21 and Tn7 mutually stimulate transposition of each other in *trans* (104).

The process of IS21 tandem repeat formation (step 3 in Figure 1) is poorly understood. A *recA*-dependent amplification mechanism (3,89) is unlikely because IS21 does not have directly repeated sequences at its ends. We favour models (32,92) which propose that tandem duplications may be formed within the replication fork by a slippage error or by a one-ended transposition (10,25) of one transposable element to the end of the other transposable element on the other branch of the replication fork. The generation of a tandem repeat would then be completed by repair and a further round of replication. There is evidence from several different experiments that IS21 has a tendency to transpose into sites at or very near the termini of another IS21 element (27,94).

The chances of some replicon fusion process taking place may be improved if additional transposable elements are loaded onto IncP plasmids. For example, the mini-Mu transposon (8.0 kb) on pULB113 (= RP4::mini-Mu; 114) allows efficient chromosome mobilization in many bacteria (Table II). During lytic growth, Mu phage predominantly uses a replicative mode of transposition (82). The interaction between pULB113 and other replicons is therefore best explained by a replicative, mini-Mu-dependent pathway of cointegrate formation (step 1 in Figure 2A; 114). Transposons Tn501 and Tn813 (= Tn21ΔtnpR) both enhance the Cma of IncP plasmids in *Rhodobacter sphaeroides* (88,13). Both transposons are in the same subgroup of Tn3-like elements and transpose via a replicative pathway (30). Deletion of the resolvase gene in Tn813 results in the formation of stable cointegrates between R751::Tn813 and pBR322 (13). Like pME134 (93), R751::Tn813 has the potential to form stable cointegrates with the chromosome; this has been observed in *R. sphaeroides* and other bacterial species (13). An insertion sequence of *P. cepacia*, IS401 (1.3 kb), promotes replicon fusion and integration of an RP1 derivative into the chromosome of this organism (6). The *P. aeruginosa* transposon Tn2521 has proved effective in directing chromosomal integration

TABLE II. *Chromosome mobilization by IncP-1 plasmids (not stably integrated) in Gram-negative bacteria*

Bacteria	Plasmid	Approximate frequency of chromosome mobilization (= recombinants per donor)	Ref
Acinetobacter calcoaceticus	RP4	$10^{-6}-10^{-8}$	112,113
Aeromonas hydrophila	pULB113	$10^{-4}-10^{-5}$	36
Agrobacterium tumefaciens	R68.45	$10^{-4}-10^{-6}$	61,81
(chrom. Tn5)	R68.45::Tn5	$10^{-4}-10^{-7a}$	90
Alcaligenes eutrophus	pULB113	$10^{-4}-10^{-5}$	71
Azospirillum brasilense	R68.45	10^{-6}	34
Azotobacter vinelandii	R68.45	$10^{-4}-10^{-5}$	111a
Bordetella pertussis	R68.45	$10^{-6}-10^{-8}$	110
Caulobacter crescentus	RP4	$10^{-6}-10^{-7}$	5
Enterobacter cloacae	pULB113	$10^{-5}-10^{-7}$	105
Erwinia amylovora	pULB113	10^{-6}	21
Erwinia carotovora			
subsp. *atroseptica*	pULB113	$10^{-7}-10^{-8}$	21
subsp. *carotovora*	pULB113	$10^{-6}-10^{-7}$	21
	R68.45	$10^{-5}-10^{-7}$	20
	R68::Mu	10^{-5}	33
subsp. *chrysanthemi*	pULB113	10^{-3}	102
Escherichia coli	RP4	$10^{-7}-10^{-8}$	114
	R68	$5x10^{-8}$	85
	R68.45	$10^{-6}-10^{-7}$	85
	RP4'	$10^{-3}-10^{-7a}$	7,63
	pULB113	10^{-4}	114
Klebsiella pneumoniae	R68.45	10^{-5}	72
	pULB113	$10^{-6}-10^{-7}$	114
Legionella pneumophila	R68.45	10^{-8}	31
Methylobacterium AM1	R68.45	$10^{-4}-10^{-5}$	111
Methylophilus methylotrophus	R68.45	$10^{-6}-10^{-7}$	83,58
Proteus mirabilis	R772	10^{-5}	22
	pULB113	$10^{-6}-10^{-7}$	114
Proteus morganii	RP4'	$10^{-4}-10^{-6a}$	8
Pseudomonas aeruginosa PAO	R68.45	$10^{-3}-10^{-5}$	43
	RP4::D3112cts	10^{-4}	124
(chrom. Tn1)	R18 = RP4	$10^{-5}-10^{-9a}$	68
P. aeruginosa	R68	$10^{-3}-10^{-5a}$	116
	R68.45	$10^{-3}-10^{-4}$	116
Pseudomonas fluorescens	pMO47b	10^{-5}	70
	pULB113	$10^{-4}-10^{-5}$	71
Pseudomonas glycinea	R68	$10^{-6}-10^{-7}$	35

TABLE II. *continued*

Bacteria	Plasmid	Approximate frequency of chromosome mobilization (= recombinants per donor)	Ref
	R68.45	10^{-6}–10^{-8}	35
Pseudomonas putida	pMO61[b]	10^{-4}–10^{-6}	16
Rhizobium leguminosarum	R68.45	10^{-6}	11
Rhizobium melitoti	RP4	10^{-5}–10^{-6}	78
	R68.45	10^{-3}–10^{-5}	66
	RP4'	10^{-4}–10^{-7a}	63
Rhizobium trifolii	R68.45	10^{-4}–10^{-5}	80
Cowpea rhizobia	RP4	10^{-6}	77
	R68.45	10^{-6}	77
Rhodobacter capsulatus	R68.45	10^{-6}	126
	pTH10[c]	10^{-3}–10^{-6}	121
Rhodopseudomonas spharoides	RP1::Tn501	10^{-3}–10^{-7a}	88
	R751::Tn813	10^{-4}–10^{-5a}	13
	R68.45	10^{-4}–10^{-5}	169
	pBLM2[b]	10^{-3}–10^{-4}	75
Salmonella typhimurium	pULB113	10^{-6}–10^{-7}	114

Further references can be found in an earlier review (55)
[a] Polarized chromosome transfer from one or several sites
[b] pMO47. pMO61, and pBLM2 are R68.45-like plasmids
[c] pTH10 = RP4*trfA*(Ts)

of an R68 derivative in *P. aeruginosa* (86), and an unidentified insertion element of *E. coli* accounts for Cma of several R68 derivatives in this bacterium (85).

IV. STABLE INTEGRATION OF IncP PLASMIDS INTO THE CHROMOSOME AND FORMATION OF HFR DONORS

Spontaneous stable integration of IncP plasmids into the bacterial chromosome is rare, with the exception of *Myxococcus xanthus*, where RP4 is maintained integrated in the chromosome (18). In *E. coli*, strains carrying a chromosomal IncP plasmid copy can be selected in several ways.
(i) An RP4 lambda *att* recombinant plasmid (constructed *in vitro* by insertion of a lambda *att* fragment) integrates into the chromosome at the *att* lambda site and promotes Hfr chromosome transfer from this site. The integrated plasmid can be transmitted to a recipient provided that the entire chromosome

chromosome is transferred (117).

(ii) Integrative suppression of a *dnaA*(Ts) mutation by the wild-type RP1 plasmid has been obtained at 41°C and results in Hfr donor strains. Without selective pressure (at 30°C) the plasmid is excised and the Hfr properties are lost (76).

iii) IncP plasmid derivatives which are temperature-sensitive for replication can be maintained integrated in the chromosome by selection for the plasmid at nonpermissive temperature (49,28).

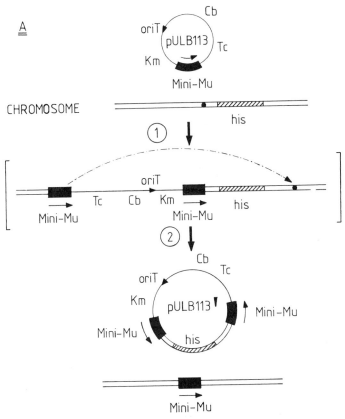

Figure 2. Models of R′ plasmid formation. **A,** pULB113 (= RP4::mini-Mu); 1, replicative mini-Mu transposition directs pULB113 integration into the chromosome; 2, mini-Mu-promoted deletion (-··-··-) excises R′ plasmid (after ref.114). **B,** R68.45; 1, conservative, IS*21*-mediated transposition of two R68.45 molecules in the same orientation into nearby targets; 2, excision of R′ plasmid via homologous recombination between two directly repeated R-plasmids or via *tnpR*-dependent resolution (----). Dots (·) designate the target sites of transposition. Structures in brackets ([]) are unstable and have not been isolated.

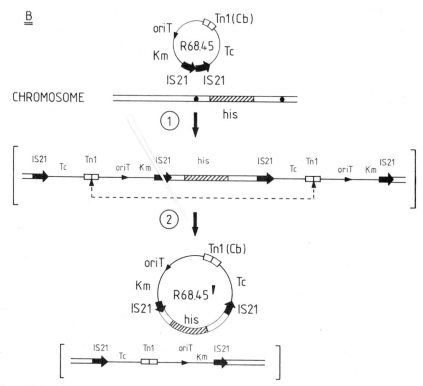

Figure 2 (continued)

(iv) Coselection for two incompatible IncP plasmids occasionally results in the chromosomal integration of one of the plasmids, and the other, autonomous plasmid may be eliminated by curing agents (107).

Studies on integrated IncP plasmids in *E. coli* have revealed the following features:

(i) Chromosome transfer and hence IncP plasmid transfer are unidirectional, as in the case of the F-plasmid (38,1). Bidirectional chromosome transfer has also been observed (49) but this is probably due to the presence of two R-plasmid copies mobilizing the chromosome in opposite direction.

(ii) An integrated IncP plasmid coexists with autonomous IncP plasmids ('integrative compatibility'), unlike F (118). The copy number of the autonomous RP4 (normally ca. 5) is reduced by a chromosomal copy of RP4 so that the total RP4 copy number remains constant (39).

(iii) Unless special measures are taken, integrated IncP plasmids usually have a marked tendency to excise, suggesting that a chromosomal IncP plasmid may be deleterious to the host (presumably because plasmid replication is initiated more often than is chromosome replication) (40). When IncP plasmids are forced to exist integrated in the chromosome without autonomous plasmid copies being present, often (but not always) the replication functions of the integrated plasmid are inactivated (observed in *E. coli* and *P. aeruginosa*; 40,93).

Outside *E. coli* and *M. xanthus*, temperature-sensitive (Ts) IncP derivatives, e. g., the *trfA*(Ts) mutants pMO190 (57), pMT1000 (= pMO190::Tn*501*; 48), pMO514 (= pMO190::Tn*2521*; 86), and pME134 (= pME301*tnpR*::Tp^R; 93,97), as well as pTH10 (49) and pMJP7 (74), have been used to select for plasmid integration into the chromosome. In this way, a variety of Hfr donors have been obtained in *P. aeruginosa* (47,57,48,86,87,93,45) and in *Rhodobacter capsulatus* (74). Unfortunately, the Ts defects of all the IncP plasmids mentioned above except pMJP7, which is Ts at 37°C (74), are manifested at temperatures above 40°C and hence unsuitable for selection of plasmid integration in bacteria that do not tolerate high incubation temperatures (60,97). An interesting alternative is provided by pAS8–121, a ColE1-RP4 hybrid with a Tn7 insertion in the RP4 *trfA* replication gene; this plasmid cannot replicate in nonenteric bacteria and has been used to create Hfr donors in *R. sphaeroides* (63a).

The molecular structures of chromosomally integrated IncP plasmids have been investigated mainly in *P. aeruginosa*. As mentioned, an integrated pME134 is flanked by directly repeated Tn*1*/Tp elements, and in 50% of all cases an IS*21* insertion has inactivated the *trfA* locus (93). Additional pME134 copies are found to be inserted adjacent to the first one, perhaps because selective antibiotic concentrations favour the integration of multiple pME134 copies (93).

Plasmid pME487 (= R68.45 delta Tn*1* *trfA*(Ts)) can also be used to construct unidirectional Hfr strains of *P. aeruginosa* (45). In its chromosomal state, this plasmid is bounded by single IS*21* elements in the same orientation and several plasmid copies may be interlocked, with concomitant *trfA* inactivation (95).

In *M. xanthus* RP4 integrates into the chromosome via a mechanism that involves neither Tn*1* nor IS*21* but a 'hot spot' between IS*21* and the Tra2 region (62). This site might be identical with a multimer resolution site located on the 6 kb *Pst*I fragment of RP4 (see Chapter 1, Figure 1;and ref. 41). The integrated plasmid, which has Cma, can be stabilized by host mutations (62); these possibly diminish the interaction at the postulated resolution site. Plasmid pME305 (96), which lacks IS*21* and the hot spot, is integrated via a different plasmid site and cannot be stabilized (62).

V. CHROMOSOME MOBILIZATION BY IncP PLASMIDS WITHOUT STABLE INTEGRATION

A survey (Table II), which is by no means exhaustive, shows that wild-type IncP plasmids mobilize the chromosome in some species of Gram-negative bacteria, albeit usually at low frequencies. Enhanced Cma may be obtained by adding transposable elements (described in III) to the IncP plasmids. The most popular plasmids with enhanced Cma are R68.45 (43) and pULB113 (114). The enhancement is due to the IS21 tandem repeat on R68.45 (120,23,46) and the mini-Mu transposon on pULB113 (114), respectively. R68.45 promotes chromosome transfer from many different insertion sites and in both orientations (44). This is a consequence of the low target-sequence specificity of the IS21 transposase (99,95); the Mu transposase has the same property (82). Since the IncP plasmid-chromosome cointegrates are generally unstable, they have not been characterized physically. In exponential cultures of *P. aeruginosa*, R68.45 will transfer only ca. 10% of the total chromosome length, as indicated by linkage analysis (44). By contrast, *P. aeruginosa* Hfr donors carrying an integrated pME134 or pME487 can transfer long chromosome segments under the same experimental conditions, and widely separated chromosomal markers show conjugational linkage (93,95). Thus, we believe that R68.45 stays integrated in the *P. aeruginosa* donor chromosome only for a short period of time. The situation in *Rhizobium* species is different; in these bacteria R68.45 is able to transfer large parts of the chromosome (66,79). Therefore, there is no universal formula that converts conjugational linkage in R68.45-mediated matings to physical distance (in kb), although the formulae of Wu (122) and Kemper (64) have been used to obtain 'map distances' from linkage values in *Rhizobium* (66,79). In matings on the plate pULB113 and R68.45 may transfer chromosomal markers in the 'usual' direction (from the donor to the recipient) but sometimes also in the opposite direction (from the recipient to the donor). This 'retrotransfer' leads to the recovery of recipient markers in the donor (80).

Polar chromosome transfer has been achieved by the use of R′ plasmids (constructed *in vitro*) in e. g. *E. coli* (7), *R. meliloti* (63) and *Proteus morganii* (8), or by the use of transposon-facilitated recombination in *A. tumefaciens* (90) and *P. aeruginosa* (68). The frequencies of chromosome mobilization by some Cma plasmids vary considerably depending on the host (Table II). The relative efficiencies of three processes contribute to these variations: interaction between the R-plasmid and the chromosome in the donor; mating pair formation; and recombination in the recipient.

Reviews on the application of IncP and other plasmids for mapping in *Rhizobium* (12) and in *Pseudomonas* (56,59) have been published recently.

VI. FORMATION OF R′ PLASMIDS

Natural IncP plasmids are too large to be practical for routine *in vitro* cloning, but their derivatives pULB113, R68.45, the R68.45-like plasmids pBLM2, pMO172, pMO960 and pMO963, as well as RP4::D3112*cts* have been found suitable for the construction of R′ plasmids *in vivo* (114,54,125,83,16,124). There are two procedures for the selection of R′ plasmids. Both methods involve a donor harboring one of the above plasmids and a recipient that cannot integrate the R' insert by homologous recombination. In the first method, a recombination-deficient (recA) mutant of the same bacterial species is used as a recipient. In the second approach (which is useful when no recA mutants are available), the recipient belongs to another species or genus which has little or no sequence homology with the donor (56). The principal advantage of R' plasmids is that they may carry large (100 kb approx.) inserts of chromosomal or megaplasmid DNA (4,106,101,21). In pULB113' plasmids the insert is flanked by mini-Mu (71); they may be generated according to the model of replicative Mu transposition (Figure 2A; 114). Prime derivatives of R68.45 and related plasmids show directly repeated, single copies of IS21 at the junctions between the insert and the R plasmid (Figure 2B; 69,127,2,16). We propose that such R' plasmids may be formed by the direct (conservative) transposition of two R68.45 molecules into nearby sites, followed by plasmid excision via homologous recombination between the two R-plasmids or via site-specific, Tn1-dependent resolution (Figure 2B).

R' plasmids have proved useful for complementation mapping in pseudomonads and methylotrophs (84,16,58) and for physical mapping of long DNA segments (106,101). Although a number of R' plasmids have been found to be stable during genetic manipulations (58), we (unpublished observations) and others (54,21) have repeatedly encountered structural instability of R' plasmids in both Rec + and Rec− backgrounds. The instability problem is probably related to the relatively high copy number of IncP plasmids (4–6 copies). F'-plasmids, which have a single copy, appear to tolerate larger inserts than do IncP R'-plasmids.

VII. CONCLUSIONS

Various IncP plasmids share extensive regions of homology in the replication and conjugation genes but differ in the resistance markers and transposable elements carried (115). We are beginning to understand what broad-host-range replication and conjugation mean at the molecular level. The host range properties of the transposable elements on IncP plasmids, however, have not been investigated systematically. Since transposition functions largely determine the interactions between IncP plasmids and the host genome, it would be interesting to study and compare the expression of some widely used transposable elements (e. g. Tn1, IS21, mini-Mu) in a variety of Gram-negative bacteria.

ACKNOWLEDGEMENTS

We thank our many colleagues who have kindly provided published and unpublished information. We apologize to all those colleagues whose work we could not quote in this review because of limitations of space. We thank Antje Hitz for secretarial assistance and Hedi Strahm for preparing the figures. Work in our laboratory has been supported by the Schweizerische Nationalfonds (project 3.620–0.84).

VIII. REFERENCES

1. Al-Doori Z, Watson M, Scaife J (1982) The orientation of transfer of the plasmid RP4. *Genet Res* **39**, 99–103

2. Allibert P, Willison JC, Vignais PM (1987) Complementation of nitrogen-regulatory (*ntr*-like) mutations in *Rhodobacter capsulatus* by an *Escherichia coli* gene: Cloning and sequencing of the gene and characterization of the gene product. *J Bacteriol* **169**, 260–71

3. Anderson P, Roth J (1981) Spontaneous tandem genetic duplications in *Salmonella typhimurium* arise by unequal recombination between rRNA (*rrn*) cistrons. *Proc Natl Acad Sci USA* **78**, 3113–7

4. Banfalvi Z, Randhawa GS, Kondorosi E, Kiss A, Kondorosi A (1983) Construction and characterization of R-prime plasmids carrying symbiotic genes of *R. meliloti*. *Mol Gen Genet* **189**, 129–35

5. Barrett JT, Rhodes CS, Ferber DM, Jenkins B, Kuhl SA, Ely B (1982) Construction of a genetic map for *Caulobacter crescentus*. *J Bacteriol* **149**, 889–96

6. Barsomian G, Lessie TG (1986) Replicon fusions promoted by insertion sequences on *Pseudomonas cepacia* plasmid pTGL6. *Mol Gen Genet* **204**, 273–80

7. Barth PT (1979) Plasmid RP4, with *Escherichia coli* DNA inserted *in vitro*, mediates chromosomal transfer. *Plasmid* **2**, 130–6

8. Beck Y, Coetzee WF, Coetzee JN (1982) *In vitro* constructed RP4-prime plasmids mediate oriented mobilization of the *Proteus morganii* chromosome. *J Gen Microbiol* **128**, 1163–9

9. Bennett PM, Grinsted J, Richmond MH (1977) Transposition of TnA does not generate deletions. *Mol Gen Genet* **154**, 205–11

10. Bennett PM, Heritage J, Comanducci A, Dodd HM (1986) Evolution of R-plasmids by replicon fusion. *J Antimicrob Chemother* **18**, Suppl C, 103–11

11. Beringer JE, Hopwood DA (1976) Chromosomal recombination and mapping in *Rhizobium leguminosarum*. *Nature* **264**, 291–3

12. Beringer JE, Johnston AWB, Kondorosi A (1987) Genetic maps of *Rhizobium meliloti* and *R. leguminosarum* biovars *phaseoli*, *trifolii* and *viciae*. *In* Genetic maps **4**, (O'Brien SJ ed), Cold Spring Harbor Lab, pp 245–251

13. Bowen ARSG, Pemberton JM (1985) Mercury resistance transposon Tn*813* mediates chromosome transfer in *Rhodopseudomonas sphaeroides* and intergeneric transfer of pBR322. *In* Plasmids in Bacteria, (Helinski DR, Cohen

SN, Clewell DB, Jackson DB, Hollaender A eds), Plenum Press, NY, pp 105–15

14. Boyd AC, Sherratt DJ (1986) Polar mobilization of the *Escherichia coli* chromosome by the ColE1 transfer origin. *Mol Gen Genet* **203**, 496–504

15. Bradley, D.E. and Taylor, D.E. (1987). Transposition from RP4 to other replicons of a tellurite-resistance determinant not normally expressed by IncP plasmids. *FEMS Microbiol Lett* **41**, 237–40

16. Bray R, Strom D, Barton J, Dean HF, Morgan AF (1987) Isolation and characterization of *Pseudomonas putida* R-prime plasmids. *J Gen Microbiol* **133**, 683–90

17. Bresler SE, Krivonogov SV, Lanzov VA (1979) Genetic determination of the donor properties in *Escherichia coli* K12. *Mol Gen Genet* **177**, 177–84

18. Breton AM, Jaoua S, Guespin-Michel J (1985) Transfer of plasmid RP4 to *Myxococcus xanthus* and evidence for its integration into the chromosome. *J Bacteriol* **161**, 523–8

19. Bukhari AI, Shapiro JA, Adhya SL (1977) DNA insertion elements, plasmids and episomes, Cold Spring Harbor Lab, 639–40.

20. Chatterjee AK, Starr MP (1980) Genetics of *Erwinia* species. *Ann Rev Microbiol* **34**, 645–76

21. Chatterjee AK, Ross LM, McEvoy JL, Thurn KK (1985) pULB113, an RP4::mini-Mu plasmid, mediates chromosomal mobilization and R-prime formation in *Erwinia amylovora*, *Erwinia chrysanthemi*, and subspecies of *Erwinia carotovora*. *Appl Environ Microbiol* **50**, 1–9

22. Coetzee JN (1978) Mobilization of the *Proteus mirabilis* chromosome by R-plasmid R772. *J Gen Microbiol* **108**, 103–9

23. Currier TC, Morgan MK (1982) Direct DNA repeat in plasmid R68.45 is associated with deletion formation and concomitant loss of chromosome mobilization ability. *J Bacteriol* **150**, 251–9

24. Curtiss III R, Stallions DR (1969) Probability of F integration and frequency of stable Hfr donors in F^+ populations of *Escherichia coli* K12. *Genetics* **63**, 27–38

25. Dalrymple B (1987) Novel rearrangements of IS*30* carrying plasmids leading to the reactivation of gene expression. *Mol Gen Genet* **207**, 413–20

26. Danilevich VN, Stephanshin YG, Volozhantsev NV, Golub EI (1978) Transposon-mediated insertion of R-factor into bacterial chromosome. *Mol Gen Genet* **161**, 337–9

27. Danilevich VN, Kostyuchenko DA (1986) Immunity to repeated insertion of IS*21* sequence. *Mol Biol* **19**, 1016–22 (English translation)

28. Danilevich VN, Kostyuchenko DA, Negrii NV (1986) Interaction (incorporation and excision) of the plasmid pRP19.6, a derivative or RP1 containing a duplicated sequence IS*21*, with the chromosome of *Escherichia* coli K12. *Mol Biol* **19**, 858–67 (English translation)

29. Depicker A, De Block M, Inze D, Van Montagu M, Schell J (1980) IS-

like element IS8 in RP4 plasmid and its involvement in cointegration. *Gene* **10**, 329–38

30. Diver WP, Grinsted J, Fritzinger DC, Brown NL, Altenbuchner J, Rogowsky P, Schmitt R (1983) DNA sequences of and complementation by the *tnpR* genes of Tn21, Tn501 and Tn1721. *Mol Gen Genet* **191**, 189–93

31. Dreyfus LA, Iglewski BH (1985) Conjugation-mediated genetic exchange in *Legionella pneumophila*. *J Bacteriol* **161**, 80–4

32. Faelen M, Toussaint A, De Lafonteyne J (1975) Model for the enhancement of lambda-*gal* integration into partially induced Mu-1 lysogens. *J Bacteriol* **121**, 873–82

33. Forbes KJ, Prombelon MCM (1985) Chromosomal mapping in *Erwinia carotovora* subsp. *carotovora* with the IncP plasmid R68::Mu. *J Bacteriol* **164**, 1110–6

34. Franche C, Canelo E, Gauthier D, Elmerich C (1981). Mobilization of the chromosome of *Azospirillum brasilense* by plasmid R68.45. *FEMS Microbiol Lett* **10**, 199–202

35. Fulbright DW, Leary JV (1978) Linkage analysis of *Pseudomonas glycinea*. *J Bacteriol* **136**, 497–500

36. Gobius KS, Pemberton JM (1986) Use of plasmid pULB113 (RP4::Mini-Mu) to construct a genomic map of *Aeromonas hydrophila*. *Current Microbiol* **13**, 111–5

37. Grindley NDF, Reed RR (1985) Transpositional recombination in prokaryotes. *Ann Rev Biochem* **54**, 863–96

38. Grinter NJ (1981) Analysis of chromosome mobilization using hybrids between plasmid RP4 and a fragment of bacteriophage lambda carrying IS1. *Plasmid* **5**, 267–76.

39. Grinter NJ (1984a) Replication control of IncP plasmids. *Plasmid* **11**, 74–81

40. Grinter NJ (1984b) Replication defective RP4 plasmids recovered after chromosomal integration. *Plasmid* **11**, 65–73

41. Grinter NJ, Barth PT (1985) Plasmid stability: involvement of a plasmid-coded primase and site-specific recombination/resolution systems. *In* Plasmids in Bacteria, (Helinski DR, Cohen SN, Clewell DB, Jackson DA, Hollaender A, eds), Plenum Press, NY, pp 858

42. Guyer MS, Clark AJ (1977) Early and late transfer of F genes by Hfr donors of *E. coli* K12. *Mol Gen Genet* **157**, 215–22

43. Haas D, Holloway BW (1976) R-factor variants with enhanced sex factor activity in *Pseudomonas aeruginosa*. *Mol Gen Genet* **144**, 243–51

44. Haas D, Holloway BW (1978) Chromosome mobilization by the R-plasmid R68.45: a tool in *Pseudomonas* genetics. *Mol Gen Genet* **158**, 229–37

45. Haas D, Jann A, Reimmann C, Lthi E, Leisinger T (1987) Chromosome organization in *Pseudomonas aeruginosa*. *Antibiot Chemother* **39**, 256–63

46. Haas D, Riess G (1983) Spontaneous deletions of the chromosome-mobilizing plasmid R68.45 in *Pseudomonas aeruginosa* PAO. *Plasmid* 9, 42–52.

47. Haas D, Watson J, Krieg R, Leisinger T (1981) Isolation of an Hfr donor of *Pseudomonas aeruginosa* PAO by insertion of the plasmid RP1 into the tryptophan synthase gene. *Mol Gen Genet* 182, 240–4.

48. Harayama S, Lehrbach PR, Tsuda M, Leppik R, Iino T, Reineke W, Knackmuss HJ, Timmis KT (1984) Genetic engineering systems for *Pseudomonas* and their use in the analysis and manipulation of metabolic pathways. *In* Transferable Antibiotic Resistance, (Mitsuhashi S, Krcmry V, eds), Springer-Verlag, Berlin, pp 361–72

49. Harayama S, Tsuda M, Iino T (1980) High frequency mobilization of the chromosome of *Escherichia coli* by a mutant of plasmid RP4 temperature-sensitive for maintenance. *Mol Gen Genet* 180, 47–56

50. Harayama S, Tsuda M, Iino T (1981) Tn*1* Insertion mutagenesis in *Escherichia coli* K12 using a temperature-sensitive mutant of plasmid RP4. *Mol Gen Genet* 184, 52–5

51. Hardy K (1986) Bacterial Plasmids; 2nd ed. Aspects of Microbiology 4. Van Nostrand Reinhold (UK) Co Ltd, Wokingham, Berkshire (GB)

52. Hedges RW, Jacob AE (1974) Transposition of ampicillin resistance from RP4 to other replicons. *Mol Gen Genet* 132, 31–40

53. Heffron F (1983) Tn*3* and its relatives. *In* Mobile Genetic Elements, (Shapiro JA, ed), Academic Press, NY, pp 223–60

54. Holloway BW (1978) Isolation and characerization of an R-plasmid in *Pseudomonas aeruginosa*. *J Bacteriol* 133, 1078–82

55. Holloway BW (1979) Plasmids that mobilize bacterial chromosome. *Plasmid* 2, 1–19

56. Holloway BW (1986) Chromsome mobilization and genomic organization in *Pseudomonas*. *In* The Bacteria, (Sokatch JR, ed) Vol. X, Academic Press, NY, pp 217–49

57. Holloway BW, Crowther C, Dean H, Hagedorn J, Holmes N, Morgan AF (1982) Integration of plasmids into the *Pseudomonas* chromosome. *In* Drug Resistance in Bacteria, (Mitsuhashi S, ed), Japan Scientific Societies Press, Tokyo, pp 231–42

58. Holloway BW, Kearney PP, Lyon BR (1987) The molecular genetics of C_1 utilizing microorganisms—an overview. Antonie van Leewenhoek 53, 47–53

59. Holloway BW, Morgan AF (1986) Genome organization in *Pseudomonas*. *Ann Rev Microbiol* 40, 79–105

60. Hooykaas PJJ, den Dulk-Ras H, Schilperoort RA (1982) Phenotypic expression of mutations in a wide-host-range R-plasmid in *Escherichia coli* and *Rhizobium meliloti*. *J Bacteriol* 150, 395–7

61. Hooykaas PJJ, Peerbolte R, Regensburg-Tuink AJG, de Vries P,

Schilperoort RA (1982) A chromosomal linkage map of *Agrobacterium tumefaciens* and a comparison with the maps of *Rhizobium* spp. *Mol Gen Genet* **188**, 12–7

62. Jaoua S, Guespin-Michel JF, Breton AM (19867) Mode of insertion of the broad-host-range plasmid RP4 and its derivatives into the chromosome of *Myxococcus xanthus*. *Plasmid* **18**, 111–9

63. Julliot JS, Boistard P (1979) Use of RP4-prime plasmids constructed in vitro to promote a polarized transfer of the chromosome in *Escherichia coli* and *Rhizobium meliloti*. *Mol Gen Genet* **173**, 289–98

63a. Kameneva SV, Polivtseva TP, Belavina NV, Shstakova SV (1986) Hfr donors of phototrophic niitrogen fixing bacterium *Rhodopseudomonas sphaeroides*. Localization of mutations in genes controlling nitrogen fixation. *Genetika* (Russ) **22**, 2664–72

64. Kemper J (1974) Gene order and cotransduction in the *leu-ara-fol-pyrA* region of the *Salmonella typhimurium* linkage map. *J Bacteriol* **117**, 94–9

65. Kleckner N, Roth J, Botstein D (1977) Genetic engineering *in vivo* using translocatable drug-resistance elements. *J Mol Biol* **116**, 125–59

66. Kondorosi A, Kiss GB, Forrai T, Vincze E, Banfalvi Z (1977) Circular linkage map of *Rhizobium meliloti* chromosome. *Nature* **268**, 525–7

67. Kretschmer PJ, Cohen SN (1977) Selected translocation of plasmid genes: frequency and regional specificity of translocation of the Tn*3* element. *J Bacteriol* **130**, 888–99

68. Krishnapillai V, Royle P, Lehrer J (1981) Insertions of the transposon Tn*1* into the *Pseudomonas aeruginosa* chromosome. *Genetics* **97**, 495–511

69. Leemans J, Villarroel R, Silva B, Van Montagu M, Schell J (1980) Direct repetition of a 1.2 Md DNA sequence is involved in site-specific recombination by the P1 plasmid R68. *Gene* **10**, 319–28

70. Lejeune P, Mergeay M (1980) R-plasmid-mediated chromosome mobilization in *Pseudomonas fluorescens* 6.2. *Arch Int Physiol Biochim* **88**, B289-B290

71. Lejeune P, Mergeay M, Van Gijsegem F, Faelen M, Gerits, J, Toussaint A (1983) Chromosome transfer and R-prime plasmid formation mediated by plasmid pULB113 (RP4::mini-Mu) in *Alcaligenes eutrophus* CH34 and *Pseudomonas fluorescens* 6.2. *J Bacteriol* **155**, 1015–26

72. Leonardo JM, Goldberg RB (1980) Regulation of nitrogen metabolism in glutamine auxotrophs of *Klebsiella pneumoniae*. *J Bacteriol* **142**, 99–110

73. Lessie TG, Gaffney T (1986) Catabolic potential of *Pseudomonas cepacia*. *In* The Bacteria, (Sokatch JR, ed) Vol. X, Academic Press, New York, pp 439–81

74. Magnin JP, Willison JC, Vignais PM (1987) Construction et utilisation d'une souche Hfr chez *Rhodobacter capsulatus* pour l'établissement de sa carte génétique. *In* Abstracts of the May 1987 meeting of the Société Française de Microbiologie, Paris, pp 28

75. Marrs B (1981) Mobilization of the genes for photosynthesis from

Rhodopseudomonas capsulata by a promiscuous plasmid. *J Bacteriol* **146**, 1003-12

76. Martin R, Thorlton CL, Unger L (1981) Formation of *Escherichia coli* Hfr strains by integrative suppression with the P-group plasmid RP1. *J Bacteriol* **145**, 713-721

77. Mc Laughlin W, Ahmad MH (1986) Transfer of plasmids RP4 and R68.45 and chromosomal mobilization in Cowpea rhizobia. *Arch Microbiol* **144**, 408-11

78. Meade HM, Signer ER (1977) Genetic mapping of *Rhizobium meliloti*. *Proc Natl Acad Sci USA* **74**, 2076-8

79. Megias M, Caviedes MA, Palomares AJ, Perez-Silva J (1982) Use of plasmid R68.45 for constructing a circular linkage map of the *Rhizobium trifolii* chromosome. *J Bacteriol* **149**, 59-64

80. Mergeay M, Lejeune P, Sadouk A, Gerits J, Fabry L (1987) Shuttle transfer (or retrotransfer) of chromosomal markers mediated by plasmid pULB113. *Mol Gen Genet* **209**, 61-70

81. Miller IS, Fox D, Saeed N, Borland PA, Miles CA, Sastry GRK (1986) Enlarged map of *Agrobacterium tumefaciens* C58 and the location of chromosomal regions which affect tumorigenicity. *Mol Gen Genet* **205**, 153-9

81a. Mills D (1985) Transposon mutagenesis and its potential for studying virulence genes in plant pathogens. *Ann Rev Phytopathol* **23**, 297-320.

82. Mizuuchi K, Craigie R (1986) Mechanism of bacteriophage Mu transposition. *Ann Rev Genet* **20**, 385-429

83. Moore AT, Nayudu M, Holloway BW (1983) Genetic mapping in *Methylophilus methylotrophus* AS1. *J Gen Microbiol* **129**, 785-99

84. Morgan AF (1982) Isolation and characterization of *Pseudomonas aeruginosa* R-plasmids constructed by interspecific mating. *J Bacteriol* **149**, 654-61

85. Nayudu M, Holloway BW (1981) Isolation and characterization of R-plasmid variants with enhanced chromosomal mobilization ability in *Escherichia coli* K12. *Plasmid* **6**, 53-66

86. O'Hoy K, Krishnapillai V (1985) Transposon mutagenesis of the *Pseudomonas aeruginosa* PAO chromosome and the isolation of high frequency of recombination donors. *FEMS Microbiol Lett* **29**, 299-303

87. O'Hoy K, Krishnapillai V (1987). Recalibration of the *Pseudomonas aeruginosa* strain PAO chromosome map in time units using high frequency of recombination donors. *Genetics* **115**, 611-8

88. Pemberton JM, Bowen ARSG (1981) High-frequency chromosome transfer in *Rhodopseudomonas sphaeroides* promoted by broad-host-range plasmid RP1 carrying mercury transposon Tn*501*. *J Bacteriol* **147**, 110-7

89. Petersen BC, Rownd RH (1985) Drug resistance gene amplification of plasmid NR1 derivatives with various amounts of resistance determinant

DNA. *J Bacteriol* **161**, 1042–8

90. Pischl DL, Farrand SK (1983) Transposon-facilitated chromosome mobilization in *Agrobacterium tumefaciens*. *J Bacteriol* **153**, 1451–60

91. Pittard J, Adelberg EA (1963) Gene transfer by F strains of *Escherichia coli* K12. II. Interaction between F-merogenote and chromosome during transfer. *J Bacteriol* **85**, 1402–8

91a. Plasota M, Piechuka E, Kauc B, Wlordarczyk M (1984) R68.45 plasmid mediated conjugation in *Thiobacillus* A2. *Microbios* **41**, 81–9

92. Read HA, Das Sarma S, Jaskunas SR (1980) Fate of donor insertion sequence IS*1* during transposition. *Proc Natl Acad Sci USA* **77**, 2514–8

93. Reimmann C, Haas D (1986) IS*21* insertion in the *trfA* replication control gene of chromosomally integrated plasmid RP1: a property of stable *Pseudomonas aeruginosa* Hfr strains. *Mol Gen Genet* **203**, 511–9

94. Reimmann C, Haas D (1987) Mode of replicon fusion mediated by the duplicated insertion sequence IS*21* in *Escherichia coli*. *Genetics* **115**, 619–25

95. Reimmann C (1987) Die transponierenden Elemente des Plasmides RP1. Ph.D. Thesis, Eidgenssische Technische Hochschule, Zrich

96. Rella M, Mercenier A, Haas D (1985) Transposon insertion mutagenesis of *Pseudomonas aeruginosa* with a Tn*5* derivative: Application to physical mapping of the *arc* gene cluster. *Gene* **33**, 293–303

97. Rella M, Watson JM, Thomas CM, Haas D (1987) Deletions in the tetracycline resistance determinant reduce the thermosensitivity of a *trfA*(Ts) derivative of plasmid RP1 in *Pseudomonas aeruginosa*. *Ann Inst Pasteur/Microbiol* **138**, 151–64

98. Riess G, Holloway BW, Phler A (1980) R68.45, a plasmid with chromosome mobilizing ability (Cma) carries a tandem duplication. *Genet Res* **36**, 99–109

99. Riess G, Masepohl B, Phler A (1983) Analysis of IS*21*-mediated mobilization of plasmid pACYC184 by R68.45 in *Escherichia coli*. *Plasmid* **10**, 111–8

100. Robinson MK, Bennett PM, Falkow S, Dodd HM (1980) Isolation of a temperature-sensitive derivative of RP1. *Plasmid* **3**, 343–7

101. Sano Y, Kageyama M (1984) Genetic determinant of pyocin AP41 as an insert in the *Pseudomonas aeruginosa* chromosome. *J Bacteriol* **158**, 562–70

102. Schoonejans E, Toussaint, A (1983) Utilization of plasmid pULB113 (RP4::mini-Mu) to construct a linkage map of *Erwinia carotovora* subsp. *chrysanthemi*. *J Bacteriol* **154**, 1489–92

103. Schurter, W. and Holloway,B.W. (1986). Genetic analysis of promoters on the insertion sequence IS*21* of plasmid R68.45. *Plasmid* **15**, 8–18

104. Schurter W, Holloway BW (1987) Interactions between the transposable element IS*21* on R68.45 and Tn*7* in *Pseudomonas aeruginosa* PAO. *Plasmid* **17**, 61–4

105. Seeberg AH, Wiedemann B (1984) Transfer of the chromosomal *bla*

gene from *Enterobacter cloacae* to *Escherichia coli* by RP4::mini-Mu. *J Bacteriol* **157**, 89–94

106. Shinomiya T, Shiga S, Kikuchi A, Kageyama M (1983) Genetic determinant of pyocin R2 in *Pseudomonas aeruginosa* PAO. II. Physical characterization of pyocin R2 genes using R-prime plasmids constructed from R68.45. *Mol Gen Genet* **189**, 382–9

107. Simon R, Priefer U, Phler A (1983) A broad host range mobilization system for in vivo genetic engineering: transposon mutagenesis in Gram-negative bacteria. *Bio Technology* **1**, 784–90

108. Simon R (1984) High frequency mobilization of Gram-negative bacterial replicons by the *in vitro* constructed Tn5-Mob transposon. *Mol Gen Genet* **196**, 413–20

109. Sistrom WR (1977) Transfer of chromosomal genes mediated by plasmid R68.45 in *Rhodopseudomonas sphaeroides*. *J Bacteriol* **131**, 526–32

110. Smith CJ, Coote JG, Parton R (1986) R-Plasmid-mediated chromosome mobilization in *Bordetella pertussis*. *J Gen Microbiol* **132**, 2685–92

111. Tatra PK, Goodwin PM (1983) R-Plasmid-mediated chromosome mobilization in the facultative methylotroph *Pseudomonas* AM1. *J Gen Microbiol* **129**, 2629–32

111a. Torolero M, Santero E, Casadesus J (1983) Plasmid transfer and mobilization of *nif* markers in *Azotobacter vinelandii*. *Microbios Lett* **22**, 31–5

112. Towner KJ, Vivian A (1976) RP4 Fertility variants in *Acinetobacter calcoaceticus*. *Genet Res* **28**, 301–6

113. Towner KJ (1978) Chromosome mapping in *Acinetobacter calcoaceticus*. *J Gen Microbiol* **104**, 175–80

114. Van Gijsegem F, Toussaint A (1982) Chromosome transfer and R-prime formation by an RP4::mini-Mu derivative in *Escherichia coli*, *Salmonella typhimurium*, *Klebsiella pneumoniae* and *Proteus mirabilis*. *Plasmid* **7**, 30–44

115. Villarroel R, Hedges RW, Maenhaut R, Leemans J, Engler G, Van Montagu M, Schell J (1983) Heteroduplex analysis of P-plasmid evolution: the role of insertion and deletion of transposable elements. *Mol Gen Genet* **189**, 390–9

116. Watson JM, Holloway BW (1978) Chromosome mapping in *Pseudomonas aeruginosa* PAT. *J Bacteriol* **133**, 1113–25

117. Watson MD, Scaife JG (1978) Chromosomal transfer promoted by the promiscuous plasmid RP4. *Plasmid* **1**, 226–37

118. Watson MD, Scaife JG (1980) Integrative compatibility: Stable coexistence of chromosomally integrated and autonomous derivatives of plasmid RP4. *J Bacteriol* **142**, 462–6

119. Willetts NS (1972) Location of the origin of transfer of the sex factor F. *J Bacteriol* **112**, 773–8

120. Willetts NS, Crowther C, Holloway BW (1981) The insertion sequence

IS21 of R68.45 and the molecular basis for mobilization of the bacterial chromosome. *Plasmid* **6**, 30–52

121. Willison JC, Ahombo G, Chabert J, Magnin JP, Vignais, PM (1985) Genetic mapping of the *Rhodopseudomonas capsulata* chromosome shows nonclustering of genes involved in nitrogen fixation. *J Gen Microbiol* **131**, 3001–15

122. Wu TT (1966) A model for three-point analysis of random general transduction. *Genetics* **54**, 405–10.

123. Yakobson EA, Guiney Jr D (1984) Conjugal transfer of bacterial chromosome mediated by the RK2 plasmid transfer origin cloned into transposon Tn5. *J Bacteriol* **160**, 451–3

124. Yanenko AS, Bekkarevitch AO, Akhverdian VZ, Krylov, VN (1986) Transfer of chromosomal genes and formation of R-derivatives mediated by RP4::D3112cts15 plasmids in *Pseudomonas aeruginosa*. *Genetika (Russ)* **22**, 2784–93

125. Youvan DC, Elder JT, Sandlin DE, Zsebo K, Alder DP, Panopoulos NJ, Marrs BL, and Hearst JE (1982) R-Prime site-directed transposon Tn7 mutagenesis of the photosynthetic apparatus in *Rhodopseudomonas capsulata*. *J Mol Biol* **162**, 17–41

126. Yu P-L, Cullum J, Drews G (1981) Conjugational transfer systems of *Rhodopseudomonas capsulata* mediated by R-plasmids. *Arch Microbiol* **128**, 390–3

127. Zsebo KM, Wu F,Hearst JE (1984) Tn5.7 Construction and physical mapping of pRPS404 containing photosynthetic genes from *Rhodopseudomonas capsulata*. *Plasmid* **11**, 182–184

CHAPTER 8

TRANSPOSON MUTAGENESIS IN NON-ENTERIC GRAM-NEGATIVE BACTERIA

Reinhard Simon

I. INTRODUCTION

Detailed genetic studies of Gram-negative bacteria were initially limited to a few strains, such as *Escherichia coli, Salmonella typhimurium* and some others. But as a consequence of the rapid improvement in genetic and molecular methodologies, the numerous characteristics displayed by non *E. coli* bacteria of biological relevance became more and more susceptible to genetic analysis and manipulation.

The interest in biologically important properties of bacteria apart from *E. coli* is not solely of academic origin. There is also an increasing involvement of commercially oriented groups which are attempting to obtain microorganisms with novel properties or to improve the efficiency of known processes.

Clearly, the powerful and sophisticated methods of *in vitro* DNA manipulation are indispensable for molecular genetics but they have not replaced totally the need for more traditional *in vivo* techniques. As a cornerstone of *in vivo* genetic engineering naturally-occurring transposable elements have become well established within the repertoire of molecular geneticists.

This Chapter reviews the modern methods developed to introduce selectable transposable elements into nonenteric Gram-negative bacteria.

II. TRANSPOSABLE ELEMENTS

Transposable elements are defined as genetic entities which are capable of promoting their own translocation from one site to another within or between edit

Promiscuous Plasmids of Gram-Negative Bacteria
ISBN 0-12-688480-3
© 1989 Academic Press Limited
All rights of reproduction in any form reserved

replicons without the need for extended DNA homology at the insertion sites.

Using physical and genetical criteria, transposable elements in bacteria are subdivided into three groups:

1) insertion sequences (IS elements) are small DNA segments (normally less than 2 kb) encoding no apparent phenotypic determinants unrelated to their transposition;

2) transposons (Tn elements) are larger in size, since they carry additional genes determining specific phenotypes e.g. antibiotic resistance;

3) certain bacteriophages (mutator phages), like the classical example Mu, which uses transposition reactions to replicate its DNA, and upon lysogenization inserts randomly into the host genome.

For detailed description of the biology of prokaryotic transposable elements (nomenclature, physical and genetic properties, ecological and evolutionary considerations, and models proposed to explain mechanisms of translocation), the reader is referred to a series of recent reviews that have covered the subject thoroughly (6,18,19,60,93,102,104).

Not long after their detection, especially the translocatable drug-resistance elements (transposons) turned from laboratory curiosities to useful tools for *in vivo* genetic engineering. These new methods in bacterial genetics were pioneered in *E. coli* and closely related organisms, but have rapidly been adapted to other Gram-negative bacteria. Kleckner *et al* (61) have summarized numerous techniques and ideas about how transposons can be used in virtually every type of genetic manipulation of bacteria. Using transposons many types of experiments can be simplified very much and a large series of other operations cannot be performed at all without them. The potential of the transposon methodology include, e.g. chromosome and plasmid mapping, strain construction, deletions at predetermined sites, fusion of replicons, *in vivo* cloning, and many others (61).

III. TRANSPOSONS AS MUTAGENIC AGENTS

The first and most obvious use of transposons results from their mutagenic activity. Obviously, the mutation caused by the insertion of a drug resistance transposon is accompanied by the acquisition of a selectable marker. Thus, after transposon mutagenesis, a cell carrying an insertional mutation is positively selectable; nonmutated cells will not survive the selection, for drug resistance, conferred by the presence of the transposon. This is in striking contrast to the classical mutagenesis procedures using chemical or physical treatment of a culture. In practice, it is not possible to apply such mutagens at doses which would result in one mutation per surviving cell: low level mutagenesis results in a high proportion of unmutated cells, the intrinsic problem of heavy mutagenic treatment is the high frequency of multiple mutational events.

Such considerations are of special importance if there is no straightforward test system available to detect the presence or absence of a specific gene function. One good example to illustrate such problems, is the group of bacteria that interact with plants. A mutation in a gene needed, e.g., for symbiotic nitrogen fixation in *Rhizobium* or in a gene responsible for plant pathogenicity of certain bacteria has no phenotypic consequence for cell growth of the mutant clone. Therefore, laborious and time consuming tests with appropriate host plants are necessary to screen for such mutants. In contrast to conventional mutagenesis procedures, an optimal transposon mutagenesis experiment results in 100% mutants, all of which carry a single site transposon insertion at different loci. This drastically reduces the number of plant infection assays to be performed in order to identify bacterial mutants with altered plant-interactive phenotypes. For such reasons alone a transposon would be the mutagenic agent of choice if there is no easily scorable phenotypic change in a potential mutant.

Transposon induced mutations offer a further series of advantages as compared to point mutations. Due to physical disruption of the DNA sequence, insertion of a transposon within a structural gene leads to complete loss of its function; there is no leakiness, i.e. low level expression or partial function of the encoded protein. Moreover, termination signals carried by the transposon, are also responsible for the strong polar effect on genes located downstream from the insertion site within the same operon.

The mutation is extremely stable; reversion by precise excision of the inserted transposon occurs at very low frequencies and can easily be recognized since it is accompanied by the loss of the transposon mediated drug resistance.

Many uses result from the direct physical association between the mutational and the drug resistance phenotype in a transposon induced mutant. For example, the mutation can be positively selected upon transfer into a new host. Or the DNA fragment containing the mutated gene can be cloned and identified simply by selection for the transposon mediated resistance. A transposon insertion provides also the target gene with a defined pattern of new restriction enzyme sites allowing e.g. precise mapping of the mutation. Until recently this could only be applied directly to mutagenesis of plasmid genomes or to precise mapping of insertions into chromosomal genes after cloning of the segment containing the transposon, because of the need for a relatively simple pattern of fragments to allow mapping. With the development of pulse field gel electrophoresis and the discovery of restriction endonucleases which cut chromosomal DNA very infrequently it has become feasible to produce a chromosomal restriction map and use it to locate new transposon insertion mutations directly to particular chromosomal segments.

The main limitation of transposon mutagenesis is of course, that insertions into essential genes cannot be isolated.

IV. GENERALIZED TRANSPOSON MUTAGENESIS

A. Theoretical and practical considerations

The three basic requirements for a transposon mutagenesis experiment are outlined in Figure 1.

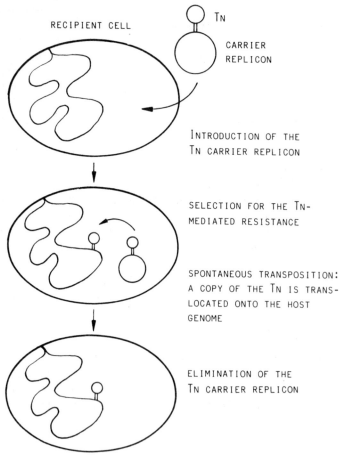

Figure 1. Requirements for a transposon mutagenesis experiment.

1. The transposon carrier replicon

The transposon of choice must be carried by a suitable vector to be propagated. There is a series of phage vectors available for this purpose. But for the nonenteric bacteria, phages are of limited application. In nonenteric bacteria, a number of different plasmids serve as transposon carriers.

2. Introduction of the transposon carrier

There must be an efficient route by which the vector carrying the transposon can be introduced into the recipient bacterium under study. A phage vector is optimal since it can be injected into every cell of a culture. Usually, the frequency of transformation with pure plasmid DNA is too low to be applicable. Especially with bacteria other than *E. coli* or *Salmonella*, the conjugative cell to cell transfer of suitable plasmids is the most feasible way to introduce a transposon into the recipient.

3. Elimination of the vector replicon

The desired translocation event from the carrier replicon onto the host genome occurs spontaneously and usually at low frequencies. It can be identified by selection for the transposon-mediated drug resistance only if the original transposon carrier vector is definitively lost from the cell. In the case of phage vectors, the derivatives used are unable to lysogenize the recipient. Plasmid vectors can be displaced from a cell e.g. by incompatibility reactions or by using temperature-sensitive replication mutants. More universally applicable systems employ transposon carrier plasmids that are *per se* not stably maintained in the recipient to be mutated.

From a practical point of view, more detailed factors must be considered when designing or choosing from the literature a transposon delivery system for a specific bacterial strain. The transposon to be used must exert a reasonably high transposition frequency combined with a low insertional specificity. Of course, the recipient strain to be mutagenized must not have intrinsic, or high-frequency of spontaneously occurring, resistance to the antibiotics for which the transposon carries the resistance genes. The transposon delivery vector must be introduced at high frequency into the recipient cell followed by a reliable loss from the out-growing daughter cells.

To fulfill these requirements, different strategies have been developed, as discussed below in detail. Unfortunately, a method shown to be suitable for a given strain need not necessarily be applicable in an other, even closely related strain. Thus, in most cases the optimal system cannot be theoretically foreseen but has to be determined by trial and error for every new recipient of interest.

B. Transposons used for mutagenesis

Numerous drug resistance transposons have been isolated from natural sources, but by far the most popular transposon used in mutagenesis experiments in non-*E. coli* strains is Tn5. This may be due partly to historical reasons: Tn5 was one of the first transposons shown to be useful for *in vivo* genetic engineering (6, 60), and the first transposon donor vector, that has gained widespread use outside of *E. coli*, was loaded with Tn5 (10). In the majority of cases, the search for alternatives has been superfluous because of the optimal or at least satisfactory properties of Tn5. To underline this, two recent papers should be mentioned: D. Haas (45) summarized a collection of different transposon delivery systems that have been used in a variety of Rhizobia and Pseudomonads : 15 out of 25 vectors listed are loaded with Tn5 or derivatives thereof. D. Mills (71) reviewed the potential of transposon mutagenesis in plant pathogenic Gram-negative bacteria: in 20 out of 29 examples mentioned Tn5 was used as a mutagen. In fact, if one follows the literature cited in these two reviews, it becomes clear that in many cases transposons other than Tn5 are of limited application.

As mentioned above, rapid progress in the genetics of the large and diverse group of nitrogen-fixing bacteria was made possible by the use of transposons; also here, Tn5 has been applied almost exclusively.

What makes Tn5 so attractive? The answer has been given in many specific research papers and general reviews (e.g. 6, 28,71) and can be summarized and discussed as follows.

The high frequency of transposition of Tn5 is of special importance, particularly when the transfer frequency of the transposon carrier vector into the recipient is far below the percentage range, as it is the case for most of the nonenteric bacteria. As a result of the generally low insertional specificity of Tn5, insertions are distributed almost randomly. Usually all possible mutants should be present in a few thousand clones selected after mutagenesis. It should be noted however, that cases of moderate (6, 9) or even high specificity (99) of Tn5 insertions have also been reported.

Tn5 encodes an aminoglycoside-3'-phosphotransferase II (7, 56) which confers resistance to the commonly used antibiotics Nm and Km. In addition, there is a gene coding for resistance to metalloglycopeptide antibiotics like bleomycin (26, 42, 68), which is followed by a streptomycin phosphotransferase gene that is expressed to various degrees in several non-*E. coli* strains (29,69,76,79,91). Thus, if selection for resistance to Nm or Km is not sufficient (e.g. because of high frequency of spontaneous resistance in the recipient), the presence of Tn5 can be verified unambiguously by using one of the other resistances instead of or in addition to Km/Nm.

Other noteworthy properties of Tn5 (as reviewed in 6) are the low probability of genome rearrangements and deletions upon transposition, the insertion stability of Tn5 once established in a genome, and, although a few exceptions have been reported (6, 28, 34), the strong polarity of a Tn5 induced mutation.

Finally, Tn5 is best characterized at the molecular level: the complete sequence (5818 bp) is published (3,4,68). There are convenient sites for several conventionally used restriction enzymes to map the position of Tn5 insertion; some sites are absent so that cloning of DNA fragments containing Tn5 is possible with commonly used enzymes (e.g. *Eco*RI); and there are restriction sites allowing the construction of new derivatives by cloning DNA fragments into Tn5 without affecting transposition. Thus, a number of different Tn5 derivatives have been constructed which combine the advantageous transposition properties with other useful features.

In Tn5–132 the Nm resistance is replaced by the Tc resistance gene from Tn10 (56). Tn5-Tc contains in addition to Km/Nm the Tc resistance gene from plasmid RP4 (97). The central part of Tn5 has been exchanged for the drug resistance determinants from Tn7 (Tp, Sm, Sp) to construct Tn5.7 (116), or from Tn1697 to yield Tn5–GmSpSm (51). Similarly, Tn5–233 carries the Gm/Km and Sp/Sm resistance markers from plasmid pSa (30). In Tn5–751 the Tp resistance gene from plasmid R751 is inserted as an additional marker (81). Tn5–235 (Km) carried the lacZ gene from which β–galactosidase is expressed constitutively (30). Tn5–oriT (114), Tn5–A1 (92) and Tn5–Mob (95) provide the target replicon with the ability of being efficiently mobilized between different Gram-negative bacteria by the broad-host-range transfer functions of RP4. Of special value are the promoter probe transposons constructed on the basis of Tn5. Tn5–VB32 contains a promoterless Nm resistance gene (5), and Tn5–lac carries the E. coli lac operon lacking the transcriptional start signals (63). With these transposons genes can not only be mutated, but the level and conditions of their expression can be monitored by the inserted indicator genes. Finally, Tn5phoA contains an alkaline phosphatase gene without its signal peptide allowing identification by insertional mutagenesis genes whose products are secreted (67).

As mentioned above, transposons other than Tn5 have not gained such widespread use. In many cases when other transposons have been used, clear-cut disadvantages can be deduced from the published data as far as generalized mutagenesis is concerned. A few examples should illustrate this point. Various degrees of undesired insertional specificity, ranging from regional preference to single insertion sites, have been reported for transposons Tn1(Ap) (21, 62) and Tn7 (12, 20, 37, 38, 103, 106). Tn10(Tc) (60) and its new derivatives (carrying e.g. other resistance markers or the indicator gene lacZ (110) are extremely useful tools in E. coli and Salmonella. Unfortunately, only very low or even no detectable transposition frequencies are reported for several other organisms (38, 92, 96). Tn501(Hg) has successfully been used for insertion mutagenesis e.g. in Pseudomonas aeruginosa (78, 105) whereas in other bacterial sites its transposition frequency was found to be too low (38). Closely related Tn501 is transposon Tn1721(Tc) (1, 43, 90), and there are interesting derivatives of it available carrying e.g. other resistance markers (107) or the promoterless luciferase operon (94); but again these transposons have been tested only in a limited number of recipient species so far (94, 107).

From the published data it is clear that Tn5 has gained the widest acceptance in work with non-E. coli strains. But to date there is no systematic overview about the properties of a representative series of transposons in many different Gram-negative strains available. Thus, other transposons may well be useful for specific purposes or at least worth testing in a particular recipient of interest.

V. VECTOR SYSTEMS FOR TRANSPOSON MUTAGENESIS

A. General aspects

To carry out transposon mutagenesis, a vehicle is needed which as a carrier introduces the transposon at high frequency into the target cells but which fails to become established in the recipient.

A few successful attempts have been made to use bacteriophage vectors constructed for enteric bacteria outside their natural host range. For example, *E. coli* phage P1 also infects *Myxococcus* bacteria but does not replicate therein, i.e. it can be used as a transposon delivery system for this organism (63, 64). Plasmids have been constructed that express the *E. coli* *lam*B gene, encoding the lambda receptor, even in naturally lambda-resistant hosts (25, 31, 49). By introducing one of these vectors, pHCP2, strains of *Erwinia* were made susceptible to phage lambda infection, which subsequently allowed the use of derivatives carrying Tn*5* or Tn*10* for transposon mutagenesis (87).

However, for the vast majority of recipient strains, no phage vector system has been developed so far. Thus the conjugative cell-to-cell transfer of appropriate vector plasmids is the most commonly used way to introduce carrier replicon. Figure 2 summarizes schematically the basic features of such a system.

The transposon is carried by a plasmid which stably replicates in the *E. coli* donor strain. The plasmid must either exhibit its own broad-host-range transfer system or must be mobilizable by such a system, so that it can be introduced into the non-*E. coli* recipient by conjugation. In the recipient, the stable replication and maintenance of the transposon carrier replicon must be prevented, so as to identify the cells in which the transposon has been rescued by transposition onto the host genome. The original carrier plasmid is lost by segregation and a stable drug resistant clone ideally results from a single transposition event.

The vehicles displaying such properties are often referred to as 'suicide plasmids'. On the basis of their suicide mechanism the donor plasmids currently being used may be divided into three classes:

(i) The classical suicide plasmid is a IncP type resistance plasmid harboring a Mu prophage that prevents vector establishment in many non-*E. coli* recipients;

(ii) There are also temperature-sensitive promiscuous replicons, the use of which is of course limited to bacteria able to grow at elevated temperatures;

(iii) More universally applicable donor plasmids combine broad-host-range transfer or mobilization functions with a narrow host range of replication. This diverse group of vehicles consists mainly of *in vitro* constructions but also naturally-occurring plasmids have been found which exhibit these properties.

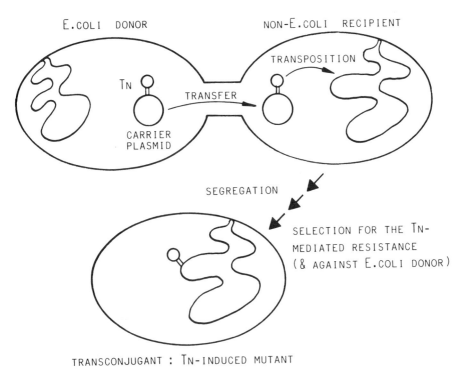

Figure 2. Basic features of plasmid-mediate transfer of transposons from cell to cell.

B. Mu-mediated suicide of vector plasmid

The first suicide plasmids were found more or less by chance. The initial intention of inserting bacteriophage Mu into the promiscuous IncP plasmid was to introduce the phage conjugally into naturally Mu-resistant soil bacteria like *Rhizobium*, *Agrobacterium* or *Pseudomonas* and to take advantage of its properties in the new hosts. As a result of these experiments, it was found that the Mu prophage carried by the plasmid severely reduced its ability to become established in some bacteria (13, 32, 108). This phenomenon was explained by Mu-specific gene function(s) as well as by restriction mechanisms that primarily affect the Mu but not the plasmid DNA (108).

Consequently, this Mu dependent suicide effect has been utilized to construct a transposon delivery system of wide applicability: The IncP type resistance plasmid pPH1JI (Gm,Sp) was loaded with phage Mu and transposon Tn5 to yield plasmid pJB4JI (10, 50). Using pJB4JI as a Tn5 donor vehicle, many new research activities have been initiated primarily in *Rhizobium* genetics by the isolation of symbiotic mutants. In the meantime there exists a long and diverse list of other organisms that have been successfully mutagenized by pJB4JI transfer (see in 6, 71). However, a substantial amount of published data has also accumulated showing disadvantages or failure of Mu-based plasmid suicide. For example pJB4JI showed high frequency of survival in subspecies in *Pseudomona solanacearum* (14, 72), *Xanthomonas campestris* (106), *Rhodopseudomonas capsulata* (57), *Azospirillum brasilense* (100), *P. aeruginosa* (81), or *Erwinia carotovora* (115). Also stably replicating deletion derivatives of the suicide plasmids have been identified in several *Rhizobium* strains (22, 74). Another type of complication has been observed after pJB4JI transfer to *Rhizobium meliloti*; in a high proportion of clones, Mu sequences have transposed together with Tn5 onto the recipient genome making further analyses of the mutants difficult (40, 70).

C. Temperature-sensitive replicons

The idea to use Rep-ts mutants of RP4 as transposon delivery vehicles for a broad range of Gram-negative recipients was already proposed in the first review about transposon work in enteric bacteria (61). Meanwhile several derivatives of RP4 (= RP1, R68, RK2) that are not autonomously maintained at 43°C, have been constructed (27, 47, 54, 82, 109) and used in a variety of experiments. For example, new Hfr strains of *E. coli* (48, 82) or *P. aeruginosa* (46, 54, 80, 105) were isolated by transposon Tn*1* mediated chromosomal integration of the Rep-ts plasmid. Similarly, Tn*2521* (Cb,Sm,Sp,Su) isolated from *P. aeruginosa* and inserted into the Rep-ts plasmid pMO190 (54) produced new Hfr strains at high frequency by vector integration (75). Plasmid pMO190 has also been loaded with Tn*5* (pMO194) or Tn*501* (pMT1000) and used as transposon donor vehicle in *P. aeruginosa* (75, 78, 105). Tn*5* mutants of *Legionella pneumophila* have been isolated after transfer of pRK340 (58) and elevated temperatures. Transposon Tn*1* was introduced into the *P. aeruginosa* chromosome with pME319 (46, 73) and pME9 served as a donor vehicle for the *in vitro* constructed transposon Tn*5−751* (81).

Unfortunately, the nonpermissive temperature for all these vectors is above 40°C and this limits their possible application to a minority of Gram-negative bacteria able to grow at high temperature.

D. Combination of broad host range transfer with narrow host range replication systems

1. Replication defective mutants of RP4

During the course of early experiments performed to map replication genes of RP4, the hybrid plasmid pAS8 was constructed which is composed of RP4 and ColE1 fused at their unique *Eco*RI sites (86). A derivative of this hybrid, pAS8 Rep-1, carries a Tn7 insertion in a broad host range maintenance function of RP4 and is therefore dependent on the narrow-host-range ColE1 replication system. The fertility functions of the parent RP4 still allow the transfer of pAS8 Rep-1 from *E. coli* into other Gram-negative bacteria in which it is not stably replicated. These properties have been used to deliver transposon Tn7 into subspecies of *Pseudomonas syringae* (89). The derivative pUW942 served as donor vehicle for Tn501 mutagenesis of *Bordetella pertussis* (111).

2. In vitro combinations of promiscuous transfer genes and narrow host range replication functions

The basic idea to combine in one replicon both narrow-host-range maintenance functions and interspecific transfer proficiency has also been realized by cloning transfer genes from different sources into *E. coli* specific vector plasmids.

(i) IncP type transfer genes. Plasmid pRK2013 resulted from work undertaken to study replication functions of RK2. It consists of the entire set of the broad-host-range transfer genes of RK2 inserted, together with a Km resistance determinant into ColE1 (39). The selection marker is located on Tn903, and this transposon has been introduced into a number of plant-associated Pseudomonads using pRK2013 directly as a suicide vector (65). Derivatives of pRK2013 have also been used as suicidal carrier replicons for other transposons. In *R. meliloti* (12) and *X. campestris* (106) site specific transpositon of Tn7 has been demonstrated after pRK2013::Tn7 transfer. Other pRK2013 based delivery vectors are pBEE7 (Tn7), pBEE10 (Tn10), pBEE104 (Tn10—HH104, overproducing transposase (41)), and pBEE132 (Tn5—132), which have been tested in *Caulobacter crescentus* and *Acinetobacter calcoaceticus* (38). With pMD100, Tn501 has been transferred into *Rhizobium japonicum* strains (17) and with pUW964, Tn5 mutants of *B. pertussis* (112) as well as subspecies of *P. syringae* (66,77) have been obtained. Finally, the Tn5 derivatives Tn5—233 and Tn5—235 have been inserted into the *R. meliloti* genome via pRK2013 transfer (30).

(ii) IncN type transfer genes. By cloning the promiscuous transfer genes from the IncN plasmid pCU1 into the *E. coli* p15A derived plasmids pACYC184 and pACYC177 (24), a further series of suicidal vectors have been constructed: pGS9 (Tn5), pGS18 (Tn1), pGS27 (Tn9), and pGS16 (Tn10) (92). Because of the advantageous properties of Tn5, pGS9 in particular has been used successfully to isolate transposon insertion mutants of various strains of *R. meliloti* and *R. leguminosarum* (92), slow and fast growing strains of *R. japonicum* (83) and *Azotobacter vinelandii* (59).

(iii) IncW type transfer genes. A DNA fragment from the IncW plasmid R388 containing the genes for conjugal transfer has been cloned into the *E. coli* vector pBR322 (11) and the resulting hybrid was loaded with Tn5 to construct pWI281 (72). As expected from results of transfer experiments with the parent plasmid R388, pWI281 is transmissible from *E. coli* to strains of *P. solanacearum* and suitable as suicidal donor for for Tn5 (72).

3. Naturally occurring suicide plasmids

Some transposon delivery systems are based on plasmids whose transfer but not replication systems are naturally promiscuous. One example is the IncIα enteric bacterial plasmid ColIb. The transfer-derepressed derivative ColIb-drd1 was loaded with transposons Tn5 (pLG221) and Tn10 (pLG223) (15). Especially pLG221 proved to be an efficient suicidal donor vehicle in Tn5 mutagenesis experiments with a variety of *Pseudomonas* strains (16, 36). Another example is the transfer derepressed IncP10 plasmid R91–5, identified in *P. aeruginosa* (23). Its Tn5-carrying derivative, pMO75, has been shown to be transmissible to but not maintained in *E. coli*, *Methylobacterium* and other Pseudomonads, so that it can be used as a transposon donor vehicle for these organisms (113).

4. Mobilization of transposon carrier vehicles

In all methods discussed so far, the transposon donor plasmids harbour a complete set of transfer genes, i. e. they are self-transmissible. But the gene products of a broad-host-range system can also act in *trans* and promote the mobilization of plasmids that carry the suitable origin of transfer (*oriT*) as has been shown during analysis of RK2 conjugation (44).

By cloning into usual *E. coli* vectors like pACYC184 and 177 (24), or pBR325 (11) the IncP-type specific recognition site (Mob-site = *oriT*) for mobilization, the series of pSUP vectors was constructed (95–98). These small multicopy plasmids are mobilized at very high frequencies by the transfer functions of IncP plasmids.

For optimal use of the pSUP vectors, special mobilizing *E. coli* strains, that carry derivatives of RP4 integrated into their chromosomes have been constructed. From these donors, the mobilizable vectors are transferred to a recipient cell without concomitant self-transmission of the helper plasmid that provides the transfer functions. The pSUP vectors can be introduced into all bacteria that can act as recipients for RP4, but their replication host range is restricted to *E. coli* and some close relatives. Consequently, they have been used as suicidal carrier vehicles for transposon Tn5 mutagenesis in numerous different Gram-negative bacterial strains (71, 96, 97).

With pSUP102(= pACYC184-Mob) and pSUP202(= pBR325-Mob) as carriers it has been shown, that the insertion site of Tn5 in the vector molecules has significant influence on the transposition frequency in a given recipient; transposition frequencies differing more than a hundred-fold have been found in a series of different Tn5 insertion sites in a vector (96). Furthermore, a Tn5 carrier such as pSUP1021, demonstrated in numerous experiments to be optimal for various *Rhizobium* strains, need not necessarily be optimal for other recipients. For example, another insertion site of Tn5 in pSUP102 resulted in much higher yield of Tn5-induced mutants with *X. campestris*, but lower transposition frequency in *R. meliloti*, when compared with pSUP1021 under identical conditions (98a). A reason for the marked differences in the transposition frequencies could be the promoter activity in the border sequences of the Tn5 insertion site (88), since the rate of transcription is of course not equally distributed within a replicon and most probably is also strain dependent.

E. Conclusions

Transposon mutagenesis has become a standard method since it offers tremendous potentials for genetic analyses and manipulation of Gram-negative bacteria. Although the various techniques using transposons in nonenteric strains have been developed within the last several years, it is now almost impossible to compile a complete list of references that would cover all important results of transposon mutagenesis. Nevertheless, the knowledge about the behaviour of the different suicide vectors and transposons within the huge variety of Gram-negative bacteria is very incomplete. Thus, as stated above, the optimal transposon delivery system for a new recipient strain of interest, in most cases still has to be experimentally elaborated. Most obvious possibilities of recipient strain-dependent variations are the transfer frequency and the efficiency of the suicide mechanism. But also the frequency of transposition of a given transposon may vary significantly from strain to strain. Furthermore, as shown for the Tn5 carrying pSUP vectors, the insertion site of the transposon within the plasmid has an influence on the frequency of its transposition in a target cell. Therefore, well established suicide vector systems may also still be optimized simply by testing derivatives of a given vector having the transposon inserted at different sites.

But even if a chosen vector for Tn5 mutagenesis seems to work satisfactorily in a bacterial strain under study, experience has shown, that it is advisable to check the mutant clones for possible anomalies. One of the 1.5 kb terminal repeats of Tn5, called IS50, may have transposed independently from a genuine Tn5 insertion, or the whole vector may have been inserted together with Tn5 into the host genome (6,8,52,53,55). Such events have been reported to take place in mutagenesis experiments with pSUP::Tn5 vectors, but since dependent on the transposon, these anomalies are likely to occur also with other suicide vectors. Rare cases of independent IS50 transpositions have been found by hybridization in *P. syringae* (2) or in *R. meliloti* (96). Tn5-mediated vector insertion has been found to occur with widely varying frequencies. For example, there was no significant frequency of vector insertion in species of *P. syringae* (2) or *P. putida* (98a), whereas about 5 to 30% of Tn5-induced mutants of *R. meliloti* (96) or two species of *R. japonicum* (101) also have the vector molecule inserted. In extreme situations, like with one *Rhizobium* species (35) or *X. campestris* (98a), all mutant clones result from concomitant insertion of Tn5 and vector DNA. This again reflects the strain-dependent variations inherent in a Tn5 mutagenesis system and emphasizes the necessity of careful tests with any new recipient to be mutagenized.

VI. TRANSPOSON MUTAGENESIS OF CLONED DNA

An extremely useful method for fine structure genetic analysis results from combining *in vitro* cloning with *in vivo* transposon mutagenesis. This technique, termed site-specific transposon mutagenesis, homogenotization, marker exchange or gene replacement, was first applied to the study of symbiotic genes of *Rhizobium* (85,85) and is now being widely used (for review see 28, 33). The principle of site-directed transposon mutagenesis can be briefly summarized as follows. A cloned DNA fragment is mutagenized by transposon insertion in *E. coli*, using one of the well established standard procedures, e.g. with a ::Tn5 vector (28). The insertion site can be easily mapped by restriction analysis of the isolated plasmid DNA. The mutagenized DNA fragment is transferred back to its original host, where sequence homology allows recombination on both sides of the transposon insertion. This double cross-over finally leads to the exchange of the wild-type gene for the transposon-mutated one. This technique not only enables the insertion of a transposon at a predetermined site in a target cell, but also allows very accurate analyses of genes and operons by comparing the physical map with the phenotype resulting from the gene replacement.

Mainly two vector systems are currently being used for such experiments. There is a series of mobilizable broad-host-range cloning vectors, e.g. pRK290 (33, 33a), which after introduction into a recipient have to be replaced by an incoming incompatible plasmid to identify the homogenotization event. Alternatively, the above mentioned mobilizable pSUP vectors which do not replicate in the non-*E. coli* recipient can be used.

VII. REFERENCES

1. Altenbucher J, Choi CL, Grinsted J, Schmitt R, Richmond MH (1981) The transposons Tn501(Hg) and Tn1721(Tc) are related. *Genet Res* **37**, 285–9
2. Anderson DM, Mills D (1985) The use of transposon mutagenesis in the isolation of nutritional and virulence mutants in two pathovars of *Pseudomonas syringae*. *Phytopathology* **75**, 104–8
3. Auerswald EA, Ludwig G, Schaller H (1980) Structural analysis of Tn5. *Cold Spring Harbor Symp Quant Biol* **45**, 107–13
4. Beck E, Ludwig G, Aeuerswald EA, Reiss B, Schaller, H. (1982) Nucleotide sequence and exact location of the neomycin phosphotransferase gene from transposon Tn5. *Gene* **19**, 327–36
5. Bellofatto V, Shapiro L, Hodgson DA (1984) Generation of a Tn5 promoter probe and its use in the study of gene expression in *Caulobacter crescentus*. *Proc Natl Acad Sci USA* **81**, 1035–9

6. Berg DE, Berg CM (1983) The prokaryotic transposable element Tn5. *Biotechnology* **1**, 417–35

7. Berg DE, Davies J, Allet B, Rochaix JD (1975) Transposition of R-factor genes to bacteriophage lambda. *Proc Natl Acad Sci USA* **72**, 3628–32

8. Berg DE, Johnsrud L, McDivitt L, Ramabhadran R, Hirschel BJ (1982) The inverted repeats of Tn5 are transposable elements. *Proc Natl Acad Sci USA* **79**, 2632–5

9. Berg DE, Schmandt MA, Lowe JB (1983) Specificity of transposon Tn5 insertion. *Genetics* **105**, 813–28

10. Beringer JE, Benynon JL, Buchanan-Wollaston AV, Johnston AWB (1978) Transfer of the drug-resistance transposon Tn5 to *Rhizobium*. *Nature* **276**, 633–4

11. Bolivar F (1978) Construction and characterization of new cloning vehicles III. Derivatives of plasmid pBR322 carrying unique *Eco*RI sites for selection of *Eco*RI generated recombinant molecules. *Gene* **4**, 121–36

12. Bolton E, Glynn P, O'Gara F (1984) Site specific transposition of Tn7 into a *Rhizobium meliloti* megaplasmid. *Mol Gen Genet* **193**, 153–7

13. Boucher C, Bergeron B, Barate de Bertalmio M, Denarie J (1977) Introduction of bacteriophage Mu into *Pseudomonas solancearum* and *Rhizobium meliloti* using the R-factor RP4. *J Gen Microbiol* **98**, 253–263

14. Boucher C, Message B, Debieu D, Zischek C (1981) Use of P-1 incompatibility group plasmids to introduce transposons into *Pseudomonas solanacearum*. *Phytopathology* **71**, 639–42

15. Boulnois GJ (1981) Colicin Ib does not cause plasmid-promoted abortive phage infection of *Escherichia coli* K12. *Mol Gen Genet* **182**, 508–10

16. Boulnois GJ, Varley JM, Sharpe GS, Franklin FCH (1985) Transposon donor plasmids based on ColIb-P9 for use in *Pseudomonas putida* and a variety of other Gram-negative bacteria. *Mol Gen Genet* **200**, 65–7

17. Bullerjahn GS, Benzinger RH (1984) Introduction of the mercury transposon Tn501 into *Rhizobium japonicum* strains 31 and 110. *FEMS Microbiol Lett* **22**, 183–7

18. Calos MP, Miller JH (1980) Transposable elements. *Cell* **20**, 579–95

19. Campbell A, Berg DE, Botstein D, Lederberg EM, Novick RP et al (1979) Nomenclature of transposable elements in bacteria. *Gene* **5**, 197–206

20. Caruso M, Shapiro JA (1982) Interaction of Tn7 and temperate phage F116L of *Pseudomonas aeruginosa*. *Mol Gen Genet* **188**, 292–8

21. Casadesus J, Janez E, Olivares J (1980) Transposition of Tn1 to the *Rhizobium meliloti* genome. *Mol Gen Genet* **180**, 405–10

22. Casey C, Bolton E, O'Gara F (1983) Behaviour of bacteriophage Mu-based IncP suicide vector plasmids in *Rhizobium* spp. *FEMS Microbiol Lett* **20**, 217–23

23. Chandler PM, Krishnapillai V (1977) Characterization of *Pseudomonas aeruginosa* derepressed R-plasmids. *J Bacteriol* **130**, 596–603

24. Chang ACY, Cohen SN (1978) Construction and characterization of amplifiable multicopy DNA cloning vehicles derived from the pA15 cryptic miniplasmid. *J Bacteriol* **134**, 1141–56

25. Clement JM, Perin D, Hedgepeth J (1982) Analysis of receptor and β-lactamase synthesis and export using cloned genes in a minicell system. *Mol Gen Genet* **185**, 302–10

26. Collins CM, Hall RM (1985) Identification of a Tn*5* determinant conferring resistance to phleomycins, bleomycins and tallysomycin. *Plasmid* **4**, 143–51

27. Danilevich VN, Stephanshin YG, Volzhantsev NV, Golub, EI (1978) Transposon-mediated insertion of R-factor into bacterial chromosome. *Mol Gen Genet* **161**, 337–9

28. De Bruijin FJ, Lupski JR (1984) The use of transposon Tn*5* mutagenesis in the rapid generation of correlated physical and genetic maps of DNA segments cloned into multicopy plasmids—A review. *Gene* **27**, 131–49

29. De Vos GF, Finan TM, Signer ER, Walker GC (1984) Host dependent transposon Tn*5*-mediated streptomycin resistance. *J Bacteriol* **159**, 395–8

30. De Vos GF, Walker GC, Signer ER (1986) Genetic manipulations in *Rhizobium meliloti* utilizing two new transposon Tn*5* derivatives. *Mol Gen Genet* **204**, 485–91

31. De Vries GE, Raymond CK, Ludwig RA (1984) Extension of bacteriophage lambda host range: selection, cloning and characterization of a constitutive receptor gene. *Proc Natl Acad Sci USA* **81**, 6080–4

32. Denarie J, Rosenberg C, Bergeron B, Boucher C, Michel M, Borate de Bertalmio M (1977) Potential of RP4::Mu plasmids for *in vivo* genetic engineering of Gram-negative bacteria. *In* DNA Insertion Elements, Plasmids and Episomes, (Bukhari AI, Shapiro JA, Adhya SL eds.) Cold Spring Harbor, NY: Cold Spring Harbor Biol Labs pp. 507–20

33 Ditta G (1986) Tn*5* mapping of *Rhizobium* nitrogen fixation genes. *In* Plant Molecular Biology, (Weisbach A, Weisbach H eds) *Methods in Enzymology* **118**, 519–28

33a. Ditta G, Stanfield S, Corbin D, Helinski DR (1980) Broad-host-range DNA cloning system for Gram-negative bacteria. Construction of a gene bank of *Rhizobium meliloti*. *Proc Natl Acad Sci USA* **77**, 7347–51

34. Don RH, Weightman AJ, Knackmus JH, Timmis KN (1985) Transposon mutagenesis and cloning analysis of the pathways for degradation of 2,4-dichlorophenoxyacetic acid and 3-chlorobenzoate in *Alcaligenes eutrophus* JMP134 (pJP4). *J Bacteriol* **161**, 85–90

35. Donald RG, Raymond CK, Ludwig RA (1985) Vector insertion mutagenesis of *Rhizobium* sp.strain ORS571: direct cloning of mutagenized DNA sequences. *J Bacteriol* **162**, 317–23

36. Eaton RW, Timmis KN (1986) Characterization of a plasmid-specified pathway for catabolism of isopropylbenzene in *Pseudomonas putida* RE204.

J Bacteriol **168**, 123–31

37. Ely B, Croft RH (1982) Transposon mutagenesis in *Caulobacter crescentus*. *J Bacteriol* **149**, 620–5

38. Ely B (1985) Vectors for transposon mutagenesis of nonenteric bacteria. *Mol Gen Genet* **200**, 302–4

39. Figurski DH, Helinski DR (1979) Replication of an origin-containing derivative of plasmid RK2 dependent on a plasmid function provided in *trans*. *Proc Natl Acad Sci USA* **76**, 1648–52

40. Forrai T, Vincze E, Banfalvi Z, Kiss GB, Randhawa GS, Kondorosi A (1983) Localization of symbiotic mutations in *Rhizobium meliloti*. *J Bacteriol* **153**, 635–43

41. Forster TJ, Davis MA, Roberts DE, Takerhita K, Kleckner N (1981) Genetic organization of transposon Tn*10*. *Cell* **23**, 201–13

42. Genilloud, O., Garrido, M.C. and Moreno, F. (1984). The transposon Tn*5* carries a bleomycin-resistance determinant. *Gene* **32**, 225–33

43. Grinsted J, de la Cruz F, Altenbuchner J, Schmitt R (1982) Complementation of *tnpA* mutants of Tn*3*, Tn*21*, Tn*501* and Tn*1721*. *Plasmid* **8**, 276–86

44. Guiney DG, Helinski DR (1979) The DNA-protein relaxation complex of the plasmid RK2: Location of the site-specific nick in the region of the proposed origin of transfer. *Mol Gen Genet* **176**, 183–9

45. Haas D (1986) Jumping genes as tools in the analysis of bacterial metabolism. *Swiss Biotech* **4**, 20–2

46. Haas D, Watson J, Krieg R, Leisinger T (1981) Isolation of an Hfr donor of *Pseudomonas aeruginosa* PAO by insertion of the plasmid RP1 into the tryptophan synthase gene. *Mol Gen Genet* **182**, 240–244

47. Harayama S, Tsuda M, Iino T (1980) High frequency mobilization of chromosome of *Escherichia coli* by a mutant of a plasmid RP4 temperature-sensitive for maintenance. *Mol Gen Genet* **180**, 47–56

48. Harayama S, Tsuda M, Iino T (1981) Tn*1* insertion mutagenesis in *Escherichia coli* K12 using a temperature-sensitive mutant of plasmid RP4. *Mol Gen Genet* **184**, 52–5

49. Harkki A, Palva ET (1985) A *lamB* expression plasmid for extending the host range of phage to other bacteria. *FEMS Microbiol Lett* **27**, 183–87

50. Hirsch PR, Beringer JE (1984) A physical map of pH1JI and pJB4JI. *Plasmid* **12**, 133–41

51. Hirsch PR, Wang CL, Woodward MJ (1986). Construction of a Tn*5* derivative determining resistance to gentamycin and spectinomycin using a fragment cloned from R1033. *Gene* **4**, 203–9

52. Hirschel BJ, Berg DE (1982) A derivative of Tn*5* with direct terminal repeats can transpose. *J Mol Biol* **155**, 105–20

53. Hirschel BJ, Galas DJ, Berg DE, Chandler M (1982) Structure and stability of transposon 5-mediated cointegrates. *J Mol Biol* **159**, 557–80

54. Holloway BW, Crowther C, Dean H, Hagedorn J, Holmes N, Morgan AF (1982) Integration of plasmids into the *Pseudomonas* chromosome. *In* Drug resistance in bacteria, (Mitsuhashi S, ed). Japan Scientific Society Press, Tokyo. pp 231–42

55. Isberg RR, Syvanen M (1981) Replicon fusions promoted by the inverted repeats of Tn5: The right repeat is an insertion sequence. *J Mol Biol* **150**, 15–32

56. Jorgensen RA, Rothstein SJ, Reznikoff WS (1979) A restriction enzyme cleavage map of Tn5 and localization of a region encoding neomycin resistance. *Mol Gen Genet* **177**, 65–72

57. Kaufman N, Hudig H, Drews G (1984) Transposon Tn5 mutagenesis of genes for the photosynthetic apparatus in *Rhodopseudomonas capsulata*. *Mol Gen Genet* **198**, 153–8

58. Keen MG, Street ED, Hoffman PS (1985) Broad-host-range plasmid pRK340 delivers Tn5 into the *Legionella pneumophila* chromosome. *J Bacteriol* **162**, 1332–5

59. Kennedy C, Gamal R, Humphrey R, Ramos I, Brigle K, Dean D (1986) The *nifH* and *nifN* genes of *Azotobacter vinelandii*: characterization by Tn5 mutagenesis and isolation from pLAFR1 gene banks. *Mol Gen Genet* **205**, 318–25

60. Kleckner N (1981) Transposable elements in prokaryotes. *Ann Rev Genet* **15**, 341–404

61. Kleckner N, Roth J, Botstein D (1977) Genetic engineering *in vivo* using translocatable drug resistance elements: New methods in bacterial genetics. *J Mol Biol* **116**, 125–59

62. Krishnapillai V, Royle P, Lehrer J (1981) Insertion of the transposon Tn1 into the *Pseudomonas aeruginosa* chromosome. *Genetics* **97**, 295–311

63. Kroos L, Kaiser D (1984) Construction of Tn5–*lac*, a transposon that fuses *lacZ* expression to exogenous promoters, and its introduction into *Myxococcus xanthus*. *Proc Natl Acad Sci USA* **81**, 5817–20

64. Kuner JM, Kaiser D (1981) Introduction of transposon Tn5 into *Myxococcus* for analysis of developmental and other nonselectable mutants. *Proc Natl Acad Sci USA* **78**, 425–29

65. Lam ST, Lam BS, Strobel G (1985) A vehicle for the introduction of transposons into plant-associated Pseudomonads. *Plasmid* **13**, 200–4

66. Lindgren PB, Peet RC, Panopulos NI (1986) Gene cluster of *Pseudomonas syringae* pv. *phaseolicola* controls pathogenicity of bean plants and hypersensitivity on non-host plants. *J Bacteriol* **168**, 512–522

67. Manoil C, Beckwith J (1985) Tn*phoA*: A transposon probe for protein export signals. *Proc Natl Acad Sci USA* **82**, 8129–33

68. Mazodier P, Cossart P, Giraud E, Gasser F (1985) Completion of the nucleotide sequence of the central region of Tn5 confirms the presence of three resistance genes. *Nucl Acids Res* **13**, 195–205

69. Mazodier P, Giraud E, Gasser F (1983) Genetic analysis of the streptomycin resistance encoded by Tn5. *Mol Gen Genet* **192**, 155–62

70. Meade HM, Long SR, Ruvkun GB, Brown SE, Ausubel FM (1982) Physical and genetic characterization of symbiotic and auxotrophic mutants of *Rhizobium meliloti* induced by transposon Tn5 mutagenesis. *J Bacteriol* **149**, 114–22.

71. Mills D (1985) Transposon mutagenesis and its potential for studying virulence genes in plant pathogens. *Ann Rev Phytopathol* **23**, 297–320

72. Morales VM, Sequeira L (1985) Suicide vector for transposon mutagenesis in *Pseudomonas solanacearum*. *J Bacteriol* **163**, 1263–4

73. Nikas TI, Iglewski BH (1984) Isolation and characterization of transposon induced mutants of *Pseudomonas aeruginosa* deficient in production of exoenzyme S. *Infect Immunol* **45**, 470–4

74. Noel KD, Sanchez A, Fernandez L, Leemans J, Cevallos M (1984) *Rhizobium phaseoli* symbiotic mutants with transposon Tn5 insertions. *J Bacteriol* **158**, 148–55

75. O'Hoy K, Krishnapillai V (1985) Transposon mutagenesis of the *Pseudomonas aeruginosa* PAO chromosome and the isolation of high frequency of recombination donors. *FEMS Microbiol Lett* **29**, 299–303

76. O'Neill EA, Kiely GM, Bender RA (1984) Transposon Tn5 encodes streptomycin resistance in nonenteric bacteria. *J Bacteriol* **159**, 388–9

77. Peet RC, Lindgren PB, Willis DK, Panopulos NI (1986) Identification and cloning of genes involved in phaseolotoxin production by *Pseudomonas syringae* pv. *phaseolicola*. *J Bacteriol* **166**, 1096–105

78. Poole K, Hancock REW (1986) Isolation of a Tn501 insertion mutant lacking porin protein P-of *Pseudomonas aeruginosa* PAO. *Mol Gen Genet* **202**, 403–9

79. Putnoky P, Kiss GB, Ott I, Kondorosi A (1983) Tn5 carries a streptomycin resistance determinant downstream from the kanamycin resistance gene. *Mol Gen Genet* **191**, 288–94

80. Reimmann C, Haas D (1986) IS21 insertion in the *trfA* replication control gene of chromosomally integrated plasmid RP1: a property of stable *Pseudomonas aeruginosa* Hfr strain. *Mol Gen Genet* **203**, 511–9

81. Rella M, Mercenier A, Haas D (1985) Transposon insertion mutagenesis of *Pseudomonas aeruginosa* with a Tn5 derivative: Application to physical mapping of the *arc* gene cluster. *Gene* **33**, 293–303

82. Robinson MK, Bennet PM, Falkow S, Dodd HM (1980) Isolation of a temperature-sensitive derivative of RP1. *Plasmid* **3**, 343–7

83. Rostas K, Sista PR, Stanley J, Verma DPS (1984) Transposon mutagenesis of *Rhizobium japonicum*. *Mol Gen Genet* **197**, 230–5

84. Ruvkun GB, Ausubel FM (1981) A general method for site-directed mutagenesis in prokaryotes. *Nature* **289**, 85–8

85. Ruvkun GB, Sundaresan V, Ausubel FM (1982) Directed transposon

mutagenesis and complementation analysis of *Rhizobium meliloti* symbiotic nitrogen fixation genes. *Cell* **29**, 551–9

86. Sakanyan VA, Yakubov LZ, Alikhanian SI, Stepanov AI (1978) Mapping of RP4 plasmids using deletion mutants of pAS8 hybrid (RP4–ColE1). *Mol Gen Genet* **165**, 331–41

87. Salmond GPC, Hinton JCD, Gill DR, Perombelon MCM (1986) Transposon mutagenesis of *Erwinia* using phage vectors. Mol Gen Genet **203**, 524–8

88. Sasakawa C, Lowe JB, McDivitt L, Berg DE (1982) Control of transposon Tn5 transposition in *Escherichia coli*. *Proc Natl Acad Sci USA* **79**, 7450–4

89. Sato M, Staskawicz BI, Panopulos NI, Peters S, Honma M (1981) A host-dependent hybrid plasmid suitable as a suicidal carrier for transposable elements. *Plasmid* **6**, 325–31

90. Schmitt R, Altenbuchner J, Weibauer K, Arnold W, Puhler A, Schoffl F (1981) Basis of transposition and gene amplification by Tn*1721* and related tetracycline resistance transposons. *Cold Spring Harbor Symp Quant Biol* **45**, 59–65

91. Selveraj G, Iyer VN (1984) Transposon Tn5 specifies streptomycin resistance in *Rhizobium* spp. J Bacteriol **158**, 580–9

92. Selveraj G, Iyer VN (1983) Suicide plasmid vehicles for insertion mutagenesis in *Rhizobium meliloti* and related bacteria. *J Bacteriol* **156**, 1292–1300

93. Shapiro JA (ed.) (1983). Mobile Genetic Elements, New York, Academic Press

94. Shaw JJ, Kado CI (1987) Direct analysis of the invasiveness of *Xanthomonas campestris* mutants generated by Tn*4431*, a transposon containing a promoterless luciferase cassette for monitoring gene expression. *In* Molecular Genetics of Plant-Microbe Interaction, (Verma DPS, Brisson N, eds) Martinus Nijhoff Publishers, pp. 57–60

95. Simon R (1984) High frequency mobilization of Gram-negative bacterial replicons by the *in vitro* constructed Tn5–Mob transposon. *Mol Gen Genet* **196**, 413–20

96. Simon R, O'Connell M, Labes M, Puhler A (1986) Plasmid vectors for the genetic analysis and manipulation of Rhizobia and other Gram-negative bacteria. *In* Plant Molecular Biology, (Weisbach A, Weisbach H, eds) *Methods in Enzymology* **118**, 640–59

97. Simon R, Priefer U, Puhler A (1983) A broad-host-range mobilization system for *in vivo* genetic engineering: Transposon mutagenesis in Gram-negative bacteria. *BioTechnology* **1**: 784–791

98. Simon R, Priefer U, Puhler A (1983) Vector plasmids for *in vivo* and *in vitro* manipulations of Gram-negative bacteria. *In* Molecular Genetics of the Bacteria-Plant Interaction, (Puhler A, ed). Springer Verlag, Heidelberg. pp. 98–106

98a. Simon R, unpublished results

99. Singer JT, Finnerty WR (1984) Insertional specificity of transposon Tn5 in *Acinetobacter* sp. *J Bacteriol* **157**, 607–11

100. Singh M, Klingmuller W (1986) Transposon mutagenesis in *Azospirillum brasiliense*: Isolation of auxotrophic and Nif mutants and molecular cloning of the mutagenized *nif* DNA. *Mol Gen Genet* **202**, 136–142

101. So JS, Hodgson ALM, Haugland R, Leavitt M, Banfalvi Z, Nieuwkoop A, Stacey G (1987) Transposon-induced symbiotic mutants of *Bradyrhizobium japonicum*: Isolation of two gene regions essential for nodulation. *Mol Gen Genet* **207**, 15–23

102. Starlinger P (1980) IS elements and transposons. *Plasmid* **3**, 242–59

103. Thomson JA, Hendson M, Magnes RM (1981) Mutagenesis by insertion of drug resistance transposon Tn7 into *Vibrio* species. *J Bacteriol* **148**, 374–8

104. Toussaint A, Resibois A (1983) Phage Mu: transposition as a life-style. *In* Mobile Genetic Elements, (Shapiro JA ed). New York: Academic Press. pp. 105–158

105. Tsuda M, Harayama S, Iino T (1984) Tn*501* insertion mutagenesis in *Pseudomonas aeruginosa* PAO. *Mol Gen Genet* **196**, 494–500

106. Turner P, Barber C, Daniels M (1984) Behaviour of the transposons Tn5 and Tn7 in *Xanthomonas campestris* pv. *campestris*. *Mol Gen Genet* **195**, 101–7

107. Ubben D, Schmitt R (1986) Tn*1721* derivatives for transposon mutagenesis, restriction mapping and nucleotide sequence analysis. *Gene* **41**, 145–52

108. Van Vliet F, Silva B, Van Montagu M, Schell J (1978) Transfer of RP4::Mu plasmids to *Agrobacterium tumefaciens*. *Plasmid* **1**, 446–55

109. Watson J (1980) Replication mutants of the IncP-1 plasmid RP1. *Experientia* **36**, 1451

110. Way JC, Davis MA, Morisato D, Roberts DE, Kleckner N (1984) New Tn*10* derivatives for transposon mutagenesis and for construction of *lacZ* operon fusions by transposition. *Gene* **32**, 369–79

111. Weiss A, Falkow S (1983) Transposon insertion and subsequent donor formation promoted by Tn*501* in *Bordetella pertussis*. *J Bacteriol* **153**, 304–9

112. Weiss A, Hewlett EL, Myers GA, Falkow S (1983) Transposon Tn5-induced mutations affecting virulence factors of *Bordetella pertussis*. *Infect Immunol* **42**, 33–41

113. Whitta S, Sinclair MI, Holloway BW (1985) Transposon mutagenesis in *Methylobacterium* AM1 (*Pseudomonas* AM1). *J Gen Microbiol* **131**, 1547–9

114. Yakobson EA, Guiney DG (1984) Conjugal transfer of bacterial chromosomes mediated by the RK2 plasmid transfer origin cloned into transposon Tn5. *J Bacteriol* **160**, 451–3

115. Zink RT, Kemble RJ, Chatterjee AK (1984) Transposon Tn*5* mutagenesis in *Erwinia carotovora* subsp. *carotovora* and *E. carotovora* subsp. *atroseptica*. *J Bacteriol* **157**, 809–14

116. Zsebo KM, Wu F, Hearst JE (1984) Tn*5.7* construction and physical mapping of pRPS404 containing photosynthetic genes from *Rhodopseudomonas capsulata*. *Plasmid* **11**, 182–4

CHAPTER 9

GENE EXPRESSION SIGNALS IN GRAM-NEGATIVE BACTERIA

Christopher M. Thomas and F. Christopher H. Franklin

I. INTRODUCTION

The subject of this book is the molecular biology and exploitation of broad-host-range plasmids. The nature of gene expression signals in Gram-negative bacteria is clearly of interest both to understanding the promiscuity of these plasmids and to achieving successful expression of cloned genes in a variety of species. Barriers to expression may occur at many levels: initiation, elongation and termination of transcription; stability of mRNA; translational efficiency; and protein product stability. Unfortunately studies on most aspects of gene expression in Gram-negatives other than *Escherichia coli* are in their infancy. Consequently this review will concentrate mainly on the transcription initiation signals.

The process of transcription initiation has been reviewed extensively elsewhere (59,73,83) and it is not our intention here to duplicate such efforts. Thus many background references can be found in the reviews cited above. Transcription initiation signals can vary considerably. While some bacteriophage such as T7 encode a complete RNA polymerase, most variation in promoter sequence reflects either the association of different sigma factors (σ) with a core polymerase or the requirement for additional positive regulators. The fact that there seems to be evolutionary conservation of RNA polymerases suggests that all bacteria are likely to possess related transcription apparatus and indeed the transcriptional signals required for the major form of RNA polymerase in *E. coli* and *Bacillus subtilis*, representatives of

Promiscuous Plasmids of Gram-Negative Bacteria   © 1989 Academic Press Limited
ISBN 0-12-688480-3   All rights of reproduction in any form reserved

Gram-negative and Gram-positive bacteria respectively, are almost identical. On the other hand, since there has been both an evolution of homologous components of the transcription apparatus and also a diversification of sigma factors (see below), it is quite possible that the activity of a given promoter may vary from species to species. It is also possible that in other Gram-negative species the dominant or most frequently used forms of RNA polymerase may differ from that in *E. coli* and in some cases it is possible that no equivalent of the major sigma factor of *E. coli* exists. This would then create a barrier to plasmid promiscuity which may have been overcome in certain broad-host-range plasmids. It could also create a barrier to the expression, in diverse species, of genes which have been manipulated by standard techniques which are optimized for *E. coli*. It would therefore be helpful to know what expression signals are used in particular species since this not only would allow rational manipulation of a gene of interest but would also facilitate the identification of putative gene expression signals in the analysis of newly isolated genes. At present the information which would be needed to make valid generalizations about promoter usage in other species does not exist. Nevertheless we feel it is worthwhile to compile the information which is currently available in the hope that this may stimulate more comprehensive studies from which a useful picture may emerge.

II. TRANSCRIPTION IN *E. coli* USING THE MAJOR SIGMA FACTOR

Studies on transcription in *E. coli* have been largely confined to the major form of RNA polymerase which consists of the core polymerase in association with the 70 kDa sigma factor (σ^{70}) forming what is referred to as the holoenzyme ($E\sigma^{70}$). Initiation of transcription occurs in a series of steps that can be characterized kinetically and which have been well reviewed elsewhere (59,83). Initial binding of the RNA polymerase to the DNA in the promoter region is characterized by a binding constant, K_b, and results in the reversible formation of a so-called 'closed complex'. An irreversible isomerization step (with rate constant K_f) then leads to the separation of about 10–12 bp of the DNA strand and the formation of the so-called 'open complex'. In the presence of NTP substrates polymerization then ocurs. After the addition of 8–9 bases to the nascent transcript the RNA polymerase is thought to have cleared the promoter. The sigma factor then dissociates and appears to be replaced by the NusA protein (76).

The RNA polymerase bound at a promoter can occupy as much as 70 bp of DNA stretching from +20 (where +1 is the base at which transcription is initiated) to −50. It appears that sequences throughout this region can be important for promoter function. The universal features of $E\sigma^{70}$ promoters have been reaffirmed by a recent compilation of such promoters (36). These consist of the '−35' (TTGACA) and '−10' (TATAAT) consensus hexamers with an optimal spacing of 17 ± 1 bp between these two regions. While there appears to be a preference around the transcription start for CAT with the A at the $+1$ position (purines are favoured but not exclusively used for the initiating base) there is considerable flexibility in this region, the transcription start being governed more by the distance from the -10 region than by the exact sequence (5). However sequence variations in this region can affect a promoter's response to factors such as ionic strength and temperature (5). While in general, mutations which reduce the homology of a given promoter to the various components of the consensus will reduce its strength, possession of all these consensus features does not guarantee maximum promoter strength. Sequences both within the $+1$ to $+20$ transcribed region and upstream from the -35 region can influence promoter activity more than ten-fold (21,47).

Promoter strength depends on the frequency with which RNA polymerase molecules clear the promoter and this depends on a combination of the rates associated with the different steps of initiation outlined above. While in extremely strong promoters all steps may proceed efficiently, most promoters possess a combination of efficient and not so efficient steps and often recombination of the better parts from two or more promoters may result in a more efficient promoter than either of the parent promoters. This is well illustrated by the formation of the *tac* promoters from the *trp* (in which the K_b step is efficient) and *lac* (in which the K_f step is efficient) promoters (18). A given promoter strength can thus be achieved in a variety of ways depending on what is the rate limiting step (21).

In addition to the standard $E\sigma^{70}$ promoters which approximate to the consensus described above, it has also become apparent (49) that promoters which do not possess a standard -35 region can function efficiently through the possession of important sequences upstream of the -10 region (specifically TG separated by 1 bp from the -10 box). This suggests that there may also be other combinations of bases at nonconsensus positions which can overcome the need for homology to the consensus at all regions of the promoter seqence. The nature of promoter signals, even for a given holoenzyme, can therefore vary considerably, making the task of identifying putative gene expression signals or assigning them to a particular RNA polymerase on the basis of DNA sequence alone very difficult unless the promoter conforms very closely to one of the standard consensuses.

III. OTHER SIGMA FACTORS

Over the last few years it has become clear that there are a number of additional sigma factors present in *E. coli*. Both σ^{70} and σ^{32} belong to a group of sigma factors which have a common evolutionary origin and recent studies using antibodies raised to the most highly conserved oligopeptide (a tetradecamer) have not only confirmed the presence of both of these proteins in *E. coli* but have also indicted the existence of at least three other proteins, of 75 kDa, 27 kDa and 23 kDa, which crossreact with the antibody (27). While one of these may correspond to the σ^{28} referred to below, these results suggest that there may be at least two members of this family of sigma factors which have yet to be discovered. Only those which have been studied in some detail are described in more detail below.

A. Sigma 32—the HtɒR protein

The product of the *htpR* (*rpoH*) gene has been shown to be an alternative sigma factor (σ^{32}) in *E. coli* (32). It shows considerable homology to the C-terminal section of σ^{70}, as do all sigma factors which have so far been studied in detail. Purified σ^{32} can be reconstituted with RNA polymerase core enzyme to form a new holoenzyme which is specific for the promoters similar to those of genes (for example *rpoD*, *dnaK* and *groE*) that are induced by the heat shock response and will not initiate transcription from standard $E\sigma^{70}$ promoters. The consensus (16) for the $E\sigma^{32}$ promoters is:

$$\text{T-t C-Cc CTTGAA } N_{13-15} \text{ CCCCATt Ta}$$
$$\qquad\quad -35 \qquad\qquad\qquad\quad -10$$

This sort of promoter is not recognized by $E\sigma^{70}$. Since the heat-shock response appears to be virtually universal it seems likely that such promoters will be utilized in other Gram-negatives as well.

B. Sigma 54—the NtrA protein

This sigma factor was discovered as a result of studies on the nitrogen fixation genes which exist in several species of Gram-negative bacteria, including *Klebsiella pneumoniae*, *Azotobacter* spp. and *Rhizobium* spp. (58). These *nif* genes together with other genes involved in nitrogen assimilation (for example, *glnA*) are regulated by the nitrogen regulation system (*ntr*). Nucleotide sequence analysis of the promoter region of the *nif* and *gln* genes (22,39,72) revealed a consensus sequence:

$$\text{CTGGPy APy Pu } N_4 \text{TTGCA}$$
$$\qquad -26 \qquad\qquad\qquad -10$$

This is clearly different from the accepted $E\sigma^{70}$ consensus and furthermore, analysis of the regulation system has identified that one of its components, the product of the *ntrA* gene is an alternative sigma factor, RpoN (σ^{54}), which is specific for the *nif* consensus (39,40). This sigma factor is not related to the other sigma factors so far discovered but there is substantial homology between the NtrA proteins of *A. vinelandii*, *K. pneumoniae* and *R. meliloti* (61,62).

Regulation via these controlling factors and specifically Rpo N appears to be more widespread than the nitrogen-fixing bacteria, being found not only in the enteric bacteria such as *E. coli* and *Salmonella* spp. but apparently also in a number of *Pseudomonas* species. Thus the promoter region of the *xylCAB* operon from the TOL plasmid of *Pseudomonas putida* has a perfect *nif* consensus which can be transcriptionally activated by the *ntr* system (23). Another gene from the TOL pathway, *xylS*, also has a *nif*-like promoter region (80). Both *xylCAB* and *xylS* are positively activated by the product of the *xylR* gene, which mechanistically at least appears very similar to the products of the positively acting genes *nifA* and *ntrC* which act in concert with Ntr A in other systems (*nif* and *gln* respectively). A putative *ntrA* homologue has been identified in *P. putida* (FCHF, unpublished). Thus we have two clearly different pathways one involved in nitrogen assimilation and the other in hydrocarbon catabolism which have close similarities in their expression and regulation. At present it is unclear how and why this situation has evolved but the identification of *nif*-like promoters upstream of the *Pseudomonas* carboxypeptidase G2 (63) and *P. putida* chloroproprionic acid dehalogenase (Tobin and PT Barth, personal communication) genes provides a clear indication that a significant number of genes of unrelated function might fall under the transcriptional control of this class of promoter.

C. Sigma 28

While the promoter regions of the coordinately regulated operons of enteric bacteria that contain genes responsible for flagellar, chemotaxis and motility functions do not contain plausible $E\sigma^{70}$ promoter sequences they all contain homology to *B. subtilis* σ^{28} promoters (consensus: TAAA N_{16} CCGATAT). In *B. subtilis* this sigma factor is also responsible for expression of genes required for flagellar synthesis. This suggests that enteric bacteria may contain a minor sigma factor equivalent to the σ^{28} of *B. subtilis* (38) and again emphasizes the possibly surprisingly close evolutionary relationship between these species.

IV. HETEROLOGOUS GENE EXPRESSION IN GRAM-NEGATIVES

Early experiments to investigate the expression of genes from *E. coli* in heterologous Gram-negative hosts showed that the *trp* genes are quite well expressed in *Pseudomonas* and *Rhizobium* but that these species do not regulate their *trp* genes in the same way as *E. coli* resulting in constitutive expression of the *E. coli* genes (64,65). By contrast the *trp* genes of *Rhizobium leguminosarum* are not expressed at all in *E. coli* and only poorly in *Pseudomonas aeruginosa* as indicated by complementation of auxotrophic mutations (46). Similarly the *trpAB* genes from *P. aeruginosa* are expressed only very poorly in *E. coli* although mutations which increase the level of expression can readily be isolated by selecting for improved complementation of auxotrophic mutations (37). Since these early experiments there have been numerous reports, a representative selection of which are summarized in Table I, of heterologous genes not being expressed well in *E. coli* although this is not always found to be the case. These observations may be explained if all the Gram-negative species studied have a core RNA polymerase and sigma factor with similar recognition properties to that of the major *E. coli* enzyme ($E\sigma^{70}$) but that many of the genes from these species that have been transferred to *E. coli* normally use a different form of RNA polymerase or require positive activation and therefore lack $E\sigma^{70}$ promoter signals. Such a proposal is consistent with the results of more comprehensive screening of gene libraries for complementation of a range of amino acid biosynthetic genes in heterologous species. For example, a gene bank from *Methylophilus viscogenes* complemented 32 out of 41 *P. aeruginosa* markers (78%) tested but only 8 out of 27 *E. coli* markers (30%) while a *Methylophilus methylotrophus* library complemented 24 out of 37 *P. aeruginosa* markers (65%) and 19 out of 28 *E. coli* markers (68%) (57).

Genes which can either fully or partially complement *recA* mutations in *E. coli* K12 have been cloned from a variety of strains or species including *E. coli* B/r, *Erwinia carotovora*, *Shigella flexneri*, *Proteus vulgaris* (48), *Proteus mirabilis* (25), *Vibrio cholerae* (29,35), *P. aeruginosa* (68,77), *Rhizobium meliloti* (9) and even *Bacteroides fragilis* (30). For the latter species this result is very surprising both because the cloned gene shows no detectable homology to the *E. coli recA* gene on the basis of Southern blotting (30) and because in the past significant barriers have appeared to exist which prevent the expression of other functions such as antibiotic resistance, plasmid replication, etc in the heterologous species (34). In the case of the *P. aeruginosa recA* gene it is interesting that while the coding region and the SOS box transcription regulatory region are highly conserved relative to the *E. coli* gene there appears to be no $E\sigma^{70}$ promoter sequence in the expected position raising the question as to the basis for its expression in *E. coli* (77).

In conclusion therefore, while from a practical point of view it is interesting to know what are the chances of obtaining heterologous gene expression, predicting whether a gene is likely to be expressed may be rather difficult.

V. NON-*E. coli* TRANSCRIPTION ENZYMES

Although radioimmunoassay analysis indicates that significant variation exists in the antigenic structure of RNA polymerases from diverse Gram-negative bacterial species, the β and β' subunits of RNA polymerase from species such as *V. cholerae*, *P. putida*, *Alcaligenes faecalis* and *Agrobacterium tumefaciens* showing as little as between 10% and 30% antigenic similarity to *E. coli* (67),

TABLE I. *Expression of Gram-negative genes in heterologous hosts*

Gene	Natural host	Heterologous host	Level of expression	Ref
rbc	*A. eutrophus*	*P. aeruginosa*	+	4
argA	*E. coli*	*P. aeruginosa*	5%	45
argF	*E. coli*	*P. aeruginosa*	4%	45
ntr	*E. coli*	*R. capsulatus*	+	2
proC	*E. coli*	*P. aeruginosa*	40%	45
proB	*E. coli*	*P. aeruginosa*	50%	45
trpA	*E. coli*	*P. aeruginosa*	23%	45
		R. leguminosarum	25%	64
trpB	*E. coli*	*P. aeruginosa*	10%	65
		R. leguminosarum	100%	64
argA	*P. aeruginosa*	*E. coli*	0.5%	45
argF	*P. aeruginosa*	*E. coli*	5%	45
lasA	*P. aeruginosa*	*E. coli*	+	78
proC	*P. aeruginosa*	*E. coli*	0.3%	45
recA	*P. aeruginosa*	*E. coli*	+	50
trpAB	*P. aeruginosa*	*E. coli*	−	37
toxA	*P. aeruginosa*	*E. coli*	−	55
xyl	*P. putida*	*C. crescentus*	+	12
recA	*P. mirabilis*	*E. coli*	+	25
trpA	*R. leguminosarum*	*E. coli*	−	46
		P. aeruginosa	+	46
trpB	*R. leguminosarum*	*E. coli*	−	46
		P. aeruginosa	+	46
trpC	*R. leguminosarum*	*E. coli*	−	46
		P. aeruginosa	+	46
trpD	*R. leguminosarum*	*E. coli*	−	46
		P. aeruginosa	+	46
trpE	*R. leguminosarum*	*E. coli*	−	46
		P. aeruginosa	+	46
trpF	*R. leguminosarum*	*E. coli*	−	46
		P. aeruginosa	+	46
recA	*R. meliloti*	*E. coli*	+	9
adhB	*Z. mobilis*	*E. coli*	+	15
bxn	*K. ozenae*	*E. coli*	+	82

there is considerable evidence to support the view that most Gram-negative species contain an RNA polymerase with similar recognition properties to the *E. coli* $E\sigma^{70}$.

First, enzyme purification and subsequent biochemical experiments indicate that despite minor differences in properties such as apparent subunit molecular weight, optimum ionic and pH conditions, elongation rate and efficiency of template utilization the major RNA polymerases from species other than *E. coli* seem able to recognize the same promoters as the *E. coli* enzyme. Thus *Myxococcus xanthus*, a species which is taxonomically distant from *E. coli* (86), has an RNA polymerase which is very similar to that of *E. coli* both in subunit structure and recognition properties (75). Similarly, the RNA polymerase holoenzyme from *Rhizobium leguminosarum* 300 apparently has four subunits: β', β, A (thought to be the sigma factor) and α of slightly different mobilities (M_r = 149 000, 146 000, 93 000 and 42 000 respectively) than the *E. coli* enzyme (M_r = 150 000, 145 000, 39 000 and 42 000 respectively) (56). This subunit structure is similar to that of the RNA polymerase of *A. tumefaciens* which is perhaps not surprising given the close relationship between these species (86). However, both *E. coli* and *R. leguminosarum* polymerases bind specifically to the same early promoters of bacteriophage T7 as estimated by electron microscopy (56). Likewise biochemical studies (methylation protection and crosslinking patterns to partially depurinized DNA) with RNA polymerase from *P. putida* suggests that on four *E. coli* promoters (P1 and P2 of pBR322, *lac*UV5 and lambda P_0) it makes contacts with the DNA which are indistinguishable from those made by the *E. coli* enzyme (31). RNA polymerase purified from *Caulobacter crescentus* is interesting (3,7). It shows different immunological crossreactivity as well as differences in the subunit sizes (specifically the α and σ subunits) as compared to the *E. coli* enzyme. In addition, while the *C. crescentus* core enzyme can associate with the *E. coli* sigma factor to give a holoenzyme with very similar specificity to the *E. coli* holoenzyme on T2 or T7 phage DNA templates, the *C. crescentus* holoenzyme does not stimulate transcription of *E. coli* DNA template (3). *C. crescentus* may thus represent an example of a species where the major form of sigma factor is significantly different from that of *E. coli*.

Second, the 5′ end of mRNA transcribed from some promoters with $E\sigma^{70}$-like promoter sequences is identical in a variety of species (Table II.A) which would not be expected if there were not common elements in promoter recognition and initiation of transcription between the species.

Third, the limited number of point mutations analysed so far in promoters that function in a variety of hosts seem to have similar effects across the range of hosts studied. Thus in the *trfA* promoter of broad-host-range plasmid RK2 a T to C transition at a conserved position in the −10 region has a similar effect on promoter strength in *E. coli*, *P. aeruginosa* and *P. putida* whille a promoter adjacent to but outside the −10 region has no apparent effect in any of these species (70). Similarly the reduced production of toxin by *Bordetella parapertussis* compared to *B. pertussis* and the nucleotide sequence differences between the toxin genes of these species, which reduce the homologoy of the *tox* promoter to the *E. coli* $E\sigma^{70}$ promoter consensus, suggest that the *tox* gene is normally transcribed by an RNA polymerase with recognition properties similar to that of the *E. coli* enzyme (66).

TABLE II. *Comparisons of promoters from Gram-negative bacteria with consensuses proposed for* E.coli *and other species*

A gap is introduced between the sequences upstream and those downstream of the + 1 position where transcription is initiated.

Species	Gene	Heterologous species	Ref

I. Promoters which are either known to direct transcription in both *E.coli* and other species or having been defined in other species show significant homology to the $E\sigma^{70}$ consensus of *E. coli*. Homology to the consensus is shown in bold.

B. pertussis   *tox*   *E. coli*   66
CGTCCGGACCGTGCTGACCCC**CCTGCCATGGTGTGATCCGTAAAAT**AGGCACC   ATCA

B. parapertussis   *tox*   *E. coli*   66
CGTCCGGACCGTGCCGACCCC**CCTGCCATGGTGTGATCCGCAAAAT**AGGCGCCACCA

M. xanthus   *vegA*   52
ACTTTTATCTCTTCCTTT**AGACA**AAACCATTTTTGGAAGG**TAA**GGGTATGGGC   AGCA

P. putida   *nahA*   79
TCAACTATGCTTT**ATTGACAAATAAAAAG**CACGCTCACCATCATCGCGAATAC   AAAT

P. putida   *nahG*   79
ATATCGAGTGGTGTGTAT**TTATCA**ATATTGTTTGCTCCGT**TATCGT**TATTAAC   AAGT

P. putida   *nahR*   79
AATAACGGAGCAAACAATA**TTGATA**AATACACCACTCGATA**TATAAT**AAATCA   TCAA

P. putida   pTN8−p1   *E. coli*   41
TTTGCTCATGATACGACTC**CACTTGAACAATGTTGTGGTACCATT**TAAAACT   ATAA

P. putida   pTN8−p2   *E. coli*   41
CACTTGAACAATGTTGTGG**TACCA**TTTAAAACTATAAAGC**TACTATA**AGGTC   AATA

P. putida   pTN8−p3   *E. coli*   41
GTTGTGGTACCATTTAAAACT**ATA**AAGCTACTATAAGGTCAA**TAGAGT**AAAGA   ATCC

RK2 (*P. aeruginosa*)   *trfA*   *E. coli, P. putida*   70
TTTAGCCGCTAAAGTT**CTTGACAGCGGAACCAATGTTTAGCTA**AACTAGAGTC   TCCT

Tn*1* (*E. coli*)   Pa   *P. aeruginosa*   10,11
TCCACGGTTTATAAAAATT**CTTGAAGACGAAAGGGCCTCGTGATA**CGCCTATT   TTTA

Tn*1* (*E. coli*)   Pb   *P. aeruginosa*   10,11
TTGAAGACGAAAGGGCCTCG**TGATACGCCTATTTTTATAGGTTAAT**GTCATGA   TAAT

Z. mobilis   *adhB*   *E. coli*   15
AATAAAGCGAACCCC**TTGATCTGATCTGATAAAACTGATAGACATA**TTGCTTT   TGCG

Z. mobilis   pLOI301   *E. coli*   13
TAAAGAAAGAATTCGC**TTGAAA**TGAAGAGATAAAGAACAAAA**GAA**CCATCATT   GCTC

Z. mobilis   pLOI302   *E. coli*   13
AAGGGCTTTTTTAACCAAA**GCCTTTCTTCTAAAAGACCTTTT**ACAATCGAGTA   AAGA

Z. mobilis   pLOI303 P−2   *E. coli*   13
TTATGTATTATTTTTAGTAAATG**AGTAATTGCATTTTATGATAAAAAT**ATAAT   ACAA

II. Non-*nif* or *gln* promoters showing similarity to *nif* (RpoN) promoters. Homology to the consensus is shown in bold.

P. putida   CPG2   63
ATCGGCGGGGGCGAAGGCACCGCAGT**GGCACTCGAATTGC**TATAAGAACCATG   GGCT

P. putida   *xylABC*   42
TCGGTATAAGCAAT**GGCATGGCGGTTGC**TAGCTATACGAG   A

P. putida   *xylS*   80
CTTAAAAAGAAGTCTTCGTTCTGCT**TGGCGT**TATTTTTGC**T**TGGAAAAGTG   GT

TABLE II. *continued*

Species	Gene	Heterologous species	Ref

III. A selection of promoters from naturally occurring non-*E. coli* genes which are not known to be expressed or are expressed at negligible levels in *E. coli*. Where the promoter has specifically been shown not to function in *E. coli* the name of this species is also shown. Homology to the σ^{70} consensus of *E. coli* is shown in bold even when in some cases it can not be significant for the mapped promoter.

Species	Gene	Heterologous species	Ref
C. crescentus	trpFBA		86

AACCCCGCGACAACGCGCCGC**TTGAGCGACTCGCCAAGTCGGCCAAAT**CTTCT GG

| *P. aeruginosa* | porF | | 24 |

AGTTGGGTAAATATTG**TCTCTCTATGCGGGAAGTTCTGATA**AAC**TT**GACCACC CA

| *P. aeruginosa* | plcA | | 71 |

AAAAACAGCGAAGACGAT**TAATCA**TCTCGAAACAAGTACGC**AGA**TTGATGGAA AT

| *P. aeruginosa* | toxA | | 33 |

CCATAAAAGCCCTCTTCCGCTCCCCGCCAGCCTCCCCGCATCCCGCACCCTAG AC

| *P. aeruginosa* | toxR | | 87 |

GATACCCCTCAACCCTGCGTGCGGGCTCCATGCCCGAGCGCC**TTGGCGAGATT** **TG**

| *P. aeruginosa* | algD | | 20 |

GCGAGCGGGACAAACGGCCG**GAACTTCCCTCGCAGAGAAAACAT**CCTATCACC GC

| *P. putida* | catBC | | 1 |

GGAGATTCATTGCATAT**TGGAC**GGCTATCAGGGTCTCGCGCA**ATCCTT**GAACA AGCA

| *P. putida* | xylDEGF | | 43 |

ATGGCTATCTCTAGAAAGGCCTACCCCT**TAGGCT**TTATGC A

| *P. putida* | xylR P−1 | | 44,80 |

GGGGATCTGCGTTGAGG**TGGA**TTTCAGTTAATCAATTGGT**TAAT**CTTTCAGGA CCAC

| *P. putida* | xylR −2 | | 44,80 |

ATCAATTGGTTAATC**TTTCA**GGACCACCTAAGCAAATGC**TAAAG**TGGCAGA TGGAT

| *R. meliloti* | d-ALA synthetase p1 | | 54 |

GCGTCGGAGCCTGTCCGGGG**TTGACCACTGA**TCGC**TTTGAAGGAA**GAAAGGCG ACAG

| *R. meliloti* | d-ALA synthetase p2 | | 54 |

GTCCTTTTTCCGCAATTGC**TTGACTTCGA**TCGATG**TTCGGGA**GAATGAAGTTT TGCC

| *Z. mobilis* | pdc | *E. coli* | 14 |

CATTTTTAAAAATGCCTA**TAGC**TAAATCCGGAACGACAC**TTTAGAG**GTTTCTG GGTC

| *Z. mobilis* | pLOI303 P−1 | *E. coli* | 13 |

TTTTTAGTAAATGAG**TTAA**TTGCATTTTATGATAAAAA**TATAAT**ACAACCTTT CTTC

In general the promoters that are active in a wide variety of species such as the *tac* conform quite closely to the $E\sigma^{70}$ consensus (6) although whether they initiate transcription at an identical position in all the species in which they function has not been determined. However, it is quite possible that this major common RNA polymerase differs sufficiently between species that a weak promoter (and therefore differing considerably from the $E\sigma^{70}$ consensus) in one species might have no activity in another species. Such a situation seems to apply to a weak promoter in the end of Tn7 which can give sufficient transcription of the RK2 *trfA* gene for replication in *P. aeruginosa* but not in *E. coli* (53). Similarly, two promoters identified in DNA from *P. putida* appear to show much greater activity in *P. putida* than *E. coli* (41).

VI. NON-*E. coli* PROMOTER SEQUENCES

Promoters from Gram-negative species other than *E. coli* which have been characterized fall into a number of groups the broader classes of which are illustrated in Table II. First, there are some promoters which resemble the standard $E\sigma^{70}$ promoter consensus of *E. coli*. Second, promoters which resemble the Ntr A-regulated promoters. Third but not shown in Table II, there are the *Rhizobium* symbiotic promoters which are turned on in response to interaction with the roots of the host plant (8). These have a consensus sequence which probably reflects the requirement for a common positive regulator. Similarly the *Agrobacterium vir* genes possess a fourth class of promoter (17). These genes are turned on by the *virG* gene product following its activation, which is in turn dependent on the interaction of the *virA* gene product with the plant wound substances acetosyringone or α-hydroxyacetosyringone. They are also turned on in response to plant stimuli (81). Finally and in fact the largest group of promoters shown in Table II, there are a growing number of promoters belonging, particularly to *Pseudomonas* species, which have not yet been linked to a particular form of RNA polymerase or general positive regulator. While homologies between different promoters from the same species or genera have been observed (28,60,41) attempts to generalize from these to create consensuses (60,41) have not yielded results which seem to be very generally applicable (19). Conclusions of general significance will probably require much more extensive genetic and biochemical analysis than has been carried out to date.

So far there is little information about the similarities or differences in the transcriptional termination signals between Gram-negative bacterial species. It is likely that rho-independent signals function efficiently in other species but since rho-dependent terminators are less well defined it is perhaps not surprising that they have not been studied in other species.

Similarly comparison of gene sequences suggests the almost all bacterial genes utilize Shine-Dalgarno sequences as part of their ribosome binding sites although in general the genes compiled in Table II appear to have rather longer sequences complementary to the 3' end of the 16S rRNA than is generally found for *E. coli* genes.

VII. CONCLUSIONS

From the previous discussion it is clear that, contrary to initial impressions, it seems likely that gene expression in most Gram-negative bacteria has basically similar properties. Most probably it is a combination of considerable diversity in sigma factors and positive control elements that results in the apparent barriers to gene expression in heterologous hosts. It is not therefore possible to predict whether an uncharacterized gene is likely to be expressed in a particular host nor whether it is likely to be regulated in its normal way. However, since the evidence suggests that most Gram-negative bacteria possess an RNA polymerase with similar recognition properties to the $E\sigma^{70}$ of *E. coli* if expression is necessary then one of a number of regulated expression vectors using efficient *E. coli* promoters can be used as discussed in Chapter 10.

VIII. REFERENCES

1. Aldrich TL, Chakrabarty AM (1988) Transcriptional regulation, nucleotide sequence, and localization of the promoter of the *catBC* operon in *Pseudomonas putida*. *J Bacteriol* **170**, 1297–1304

2. Allibert P, Wllison JC, Vignais PM (1987) Complementation of nitrogen-regulatory (*ntr*-like) mutations in *Rhodobacter capsulatus* by an *Escherichia coli* gene: Cloning and sequencing of the gene and characterization of the gene product. *J Bacteriol* **169**, 260–71

3. Amemiya K, Wu CW, Shapiro L (1977) *Caulobacter crescentus* RNA polymerase. *J Biol Chem* **252**, 4157–65

4. Andersen K, Wilke-Douglas M (1987) Genetic and physical mapping and expression in *Pseudomonas aeruginosa* of the chromosomally encoded ribulose bisphosphate carboxylase geners of *Alcaligenes eutrophus*. *J Bacteriol* **169**, 1977–2004

5. Aoyama T, Takanami M (1985) Essential structure of *E. coli* promoter II. Effect of sequences around the RNA start point on promoter function. *Nucl Acids Res* **13**, 4085–96

6. Bagdasarian MM, Amann E, Lurz R, Ruckert B, Bagdasarian M (1984) Activity of the *trp–lac* (*tac*) promoter of *Escherichia coli* in *Pseudomonas putida*. Construction of broad host range, controlled expression vectors. *Gene* **26**, 273–82

7. Bendis IK, Shapiro L (1973) Deoxyribonucleic acid-dependent ribonucleic acid polymerase of *Caulobacter crescentus*. *J Bacteriol* **115**, 848–57

8. Better M, Ditta G, Helinski DR (1985) Deletion analysis of *Rhizobium meliloti* symbiotic promoters. *EMBO J* **4**, 2419–24

9. Better M, Helinski DR (1983) Isolation and characterization of the *recA* gene of *Rhizobium meliloti*. *J Bacteriol* **155**, 311–6

10. Chen S-T, Clowes RC (1984) Two improved sequences for the β-lactamase expression arising from a single base pair substitution. *Nucl Acids Res* **12**, 3219–34

11. Chen S-T, Clowes RC (1987) Variations between the nucleotide sequences of Tn*1*, Tn*2* and Tn*3* and expression of β-lactamase in *Pseudomonas aeruginosa* and *Escherichia coli*. *J Bacteriol* **169**, 913–6

12. Chatterjee DK, Chatterjee P (1987) Expression of degradation genes of *Pseudomonas putida* in *Caulobacter crescentus*. *J Bacteriol* **169**, 2962–6

13. Conway T, Osman YA, Ingram LO (1987) Gene expression in *Zymomonas*

ticular host nor whether it is likely to be regulated in its normal way. However, since the evidence suggests that most Gram-negative bacteria possess an RNA polymerase with similar recognition properties to the $E\sigma^{70}$ of *E. coli* if expression is necessary then one of a number of regulated expression vectors using efficient *E. coli* promoters can be used as discussed in Chapter 10.

VIII. REFERENCES

1. Aldrich TL, Chakrabarty AM (1988) Transcriptional regulation, nucleotide sequence, and localization of the promoter of the *catBC* operon in *Pseudomonas putida*. *J Bacteriol* **170**, 1297–1304

2. Allibert P, Wllison JC, Vignais PM (1987) Complementation of nitrogen-regulatory (*ntr*-like) mutations in *Rhodobacter capsulatus* by an *Escherichia coli* gene: Cloning and sequencing of the gene and characterization of the gene product. *J Bacteriol* **169**, 260–71

3. Amemiya K, Wu CW, Shapiro L (1977) *Caulobacter crescentus* RNA polymerase. *J Biol Chem* **252**, 4157–65

4. Andersen K, Wilke-Douglas M (1987) Genetic and physical mapping and expression in *Pseudomonas aeruginosa* of the chromosomally encoded ribulose bisphosphate carboxylase geners of *Alcaligenes eutrophus*. *J Bacteriol* **169**, 1977–2004

5. Aoyama T, Takanami M (1985) Essential structure of *E. coli* promoter II. Effect of sequences around the RNA start point on promoter function. *Nucl Acids Res* **13**, 4085–96

6. Bagdasarian MM, Amann E, Lurz R, Ruckert B, Bagdasarian M (1984) Activity of the *trp–lac* (*tac*) promoter of *Escherichia coli* in *Pseudomonas putida*. Construction of broad host range, controlled expression vectors. *Gene* **26**, 273–82

7. Bendis IK, Shapiro L (1973) Deoxyribonucleic acid-dependent ribonucleic acid polymerase of *Caulobacter crescentus*. *J Bacteriol* **115**, 848–57

8. Better M, Ditta G, Helinski DR (1985) Deletion analysis of *Rhizobium meliloti* symbiotic promoters. *EMBO J* **4**, 2419–24

9. Better M, Helinski DR (1983) Isolation and characterization of the *recA* gene of *Rhizobium meliloti*. *J Bacteriol* **155**, 311–6

10. Chen S-T, Clowes RC (1984) Two improved sequences for the β-lactamase expression arising from a single base pair substitution. *Nucl Acids Res* **12**, 3219–34

11. Chen S-T, Clowes RC (1987) Variations between the nucleotide sequences of Tn*1*, Tn*2* and Tn*3* and expression of β-lactamase in *Pseudomonas aeruginosa* and *Escherichia coli*. *J Bacteriol* **169**, 913–6

12. Chatterjee DK, Chatterjee P (1987) Expression of degradation genes of *Pseudomonas putida* in *Caulobacter crescentus*. *J Bacteriol* **169**, 2962–6

13. Conway T, Osman YA, Ingram LO (1987) Gene expression in *Zymomonas*

mobilis: promoter structure and identification of membrane anchor sequences forming functional Lac Z' fusion proteins. *J Bacteriol* **169**, 2327–35

14. Conway T, Osman YA, Konnan JI, Hoffmann EM, Ingram LO (1987) Promoter and nucleotide sequences of the *Zymomonas mobilis* pyruvate decarboxylase. *J Bacteriol* **169**, 949–54

15. Conway T, Sewell GW, Osman YA, Ingram LO (1987) Cloning and sequencing of the alcohol dehydrogenase II gene from *Zymomonas mobilis*. *J Bacteriol* **169**, 2591–7

16. Cowing DW, Bardwell JC, Craig EA, Woolford C, Hendrix RW, Gross CA (1985) Consensus sequence for *Escherichia coli* heat-shock promoters. *Proc Natl Acad Sci USA* **82**, 2679–83

17. Das A, Stachel S, Allenza P, Montoya, Nester E (1986) Promoters of *Agrobacterium tumefaciens* Ti-plasmid virulence genes. *Nucl Acids Res* **14**, 1355–64

18. De Boer HA, Comstock LJ, Vasser M (1983) The *tac* promoter: A functional hybrid derived from the *trp* and *lac* promoters. *Proc Natl Acad Sci USA* **80**, 21–5

19. Deretic V, Gill JF, Chakrabarty AM (1987) Alginate biosynthesis: a model system for gene regulation and function in *Pseudomonas*. *Bio Technology* **5**, 469–77

20. Deretic V, Gill JF, Chakrabarty AM (1987) *Pseudomonas aeruginosa* infection in cystic fibrosis: nucleotide sequence and transcriptional regulation of the *algD* gene. *Nucl Acids Res* **15**, 4567–81

21. Deuschle U, Kammerer W, Gentz, Bujard H (1986) Promoters of *Escherichia coli*: a hierarchy of *in vivo* strength indicates alternative structures. *EMBO J* **5**, 2987–94

22. Dixon R (1984) Tandem promoters determine regulation of the *Klebsiella pneumoniae* glutamine synthetase (*glnA*) gene. *Nucl Acids Res* **12**, 7811–30

23. Dixon R (1985) The *xylABC* promoter from the *Pseudomonas putida* TOL plasmid is activated by nitrogen regulatory genes in *Escherichia coli*. *Mol Gen Genet* **203**, 129–136

24. Duchene M, Schweizer A, Lottspeich F, Krauss G, Marget M, Vogel K, von Specht B-U, Domdey H (1988) Sequence and transcriptional start site of the *Pseudomonas aeruginosa* outer membrane porin protein F gene. *J Bacteriol* **170**, 155–62

25. Eitner G, Adler B, Lanzov VA, Hofemeister J (1982) Interspecies *recA* protein substitution in *Escherichia coli* and *Proteus mirabilis*. *Mol Gen Genet* **185**, 481–6

27. Fujita N, Ishihama A, Nagasawa Y, Ueda S (1987) RNA polymerase sigma-related proteins in *Escherichia coli*: detection by antibodies against a synthetic peptide. *Mol Gen Genet* **210**, 5–9

28. Gabellin N, Sebald W (1986) Nucleotide sequence and transcription of the *fbc* operon from *Rhodopseudomonas sphaeroides*. *Eur J Biochem* **154**,

569-79

29. Goldberg I, Melalanos JJ (1986) Cloning of the *Vibrio cholerae recA* gene and construction of a *Vibrio cholerae recA* mutant. *J Bacteriol* **165**, 715-22

30. Goodman HJK, Parker JR, Southern JA, Woods DR (1987) Cloning and expression in *Escherichia coli* of a *recA*-like gene from *Bacteroides fragilis*. *Gene* **58**, 265-71

31. Gragerov AI, Chenchik AA, Aivasashvilli VA, Beabealashvilli R, Nikiforov (1984) *Escherichia coli* and *Pseudomonas putida* RNA polymerases display identical contacts with promoters. *Mol Gen Genet* **195**, 511-5

32. Grossman AD, Erickson JW, Gross CA (1984) The *htpR* gene product of *E. coli* is a sigma factor for heat-shock promoters. *Cell* **38**, 383-90

33. Grant CCR, Vasil ML (1986) Analysis of transcription of the exotoxin A gene of *Pseudomonas aeruginosa*. *J Bacteriol* **168**, 1451-6

34. Guiney DG, Hasegawa P. Davis CE (1984) Plasmid transfer from *Escherichia coli* to *Bacteroides fragilis*: differential expression of antibiotic resistance phenotypes. *Proc Natl Acad Sci USA* **81**, 7203-6

35. Hamood AN, Pettis GS, Parker CD, McIntosh MA (1986) Isolation and characterization of the *Vibrio cholerae recA* gene. *J Bacteriol* **167**, 375-8

36. Harley CB, Reynolds RP (1987) Analysis of *E. coli* promoter sequences. *Nucl Acids Res* **15**, 2343-61

37. Hedges RW, Jacob AE, Crawford IP (1977) Wide ranging plasmid bearing the *Pseudomonas aeruginosa* tryptophan synthase genes. *Nature* **267**, 283-4

38. Helman JD, Chamberlin MJ (1987) DNA sequence analysis suggest that expression of flagellar and chemotaxis genes in *Escherichia coli* and *Salmonella typhimurium* is controlled by an alternative σ factor. *Proc Natl Acad Sci USA* **84**, 6422-4

39. Hirschmann J, Wong P-K, Sei K, Keener J, Kustu S (1985) Products of the nitrogen regulatory genes *ntrA* and *ntrC* of enteric bacteria activate *glnF* transcription *in vitro*: evidence for a new sigma factor. *Proc Natl Acad Sci USA* **82**, 7525-9

40. Hunt TP, Magasanik B (1985) Transcription of *glnF* by purified *Escherichia coli* components: core RNA polymerase and the products of *glnF*, *glnG* and *glnL*. *Proc Natl Acad Sci USA* **82**, 8453-7

41. Inouye S, Asai Y, Nakazawa A, Nakazawa T (1986) Nucleotide sequence of a segment promoting transcription in *Pseudomonas putida*. *J Bacteriol* **166**, 739-45

42. Inouye S, Nakazawa A, Nakazawa T (1984) Nucleotide sequence surrounding the transcription initiation site of *xylABC* operon on TOL plasmid of *Pseudomonas putida*. *Proc Natl Acad Sci USA* **81**, 1688-91

43. Inouye S, Nakazawa A, Nakazawa T (1984) Nucleotide sequence of the promoter region of the *xylDEGF* operon on TOL plasmid of *Pseudomonas putida*. *Gene* **29**, 323-30

44. Inouye S, Nakazawa A, Nakazawa T (1985) Determination of the transcription start site and identification of the protein product of the regulatory gene *xylR* for *xyl* operons on the TOL plasmid. *J Bacteriol* **163**, 863–9

45. Jeenes DJ, Soldati L, Baur H, Watson JM, Mercenier A, Reimmann C, Leisinger T, Haas D. 1986. Expression of biosynthesis genes from *Pseudomonas aeruginosa* and *Escherichia coli* in the heterologous host. *Mol Gen Genet* **203**, 421–9.

46. Johnston AWB, Bibb MJ, Beringer JE (1978) Tryptophan genes in *Rhizobium*—their organization and their transfer to other bacterial genera. *Mol Gen Genet* **165**, 323–30

47. Kammerer W, Deuschle U, Gentz R, Bujard H (1986) Functional dissection of *Escherichia coli* promoters: information in the transcribed region is involved in late steps of the overall process. *EMBO J* **5**, 2995–3000

48. Keener SL, Mc Namee KP, Mc Entee K (1984) Cloning and characterization of *recA* genes from *Proteus vulgaris, Erwinia carotovora, Shigella flexneri* and *Escherichia coli* B/r. *J Bacteriol* **160**, 153–60

49. Keilty S, Rosenberg M (1987) Constitutive function of a positively regulated promoter reveals new sequences essential for activity. *J Biol Chem* **262**, 6389–95

50. Kokjohn TA, Miller RV (1985) Molecular cloning and characterization of the *recA* gene of *Pseudomonas aeruginosa* PAO. *J Bacteriol* **163**, 568–72

51. Kokjohn TA, Miller RV (1987) Characterization of *Pseudomonas aeruginosa recA* analog and its protein product: *rec-102* is a mutant allele of the *P. aeruginosa* PAO *recA* gene. *J Bacteriol* **169**, 1499–1508.

52. Komano T, Franceschini T, Inouye S (1987) Identification of a vegetative promoter in *Myxococcus xanthus*. A protein that has homology to histones. *J Mol Biol* **196**, 517–24

53. Krishnapillai V, Wexler M, Nash J, Figurski DH (1987) Genetic basis of a Tn7 insertion mutation in the *trfA* region of the promiscuous IncP-1 plasmid R18 which affects its host range. *Plasmid* **17**, 164–6.

54. Leong SA, Williams PH, Ditta GS. 1985. Analysis of the 5′ regulatory region of the gene for D-aminolevulinic acid of *Rhizobium meliloti*. *Nucl Acids Res* **13**, 5965–76.

55. Lory S, Strom MS, Johnson K (1988) Expression and secretion of the cloned *Pseudomonas aeruginosa* exotoxin A by *Escherichia coli*. *J Bacteriol* **170**, 714–9

56. Lotz W, Fees H, Wohlleben W, Burkardt HJ (1981) Isolation and characterization of the DNA-dependent RNA polymerase of *Rhizobium leguminosarum* 300. *J Gen Microbiol* **125**, 301–9

57. Lyon BR, Kearney PP, Sinclair MI, Holloway BW (1988) Comparative complementation mapping of *Methylophilus* spp. using cosmid gene libraries and prime plasmids. *J Gen Microbiol* **134**, 123–32

58. Magasanik B (1982) Genetic control of nitrogen assimilatioon in bacteria.

Ann Rev Genet **16**, 135–68

59. Mc Clure WR (1985) Mechanism and control of transcription initiation in prokaryotes. *Ann Rev Biochem* **54**, 171–204

60. Mermod N, Lehrbach PR, Reineke W, Timmis KN (1984) Transcription of the TOL plasmid toluate catabolic pathway operon of *Pseudomonas putida* is determined by a pair of coordinately and positively regulated promoters. *EMBO J* **3**, 2461–6

61. Merrick MJ, Gibbons JR (1985) The nucleotide sequence of the nitrogen regulation gene *ntrA* of *Klebsiella pneumoniae* and comparison with conserved features in bacterial RNA polymerase sigma factors. *Nucl Acids Res* **13**, 7607–20

62. Merrick M, Gibbins J, Toukdarian A (1987) The nucleotide sequence of the sigma factor gene *ntrA(rpoN)* of *Azotobacter vinelandii*: Analysis of conserved sequences in NtrA proteins. *Mol Gen Genet* **210**, 323–30

63. Minton NP, Clarke LE (1985) Identification of the promoter of the *Pseudomonas* gene coding for carboxypeptidase G2. *J Mol App Genet* **3**, 26–35

64. Nagahari K, Koshikawa T, Sakaguchi K (1979) Expression of *Escherichia coli* tryptophan operon in *Rhizobium leguminosarum*. *Mol Gen Genet* **171**, 115–9

65. Nagahari K, Sano Y, Sakaguchi K (1977) Derepression of *E. coli trp* operon on interfamilial transfer. *Nature* **266**, 745–6

66. Nicosia A, Rappuoli R (1987) Promoter of the pertussis toxin operon and production of pertussis toxin. *J Bacteriol* **169**, 2843–6

67. Nikiforov VG, Lebedev AN, Kalyaeva ES (1981) Antigeneic varibility of bacterial RNA polymerases. *Mol Gen Genet* **183**, 518–521

68. Ohman DE, West MA, Flynn JL, Goldberg JB (1985) Method for gene replacement in *Pseudomonas aeruginosa* used in the construction of *recA* mutants: *recA*−independent instability of alginate production. *J Bacteriol* **162**, 1068–74

70. Pinkney M, Theophilus BDM, Warne SR, Tacon WCA, Thomas CM (1987) Analysis of transcription from the trfA promoter of broad-host-range plasmid RK2 in *Esherichia coli*, *Pseudomonas putida* and *Pseudomonas aeruginosa*. *Plasmid* **17**, 222–32

71. Pritchard AE, Vasil ML. 1986. Nucleotide sequence and expression of a phosphate-regulated gene encoding a secreted hemolysin of *Pseudomonas aeruginosa*. *J Bacteriol* **167**, 291–8

72. Reitzer LJ, Magasanik B (1985) Expression of *glnA* in *Escherichia coli* is regulated by tandem promoters. *Proc Natl Acad Sci USA* **82**, 1979–83

73. Reznikoff WS, Siegele DA, Cowing DW, Gross CA (1985) The regulation of transcription initiation in bacteria. *Ann Rev Genet* **19**, 355–87

74. Ross CM, Winkler ME (1988) Structure of the *Caulobacter crescentus trpFBA* operon. *J Bacteriol* **170**, 757–68

75. Rudd KE, Zusman DR. 1982. RNA polymerase of *Myxococcus xanthus*: Purification and selective transcription *in vitro* with bacteriophage templates. *J Bacteriol* **151**, 89–105

76. Saito M, Tsugawa A, Egawa K, Nakamura Y (1986) Revised sequence of the *nusA* gene of *Escherichia coli* and identification of *nusA11*(ts) and *nusA1* mutations which cause changes in a hydrophobic amino acid cluster. *Mol Gen Genet* **205**, 380–2

77. Sano Y, Kageyama M (1987) The sequence and function of the *recA* gene and its protein in *Pseudomonas aeruginosa* PAO. *Mol Gen Genet* **208**, 412–9

78. Schad PA, Bever RA, Nicas TI, Leduc F, Hanne, LF, Iglewski BA (1987) Cloning and characterization of elastase genes from *Pseudomonas aeruginosa*. *J Bacteriol* **169**, 2691–6

79. Schell MA. 1986. Homology between nucleotide sequences of promoter regions of *nah* and *sal* operons of NAH7 plasmid of *Pseudomonas putida*. *Proc Natl Acad Sci USA* **83**, 369–73

80. Spooner RA, Lindsay K, Franklin FCH (1986) Genetic, functional and sequence analysis of the *xylR* and *xylS* regulatory genes of the TOL plasmid pWWO. *J Gen Microbiol* **132**, 1347–58

81. Stachel SE, Messens E, Van Montagu M, Zambryski P (1985) Identification of the signal molecules produced by wounded plant cells that activate T-DNA transfer in *Agrobacterium tumefaciens*. *Nature* **318**, 624–9

82. Stalker DM, McBride KE (1987) Cloning and expression in *Escherichia coli* of a *Klebsiella ozaenae* plasmid-borne gene encoding a nitrilase specific for the herbicide bromoxynil. *J Bacteriol* **169**, 955–960

83. Tybulewicz VLJ, Falk G, Walker JE (1984) *Rhodopseudomonas blastica* *atp* operon. Nucleotide sequence and transcription. *J Mol Biol* **179**, 185–214

84. Von Hippel PH, Bear DG, Moorgan WD, McSwiggen JA (1984) Protein nucleic acid interaction in transcription. *Ann Rev Biochem* **53**, 389–446

85. Winkler ME, Schoenlein PV, Ross CM, Barrett JT, Ely B (1984) Genetic and physical analyses of *Caulobacter crescentus trp* genes. *J Bacteriol* **160**, 279–287

86. Woese CR (1987) Bacterial evolution. *Micrbiol Rev* **51**, 221–71

87. Wozniak DJ, Cram DC, Daniels CJ, Galloway DR (1987) Nucleotide sequence and characterization of *toxR*: a gene involved in exotoxin A regulation in *Pseudomonas aeruginosa*. *Nucl Acids Res* **15**, 2123–35

CHAPTER 10

BROAD-HOST-RANGE CLONING VECTORS

F. Christopher H. Franklin and Robert Spooner

I. INTRODUCTION

Studies over many years have revealed that Gram-negative bacteria possess an extensive range of biochemical activities which enable them to derive their nutritional requirements from a fascinating number of sources. For example, members of the genus *Pseudomonas* can metabolize a diverse range of organic compounds (33), *Agrobacterium tumefaciens* subverts the biochemical activity of cells in dicotyledenous plant species causing them to synthesize opines which the bacterium can then catabolize (6) and *Thiobacillus* species inhabit highly acid environments (pH 1–4) where they obtain energy for growth by the oxidation of both iron and sulphur (40). During the past decade it has been appreciated that many of these abilities might be beneficially utilized by mankind. Clearly, before this can be achieved it is necessary that both the biochemical and genetic basis of the various activities are elucidated. Recombinant DNA (r DNA) technology provides a powerful analytical tool with which to investigate and ultimately manipulate these bacteria to our advantage. Central to the development of this technology has been the construction of a range of sophisticated cloning vectors with specialized functions such as promoter probe vectors, expression vectors and cosmids. Whilst these have an important role to play in the genetic analysis of Gram-negative bacteria they do have one notable limitation in that they will not replicate in nonenteric bacteria. This is a considerable handicap to both the analysis and manipulation of genes from these bacteria. One solution is to carry out analyses in a heterologous genetic background; that is, *Escherichia coli*. Whilst this can and does provide valuable data, ultimate verification of details such as gene regulation require that the studies are carried out in the natural genetic background, particularly as many of these interesting activities occur in unusual

Promiscuous Plasmids of Gram-Negative Bacteria
ISBN 0-12-688480-3
© 1989 Academic Press Limited
All rights of reproduction in any form reserved

environmental niches for which the host is specifically adapted. The availability of cloning vectors which could replicate in these bacteria would obviously solve these problems. Consequently, during the past few years significant effort has gone into the development of such vectors. In this chapter we will discuss the development of these plasmid vectors, using specific examples to demonstrate how they can be used to analyse the genetic organization of a wide range of Gram-negative bacteria. We do not propose to discuss detailed methodology of their use as this has been recently reviewed elsewhere (21).

II. BROAD-HOST-RANGE PLASMIDS

The development of cloning vectors which are suitable for use in a wide range of Gram-negative bacteria has focussed on three naturally-occurring plasmids, namely RSF1010 (28), RK2 (56) and pSa (58). Whilst these plasmids differ considerably from each other in a number of respects, they do share one important feature, that is, their ability to replicate in almost any species of Gram-negative bacteria. It is this property which has made them particularly attractive for vector development since the alternative would be to develop vectors from indigenous plasmids from each species of interest. This would obviously result in both considerable and wasteful duplication of effort.

A survey of the basic properties of plasmids RSF1010, RK2 and pSa are presented in Table I.

TABLE I. *Properties of Broad Host Range Plasmids used for Vector Development.*

Plasmid	Molecular size (kb)	Incompatibility Group	Phenotype Characteristics
RSF1010	8.9	P-4/Q	Su^R, Sm^R, Mob^+
RK2	60	P-1	Ap^R, Km^R, Tc^R, Tra^+
pSa	39	W	Km^R, Cm^R, Sp^R, Su^R, Tra^+, fertility inhibition of F phenotype

[a] Abbreviations used phenotypic characteristics: Su sulphonamide; Sm, streptomycin; Ap, ampicillin; Km, Kanamycin; Tc, tetracycline; Cm chloramphenicol; Sp, spectinomycin resistances Tra^+, conjugation proficient, Mob^+, ability to be cotransferred by a conjugative plasmid.

A. RSF1010

RSF1010 is an 8.9 kb plasmid which specifies resistance to streptomycin (Sm) and sulphonamides (Su). It belongs to the incompatibility group (Inc) Q and is almost certainly identical to three other representatives of the IncQ group, plasmids NTP2 (52), R300B and R1162 (4). Until recently it was documented as having either a copy number of about 15–20 per genome equivalent (21), or 8–12 per genome equivalent (5), but more recent experiments (Frey and Bagdasarian, unpublished) have revised this down to between 6–10. Although the plasmid itself is nonconjugative, it can be mobilized with great efficiency to a wide range of Gram-negative bacteria if appropriate transfer functions are provided in *trans* by a helper plasmid (1). Table II illustrates the range of bacterial species into which RSF1010 or one of its derivatives has been introduced and maintained. At present the molecular basis of this ability of RSF1010 (and the other plasmids mentioned in this chapter) to replicate in this diverse range of bacteria has not been clearly established (see chapter 4A,B). However, as will be apparent from reading other chapters from this book, the control of replication of broad-host-range plasmids is a major topic of investigation and with many of the components involved in replication control and transfer (or mobilization) clearly identified, it is only a question of time before these studies reveal the basis of broad host range.

B. RK2

RK2, also known as RP1, R68 and R18 (56) is a 60 kb Inc P plasmid. It encodes resistance to the antibiotics penicillin, kanamycin and tetracycline. Indeed, it is this property which first stimulated interest in this group of plasmids, following the identification of RP1 as the agent which conferred carbenicillin resistance to a strain of *Pseudomonas aeruginosa* responsible for infections in a hospital burns unit (56). In contrast to RSF1010, RK2 is self-transmissible. This characteristic requires considerable DNA coding capacity, which is reflected in the relatively large size of the plasmid. Replication control of RK2, which is reviewed in chapter 1, involves a complex interaction between a number of proteins and nucleotide sequences. However, as will become apparent, neither the complexity of the replication or conjugal transfer system, has prevented the development of several useful cloning vectors from RK2.

C. pSa

pSa is a 39 kb Inc W low copy number (1–3) plasmid originally isolated in Japan from a clinical isolate of *Shigella* sp. (56, 35). It encodes resistance to the antibiotics kanamycin, streptomycin, sulphamides and chloramphenicol. Coupled with the chloramphenicol resistance is a fertility inhibition of F-factor phenotype. The plasmid is also conjugative.

III. DEVELOPMENT OF BROAD-HOST-RANGE CLONING VECTORS

A considerable range of vectors has now been constructed from these naturally occurring broad-host-range plasmids. A representative list summarizing these

TABLE II. *Host Range of IncQ Plasmids*

Acetobacter xylinum
Acinetobacter calcoaceticus
Agrobacterium tumefaciens
Alcaligenes eutrophus
Alcaligenes faecalis
Azotobacter vinelandii
Caulobacter crescentus
Erwinia caratovora
Escherichia coli
Gluconobacter sp.
Klebsiella aerogenes
Methylophilus methylotrophus
Neisseria sicca
Neisseria subflava
Neisseria mucosa
Proteus mirabilis
Proteus morganii
Providencia sp.
Pseudomonas aeruginosa
Pseudomonas diminuia
Pseudomonas fluorescens
Pseudomonas phaseolicola
Pseudomonas putida
Rhizobium leguminosarum
Rhodopseudomonas sphaeroides
Salmonella typhimurium
Salmonella senftenberg
Salmonella dublin
Serratia marcescens
Thiobacillus A2
Xanthomonas campestris
Yersinia enterocolitica

together with their properties is presented in Table III.

In the remaining part of this chapter we aim to illustrate the development of some of these and demonstrate by reference to specific examples how they can be used as powerful tools for the genetic analysis of Gram-negative bacteria.

Development of broad-host-range vectors has followed the same pattern of development that has occurred for plasmid vectors in *E. coli*. That is, the initial objective was to produce small simple vectors, possessing a range of selectable markers and restriction endonuclease cleavage sites into which fragments of DNA could be inserted. Subsequently, more sophisticated special purpose vectors have been produced.

Of the three plasmids mentioned only RSF1010 was of small enough size

TABLE III. *Examples of broad-host-range vectors based on RSF1010, RK2 and pSA*

Vector	Size (kb)	Markers for[a] Selection	Cloning[c] Sites	Special Features	Ref.
a) Vectors based on RSF1010					
pKT210	11.8	Cm,Sm	S,H,E,Hp	Mob+	1
pKT230	11.9	Km,Sm	S,E,H,X,Xm/Sm/ Bs/B		1
pKT231	12.8	Km,Sm	S,E,H,X,Xm,C	Mob+ Mob+	1
pKT262	11.7	Deletion derivative of pKT 230		Mob−	1
pKT263	12.8	Deletion derivative of pKT 231		Mob−	11
pKT240	12.5	Ap,Sm[b]	H,X,Xm,C,S,E	promoter probe vector	2
pCF32	15.1	Km,Sm,Ap,C−230[b]	H,X,E,S,Hp	promoter probe vector	54
pIJ3100	13.2	Sm,Cm[b]	B,Sl,H,Sm/Xm	promoter probe vector	44
pMMB22/24	12.7	Ap	E,H	tac promoter expression vector	2
pMMB66 E-H/H-E	8.9	Ap	E,Sm,B,Sa,P,H	tac promoter expression vector	24
pVDtac24	12.7	Ap	Sd,Xb,B	tac promoter expression vector	16
pVDtac39	13.2	Ap	E,SL,Xb	tac promoter expression vector	16
pMMB33/34	13.75	Km	B	cosmid	23
pGSS6	12.9	Ap,Tc,Sm	SL,B,H,C,S,Ss,B	Mob+	49
pGSS15	11.5	Ap,Tc	B,H,C,E,Bs	Mob+	49
pGSS33	13.4	Ap,Tc,Cm,Sm	B,C,E,H,P,Pv,SL, S,Ss,Bs	Mob+	49

TABLE III. continued

Vector	Size (kb)	Markers for[a] Selection	Cloning[c] Sites	Special Features	Ref.
pWS3	14.0	Sm,Ap,Km	S,E,H,P,SL,Xb,B, Sm/Xm,X	lac promoter expression vector mob[+]	60
pWS6	12.0	Sm,Ap,Km	S,E,K,Sm/Xm,B,H- ,SL,X,P,Bs	Mob[+]	60
pJRD215	10.2	Kn,Sm	Polylinker cosmid	12	
pTB70	17.6	Km,Sm	E,B,SL	Mob[+]	61
b) Vectors based on RK2					
pRK290	20	Tc	E,Bg	Mob[+]	17
pRK2501	11.1	Tc,Km	X,H,SL,E,Bg	Mob[−]	17
pRK310	20.4	Tc, LacZ	H,P,B	direct identification of recombinants on X-gal[d]	18
pRK311	22.0	Tc, LacZ	H,P,B	cosmid	18
pGD500	28.1	Tc, lacZ[b]	B,Bg,Sa	promoter probe vector	18
pGD926	28.1	Tc, lacZ	H,B	translational fusions	18
pRK404	10.6	Tc, lacZ	H,P,B,E	direct identification of recombinants on X-gal	18
pLAFR1	21.6	Tc	E	cosmid	18
pLAFR3	21.6	Tc	H,P,SL,B,Sm,E	cosmid, direct identification of recombinants on X-gal	22
pVK102		Tc,Km	SL,H,X	cosmid	6 37

TABLE III. *continued*

Vector	Size (kb)	Markers for[a] Selection	Cloning[c] Sites	Special Features	Ref.
c) Vectors based on pSA					
pSA4	9.4	Sp,Cm,Km	K,B,S	Mob[+]	55
pSA151	13.3	Sp,Km	B,Pv,E,H,Ss,K,Bg	Mob[+]	55
pSA747	15.0	Sp,Km	E,K,H,Ss,Bg	cosmid	55
pGV1106	8.4	Km,Sp	E,B,P,Bg,Ss	Mob[+]	37
pGV1113	5.3	Sm,Su	B,Bg,P,H	Mob[+]	39
pGV1122	7.2	Sm,Tc	P,H,B,SL	Mob[+]	39
Amplifiable by Cm					
pGV1124	7.2	Sm,Cm	H,E,Sa,B	Mob[+]	39
Amplifiable by Cm 10					

[a] Abbreviations are the same as indicated in the footnote to Table I. Additionally C-230, catechol 2,3-dioxygenase; *lacZ*[1], α-peptide of β-galactosidase.

[b] Expression dependent on transcription initiation from cloned sequences.

[c] Abbreviations for restriction endonuclease cleavage sites E, *Eco*RI; H, *Hin*dIII; B, *Bam*HI; S, *Sst*I; Ss, *Sst*II; X, *Xho*I; Xb, *Xba*I; Xm, *Xma*I; C, *Cla*I; Sl, *Sal*I; K, *Kpn*I; Sm, *Sma*I; Hp, *Hpa*I; P, *Pst*I; Pv, *Pvu*II; Bs, *Bst*EII; Sa, *Sau*3A; Bg, *Bgl*II.

[d] X-gal, 5-bromo-4-chloro-3-indoyl-β-D-galactoside.

for immediate development into vector plasmids. Although RSF1010 itself contains four potentially useful cleavage sites, only the *Bst*EII may be used for cloning without modification to the plasmid. The other three sites (*Eco*RI, *Sst*I and *Pst*I [two adjacent sites]) are located such that introduction of a fragment into either one results in the inactivation of Sm resistance which is the only practically useful selective marker on the plasmid. By incorporating segments of DNA which specify other selectable markers into the plasmid it has been possible to solve this selection problem; moreover it has permitted further useful restriction cleavage sites to be incorporated into the molecule. For example, one of the earliest vectors derived from RSF1010, pKT231 (1) was produced by replacing the small *Pst*I fragment in RSF1010 with a 4.6 kb *Pst*I fragment derived from the R6–5 mini-plasmid pKT105. The fragment confers Km resistance and contains sites for cloning *Hind*III, *Xho*I, *Xma*I and *Cla*I fragments. Although the removal of the small *Pst*I site results in the deletion of the Sm resistance gene promoter, resistance is maintained by virtue of transcription originating in the 4.6 kb pKT105 derived fragment. Thus, fragments cloned into the *Eco*RI or *Sst*I sites may be detected by insertional inactivation of Sm. An almost identical plasmid pKT230 was constructed by ligating RSF1010 to the narrow host range vector pACYC177 (11). Using similar approaches a large number of vectors have been derived from RSF1010 in several laboratories, some of which contain multiple restriction sites for cloning purposes (Table III). All these are considerably larger in size than corresponding vectors based on narrow host range *E. coli* plasmids. Early in the development of the RSF1010 vectors numerous attempts were made to reduce the size of the RSF1010 replicon. These were not successful (M. Bagdasarian, personal communication). Subsequent studies by several groups notably Bagdasarian and co-workers (see for example 13 and chapter 5) have revealed that this failure was almost certainly due to an underestimation of the complexity of the plasmid maintenance functions, particularly replication. Few of the plasmid sequences are in fact dispensable. A major difference between RSF1010 and the other two plasmids which have been extensively developed as broad-host-range vectors is that both RK2 and pSa are conjugative and hence, considerably larger in size. Therefore, the primary task in the development of vectors from these plasmids was the deletion of regions which do not directly participate in maintenance. In both cases the most convenient sequences to delete are those encoding the *tra* genes.

Vectors derived from pSa have been constructed by two groups, namely those of Kado (55) and Schell (39). Primarily these vectors were developed for use in the genetic analysis of *A. tumefaciens* but are of course suitable for use in a wide range of Gram-negative hosts. The starting point for each group was the isolation of a 'mini-Sa' plasmid. In one case this was achieved by recircularizing the 8.7 kb *Bgl*II fragment from pSa to produce pGV1106, which encodes Km and Sm/Sp resistance but has lost Cm and Su resistance and in addition the *tra* genes (39). In other instance, pSa151 was obtained by recircularization of the 13.3 kb *Sst*II fragment from the parental plasmid (55). Both these plasmids were then modified in a similar manner to that described for RSF1010, to produce the range of plasmids presented in Table II. Examples of the most recently developed pSA based vectors are presented in Chapter 6. A number of cloning vehicles based on RK2 have now been constructed. The first plasmids suitable for use as cloning vectors were pRK290 and pRK2501 (17). Both are basically *tra* negative deletion derivatives of the parental RK2. Although both are useful vectors, indeed pRK290 was used for the construction of a gene library of *Rhizobium meliloti*, they have the disadvantage that they are both rather large, in the order of 20 kb. However, more recently a smaller derivative of RK2, pRK404 has been described (18). It has a molecular size of 10.6 kb, which is comparable with the RSF1010 and pSa derived plasmids. Additionally, a 424 bp *Hae*II fragment from pUC9 which carries the polylinker and *lacZ'* gene has been incorporated during construction of the plasmid. Hence, it has the attractive feature of permitting direct identification of recombinants on X-gal plates when transformed into a suitable *E. coli* host. Further derivatives of pRK290 are also now available (18).

During the past few years these 'general purpose' broad-host-range vectors have been used in laboratories throughout the world to analyse various activities in a wide range of Gram-negative bacteria. For example, the authors have used the RSF1010 derived vectors extensively in the genetic analysis of pathways from *Pseudomonas* spp.which specify the catabolism of aromatic compounds such as toluene/xylene, chlorobenzoate and phenolics (20, 59, 51). In addition to genetic analysis the plasmids have been used for the 'engineering' of Gram-negative bacteria. One of the first examples of this particular use is also perhaps the best known; it is the introduction of the glutamate dehydrogenase gene from *E. coli* into *Methylophilus methylotrophus* (61). The 'recruitment' of the *E. coli* gene was accomplished using a R300B (RSF1010) derived vector pTBL70, permitting the recipient glutamate synthetase negative mutant of *M. methylotrophus* to assimilate ammonia more efficiently via glutamate dehydrogenase than was previously possible via an active glutamate synthetase. The overall result was improved growth yield for the conversion of substrate (methanol) into single-cell protein.

Since obligate methylotrophs are unlikely to infect humans, they may be considered relatively safe, making them potentially very useful hosts for broad-host-range plasmids designed to express eukaryotic coding sequences. Using plasmid pGSS15, a composite plasmid with Ap and Tc resistance markers of pBR322 and the broad-host-range characteristics of R300B (RSF1010) (5), eukaryotic cDNAs encoding chicken ovalbumin and mouse dihydrofolate reductase (32) and a synthetic human $\alpha 1$ interferon gene (15) have been expressed in *M. methylotrophus*.

A further example of the potential practical exploitation of broad-host-range vectors was the recent cloning of the gene specifying parathion hydrolase into pKT230 (48). Introduction of this recombinant, pCMS55 into *Pseudomonas diminuta* resulted in a more than four-fold increase in the organism's ability to catabolize the organophosphate pesticide, parathion, a clear illustration of the important role that broad-host-range plasmids have to play in present day biotechnology.

IV. SPECIFIC PURPOSE BROAD-HOST-RANGE VECTORS

The next stage in the construction of broad-host-range vector systems was the development of specialized function vectors, analogous to those already derived from narrow-host-range replicons for use in *E. coli*.

A. Cosmids

The development of cosmid cloning by Collins and Hohn (11) has proven to be of tremendous value for the construction of gene libraries of all types of organism. In the case of prokaryotes a cosmid library need consist of only a few hundred clones. The crucial stage of course is the detection of the clone of interest. In some instances this can be achieved directly in *E. coli*. However, one of the most useful means of detection is via complementation of an appropriate mutant. Generally, this will require screening the library in its natural genetic background, hence, the requirement for broad-host-range cosmids. Ideally, it would be desirable to be able to transfect directly the host of choice following *in vitro* packaging. To date, this is limited largely to *E. coli*.

Recent experiments, however, have shown that when the *E. coli lam B* gene is introduced into cells of *Salmonella typhimurium* (29), *Vibrio cholerae* (30), *Erwinia carotovora* and *Erwinia chrysanthemi* (46) on plasmids from which it can be expressed, the resulting strains are sensitive to bacteriophage infection. One such strain of *S. typhimurium* will even support the lytic growth of *nin* derivatives, although none of the other strains and species will; nor can they, to date, be lysogenized. However, the expression and proper integration of the LamB protein in the cells' outer membranes makes these strains suitable as hosts for transduction by broad-host-range cosmids packaged *in vitro*. Indeed a number of such strains of *V. cholerae* have been successfully transduced using the broad-host-range cosmid pLAFR1 (30, 22).

Ultimately this type of strategy may permit direct transfection of other Gram-negative hosts, with cosmids packaged in lambda phage particles. However, at present a two stage process must be pursued. Firstly, the cosmid library is constructed in the standard manner using a *E. coli* host; secondly, the recombinant cosmids are then transferred to the desired host for screening purposes.

Cosmid derivatives have been produced based on each of the three parental replicons (see Table II). The first example was pFG6 (25) which was constructed by linkage of R1162 (RSF1010) to the archetypal cosmid pHC9 (11). Most other examples are similar, in that they have been produced by simply cloning a fragment containing the *cos* sequence from pHC79 into a suitable broad host range plasmid. A notable exception are the closely related RSF1010 derived cosmids pMMB33 and pMMB34 (23). With these cosmids it is possible to use the cloning procedure developed by Ish-Horowitz (36), whereby aliquots of the vector are digested with two restriction endonucleases which generate blunt-ended fragments. These cosmid arms are then digested with *Bam*HI prior to ligation with chromosomal DNA that has been partially digested with *Sau*3A or *Mbo*I to give fragments of about 35 kb. This strategy has the advantage that it precludes the formation of polycosmids (which can occur with cosmids that cannot be repared in such a manner) with the result that the cloning efficiency is considerably enhanced. A similar procedure is also possible with pLAFR3 which is an unpublished derivative of pLAFR1 (22).

A considerable range of genes from Gram-negative species has now been cloned using broad-host-range cosmids. In particular pMMB33 and pLAFR1 (or its derivative pLAFR3) have proved very useful. For example, pMMB33 has been used for the construction of libraries from a number of *Pseudomonas* species, from which genes encoding activities such as myo-inositol utilization (23), chlorobenzoate degradation (58), phenol catabolism (51) and phenazine biosynthesis (Franklin, unpublished) have been isolated. Cosmid pLAFR1 and more recently pLAFR3 have been used, amongst other things, to clone genes which specify pathogenicity determinants from veral plant pathogenic bacteria (22, 12).

B. Promoter probe vectors

Plasmid vectors which can be used to clone, detect and characterize DNA segments which contain transcription initiation signals are of tremendous value for the analysis of gene expression in bacteria. Such vectors have the same general structure, allowing DNA fragments to be inserted upstream of a 'reporter' gene, which is a promoterless gene specifying a product which is both easy to detect and assay; galactokinase (43) and galactosidase (8) are typical examples.

A range of examples of broad-host-promoter probe vehicles derived from RSF1010 and pSA are presented in Table III. Recently we have described in the construction of pCF32 which permits the fusion of sequences specifying transcriptional initiation to the TOL plasmid *xylE* gene, which encodes catechol 2,3-dioxygenase (C2,3-O) (54). This plasmid is based on an RSF1010 replicon and is in fact derived from the promoter probe vector pKT240 (2). Clones which contain sequences that activate C2,3-O can be determined by using a straightforward assay (20). This plasmid has been used to determine the nucleotide sequences required for transcription initiation of the meta-ring fission operon (OP2) that encodes genes responsible for the conversion of benzoate through to TCA cycle intermediates, from the TOL plasmid pWWO. A 400 bp fragment from the region around the transcription start site of OP2 was fused to *xylE* in pCF32 and shown to contain all the sequence information necessary for the fully regulated expression of OP2. Analysis in pCF32 of *in vitro* generated deletions of this 400 bp fragment revealed that sequences essential for regulated expression of OP2 lie within a region of 60 bp upstream of the transcriptional start (54). Recently pCF32 has been improved by the addition of a polylinker cloning region upstream of *xyl*E to produce pRLA5 (R. Allen and G. Boulnois, unpublished).

The vector pIJ3100 (44) was constructed by inserting pKK232–8 (7) between the *Pst*I sites of RSF1010. The hybrid plasmid contains a promoterless chloramphenicol acetyl transferase gene (CAT), upstream of which is a polylinker sequence. Incorporation of DNA fragments with promoter activity is readily detectable via the chloramphenicol resistant phenotype of such recombinants. A translational stop codon in all three reading frames is located between the polylinker and CAT gene to prevent translational fusion, which might adversely affect chloramphenicol resistance. Having constructed pIJ3100 Osbourn *et al* (44) used it in some very interesting studies in which they isolated promoter sequences of genes which are transcriptionally activated in the plant pathogenic bacterium *Xanthomonas campestris* during infection of susceptible plant hosts. Briefly, recombinant pIJ3100 plasmids containing fragments of *X. campestris* chromosomal DNA were transferred en mass into *X. campestris*. Individual clones were then screened for their ability to infect turnip seedlings grown in the presence of chloramphenicol, thereby identifying recombinants which contained active promoters. These isolates were then tested for expression of CAT in the absence of the host plants. This permitted the identification of a number of clones which expressed CAT *in planta* only and hence contain promoter sequences from genes which are activated during the host-pathogen interaction. A RK2 derived promoter probe vector pGD500 is also available. This vector has a *Bam*HI site upstream of a promoterless β-galactosidase gene, into which fragments can be inserted to detect and measure promoter activity (18). Most recently Greener and Helinski (unpublished) have described further RSF1010 derived vectors which permit detection and analysis of nucleotide sequences with promoter activity using either chloramphenicol acetyltransferase (pAL200), β-lactamase (pAL300) or firefly luciferase (pAL400) as reporter genes. These examples clearly demonstrate the important role that promoter probe vehicles can play in the detection, isolation and analysis of transcription initiation signals in different species of Gram-negative bacteria.

C. A vector for construction of translational fusions

In addition to promoter probe vectors, plasmids which permit the translational fusion of DNA sequences specifying transcription/translation initiation signals and 5'-coding region of a gene with those encoding *lacZ*, provide a useful means of studying various aspects of gene regulation. Such a vector has been derived from pRK290. Plasmid pGD926 permits fusions to the eighth codon of *lacZ*. It has been used for the analysis of promoter sequences from the *Rhizobium meliloti* symbiotic genes (18).

D. Regulatable expression vectors

Numerous vectors have been constructed from *E. coli* plasmids which permit the fusion of cloned genes to strong, well characterized, regulatable promoters such as *lac*UV5, *trp* and PL (41) with the result that extremely high levels of gene expression can be obtained. Elevation of gene expression in this way can make a significant contribution to the functional analysis of a gene of interest and is of course, a major step in the purification strategy for many proteins of both academic and commercial importance. In this latter instance it is probably true to say that for most purposes the existing *E. coli* based systems are satisfactory. However, there are situations where over-expression of a gene is required in the normal genetic background, notably studies of gene regulation, or if the object is improve the natural ability of an organism to perform a particular activity. An example of this might be the construction of a *Pseudomonas* strain which could convert benzene to *cis*-1,2-dihydro-1,2-dihydroxybenzene (a precursor of a useful thermotolerant polymer) at enhanced levels.

Recently, a number of RSF1010 derived expression vectors have been described (see Table III and refs. 44, 24, 16) which contain a *tac* promoter. This promoter is a fusion between the −35 region of the *E. coli trp* promoter and the −10 region of the *lac*UV5 promoter (14); it is regulated by the *lac* repressor. The first expression vectors to be constructed were pMMB22 and pMMB24 (38). These plasmids were constructed by introduction of the *tac* promoter between the *Eco*RI and *Hind*III sites of pKT240, such that transcription proceeds through the *Eco*RI site in pMMB22 and through the *Hind*III site in pMMB24. To ensure 'tight' regulation of transcription from the *tac* promoter, a fragment encoding the *lac*I^Q gene was incorporated into each plasmid. Primarily, this was aimed at preventing 'leaky' expression as a result of differences in dosage of the repressor gene and tac promoter. However, it also ensures that regulated expression would be possible in hosts which do not possess a gene equivalent to *lacI*.

The question of whether regulated expression of the *tac* promoter would be possible in Gram-negative species unrelated to *E. coli* was of considerable importance. A number of studies (see chapter 11 and Refs. 34, 42, 19) have clearly established examples where genes present in Gram-negative species use transcription initiation signals which have little homology with the accepted *E. coli* consensus promoter (31). Moreover, it is known that some of these sequences function poorly as promoters in *E. coli* (20,53). Clearly, it was conceivable that the same would be true of *E. coli* promoters in other Gram-negative species. To determine if this was true of the *tac* promoter, the *xyl*E gene from pWWO was cloned into the *Eco*RI site of pMMB22 to produce pMMB25. The level of C2,3–O specified by pMMB25 was then determined in *E. coli* and *P. putida* cells prior to and following induction of the *tac* promoter (21,2). These results demonstrated that, not only does the promoter function at comparable efficiencies in each background but also, regulation of transcription initiation is maintained in *P. putida*. These results would seem to suggest that strong promoters originating from *E. coli* or its viruses might be of general use as the basis for expression systems in Gram-negative bacteria. Supporting this is the observation that the *lac*UV5 promoter provides regulated expression of human $\alpha 1$ interferon in *M. methylotrophus* (23) when provided with the *lacI* gene in trans.

Improved broad-host-range *tac* expression vectors are now available. Plasmid pMMB66EH (24) was constructed by incorporating two DNA fragments into RSF1010. One of these contains the *tac* promoter, a polylinker cloning region and two transcription terminators and was derived from the *tac* promoter expression vector pRK233–3 (14). The other provides the *lacI* gene, permitting regulated expression from the *tac* promoter in a wide range of hosts. Plasmid pMMB66HE has the polylinker cloning sites inverted. The full potential of these plasmids, in terms of use for obtaining high level gene expression in a wide variety of Gram-negative species, has yet to be realized, although they have been used to analyse the protein products encoded by the primase region of plasmid RP4 (24). We have also used pMMB66E to boost expression of the *xylS* regulatory gene from the TOL pathway (54) enabling us to confirm data based on nucleotide sequencing that it encodes a 36 kDa peptide (53). In addition we have used it to elevate the expression of the other pathway regulatory gene *xylR*, so that it comprises about 10% of cellular protein in both *E. coli* (see Figure 1) and more importantly *P. putida* (Spooner, Bagdasarian and Franklin, unpublished). Other similar vectors, pVD*tac* 39 and pVD*tac* 24 have been derived from pMMB22 and pMMB24 by Deretic *et al* (16) and have been used for the genetic analysis of components involved in alginate biosynthesis by *P. putida*.

An additional approach to obtaining high level gene expression is the use of vectors with elevated copy number. Ideally these should be temperature-sensitive replication control mutants. Such mutants, originally described by Uhlin *et al* (57) for the plasmid R1, when grown at a permissive temperature of 30°C maintain normal copy control. However, when shifted to a nonpermissive temperature of 42°C copy number control breaks down such that extremely high plasmid copy number occurs through runaway replication. Such runaway mutants have the advantage over simple high copy number derivatives in that the switch to high copy number can be initiated towards the end of the growth phase, thereby avoiding any of the toxicity effects which sometimes occur when cloned genes are expressed at very high levels throughout the growth phase. Barth and colleagues have constructed several vectors pTB220, pTB225 and pTB234 which are based on constitutive or temperature-sensitive high copy number derivatives of RSF1010 (see Table III). One potential limitation of the vectors based on temperature-sensitive replicons is that they will not be suitable for all hosts, as many nonenteric Gram-negative bacteria have growth optima close to 30°C and do not survive at 42°C.

In certain instances, for example, the cloning of genes from pathogenic bacteria, it is important that the vectors employed are nontransmissible and that they are not mobilized by cotransfer with a conjugative plasmid. Although most plasmids listed in Table III are not self-transmissible, they can be mobilized at high frequencies.

Mobilization functions of RSF1010 are controlled by the genetic loci *mob*, *oriT* (origin of transfer) and *nic* (nick induced by the relaxation complex). These loci have been mapped on the plasmid genome and a derivative pKT261, constructed, in which the *mob* sequences have been removed by *Bal*31 deletion (1). Vectors pKT262 and pKT263 which are analogous to pKT230 and pKT231 respectively, were then constructed by appropriate modification of pKT261 (1). The RP4 mediated mobilization frequency of each plasmid was then determined and found to be less than 4×10^{-6} per donor cell. This represents a reduction of 5–6 orders of magnitude compared to pKT230 or pKT231. When the plasmids are in conjunction with the hosts *P. putida* 2440 and *P. aeruginosa* PAO1162 (1), they constitute a host-vector system which fulfils the requirements for certified HV1 systems, as specified by the US. Recombinant DNA Advisory Committee and the West German Central Commission for Biological Safety.

E. Tracer plasmids

Plasmids have been developed which permit the fate of a bacterium in the environment to be monitored. A broad-host-range tracer plasmid has been developed from pSA and is described in detail in Chapter 6.

F. Two-component transposition systems

Whilst the most convenient means of introducing genetic material into bacteria depends on the use of recombinant plasmids and phages, there are problems associated with this approach. For example, in the absence of selection, an introduced plasmid may not be maintained, and expression of a gene product from a plasmid in multiple copies may result in marked inhibition of host cell growth. Such considerations led Grinter (27) to develop a system allowing introduction of genes directly and simply into the host cell chromosome.

This system utilizes an Inc P 'carrier' plasmid, pNJ5073, with the antibiotic resistance markers (Tp, Sm) of transposon Tn7 bounded by the Tn7 termini, and an Inc Q 'helper' plasmid, pNJ9279, bearing Tn7 transposition functions. Complementation in doubly-transformed cells leads to successful transposition of the markers flanked by the Tn7 termini. After curing of pNJ9279 by displacement with another plasmid (plasmid pNJ5073 is unstable) the resistance genes were demonstrated to be stably inserted in the chromosome.

This system was recently streamlined by Barry (3), by removal of regions of homology between the Tn7 segments of the carrier and helper plasmids. The resulting helper, pMON7018 encodes transposition functions: the carrier plasmid pMON7022 contains a BamHI cloning site between the Tn7 termini. These plasmids were then used to tranpose a DNA segment containing the E. coli lacZY genes, cloned in the BamHI site of pMON7022, to the chromosome of Pseudomonas fluorescens, thereby conferring upon the pseudomonad a nutritional (and colour) selectable marker previously absent.

Since nonselected markers linked to the selected markers (in this latter case, lacZY) will also be passively inserted, the potential of this approach is enormous. One example relevant to this chapter might be the transposition of a suitably expressed E. coli lamB gene into the chromosomes of a range of Gram-negative bacteria, thus providing strains suitable for cosmid transduction (see earlier) and avoiding the need to maintain two plasmids (one expressing lamB, the other the cosmid) in one cell.

G. Transfer of broad-host-range vectors to recipient bacteria by transformation and mobilization

From a practical viewpoint the fact that most of the vectors described in this chapter can be mobilized is of considerable importance, since for most species of nonenteric Gram-negative bacteria, transformation procedures have not been established. There are exceptions: for example, P. aeruginosa PAO1162 and P. putida 2440 are restriction deficient strains which have been successfully transformed with broad-host-range vectors using a method based on a high efficiency procedure for E. coli, described by Kushner (1, 38). However, although the transformation conditions were modified and optimized for use in Pseudomonas spp. the overall frequencies obtained were at best in the region of 6×10 transformants per microgram of plasmid DNA. This represents an efficiency some two to three orders of magnitude lower than the corresponding level for E. coli.

Fortunately, mobilization does not suffer the limitations which currently exist for transformation, in that it can be used as a method to introduce broad-host-range vectors into virtually all Gram-negative bacterium; for example see Table II, for the range of bacteria into which RSF1010 or a derivative has been mobilized. In many instances the transfer is highly efficient. This latter point is particularly important, as it permits, when occasion arises, the transfer of large numbers of clones in a single conjugation experiment. Thus, in general, the strategy that is usually adopted when cloning using a broad-host-range vector of any type is to ligate vector and target DNAs then transform *E. coli* to permit maximum recovery of recombinants and follow this by mobilization into the host of choice. The actual mobilization procedure can be carried out using a so called 'triparental mating' in which the donor and recipient cells are mixed in the presence of a third strain containing a 'helper' plasmid to provide the necessary *tra* functions. Alternatively, it is possible to use an *E. coli* host which contains the 'helper' plasmid as the recipient for the transformation stage. In many cases this latter method will result in a slightly improved mobilization efficiency but has the minor drawback that if isolation of the recombinant plasmid is required for physical analysis, which is usually most conveniently accomplished from an *E. coli* host, then it is necessary to retransform the recombinant DNA into a plasmid-free *E. coli* strain. A range of useful helper plasmids are available. These include naturally-occurring plasmids such as the IncI plasmids (1) or recombinant plasmids such as the RK2 derivative pRK2013 (17) and the pSA derived pSA322 (55) which contain a ColE1 origin of replication and transfer region of the parental plasmid. Both these groups whilst promoting efficient transfer to a wide range of hosts are themselves unable to replicate in nonenteric species.

V. REFERENCES

1. Bagdasarian M, Lurz R, Ruckert B, Franklin FCH, Bagdasarian MM, Frey J, Timmis KN (1981) Specific-purpose cloning vectors II. Broad-host-range high copy number, RSF1010 derived vectors, and a host-vector system for gene cloning in *Pseudomonas*. *Gene* **16**, 237–47

2. Bagdasarian MM, Amann E, Lurz R, Ruckert B, Bagdasarian M (1984) Activity of the hybrid *trp-lac* (*tac*) promoter of *Escherichia coli* in *Pseudomonas putida*. Construction of broad host range, controlled-expressed vectors. *Gene* **26**, 273–82

3. Barry GF (1986) Permanent insertion of foreign genes into the chromosomes of soil bacteria. *BioTechnology* **4**, 446–9

4. Barth PT, Grinter NJ (1974) Comparison of the deoxyribonucleic acid molecular weights and homologies of plasmids conferring linked resistance to streptomycin and sulphonamides. *J Bacteriol* **120**, 618–30

5. Barth PT, Tobin L, Sharpe GS (1981) Development of broad host-range plasmid vectors. *In* Molecular Biology, Pathogenicity and Ecology of Bacterial

Plasmids, (Levy SB, Clowes RC, Konig EL, eds) Plenum Publishing Corporation. pp 439–48

6. Bevan MW, Chilton MD (1982) T-DNA of the *Agrobacterium* T1 and R1 plasmids. *Ann Rev Genet* **16**, 357–84

7. Brosius J (1984) Plasmid vectors for the selection of promoters. *Gene* **27**, 151–60

8. Casadaban MJ, Chou J, Cohen SN (1980) *In vitro* gene fusions that join an enzymatically active β-galactosidase segment to amino-terminal fragments of exogenous proteins: *Escherichia coli* plasmid vectors for the detection and cloning of translational initiation signals. *J Bacteriol* **143**, 971–80

9. Close TJ, Zaitlin D, Kado CI (1984) Design and development of amplifiable broad-host-range cloning vectors: Analysis of the *vir* region of *Agrobacterium tumefaciens* plasmid pTiC58. *Plasmid* **12**, 111–8

11. Collins J, Hohn B (1978) Cosmids: a type of plasmid gene-cloning vector that is packageable *in vitro* in bacteriophage lambda heads. *Proc Natl Acad Sci USA* **75**, 4242–6

12. Daniels MJ, Barber CE, Turner PC, Sawczyc MK, Byrde RJW, Fielding AH (1984) Cloning of genes involved in pathogenicity of *Xanthomonas campestris* pv. campestris showing altered pathogenicity. *EMBO J* 3, 3323–7

13. Davison J, Hensterspreute M, Chevalier N, Thi VH, Brunel F (1987) Vectors with restriction cloning sites banks V. pJRD215 a wide-host-range cosmid with multiple cloning sites. *Gene* **51**, 275–85

14. de Boer HA, Comstock LJ, Vasser M (1983) The *tac* promoter: a functional hybrid derived from the *trp* and *lac* promoters. *Proc Natl Acad Sci USA* **80**, 21–5

15. De Maeyer E, Skup D, Prasad KSN, De Maeyer-Guignard J, Williams B, Meacock P, Sharpe GS, Pioli D,Henman J, Schuch W, Atherton K (1982) Expression of a chemically synthesized human α1 interferon gene. *Proc Natl Acad Sci USA* **79**, 4256–9

16. Deretic V, Chandrasekharappa S, Gill JF, Chatterjee DK, Chakrabarty AM (1987) A set of cassettes and improved vectors for genetic and biochemical characterization of *Pseudomonas* genes. *Gene* **57**, 61–72

17. Ditta G, Stanfield S, Corbin D, Helinski DR (1980) Broad host range DNA cloning system for Gram-negative bacteria. Construction of a gene bank of *Rhizobium meliloti*. *Proc Natl Acad Sci USA* **77**, 7347–51

18. Ditta G, Schmidhauser T, Yakobson E, Lu P, Liang X-W, Finlay DR, Guiney D, Helinski DR (1985) Plasmids related to the broad host range vector, pRK290, useful for gene cloning and for monitoring gene expression. *Plasmid* **13**, 149–53

19. Drummond M, Clements J, Merrick M and Dixon R (1983) Positive control and autogenous regulation of the *nifLA* promoter in *Klebsiella pneumoniae*. *Nature* **301**, 302–7

20. Franklin FCH, Bagdasarian M, Bagdasarian MM, Timmis KN (1981)

A molecular and functional analysis of the TOL plasmid pWWO from *Pseudomonas putida* and cloning of genes for the entire regulated aromatic ring meta cleavage pathway. *Proc Natl Acad Sci USA* **78**, 7458–62

21. Franklin FCH (1985) Broad host range cloning vectors for Gram-negative bacteria. *In* DNA Cloning Vol. 1, (Glover DM ed) IRL Press, Oxford and Washington DC, pp 165–84

22. Friedman AM, Long SR, Brown SE, Buikema WJ, Ausubel FM (1982) Construction of a broad-host-range cosmid cloning vector and its use in the genetic analysis of *Rhizobium* mutants. *Gene* **18**, 289–96

23. Frey J, Bagdasarian M, Feiss D, Franklin FCH, Deshusses J (1983) Stable cosmid vectors that enable the introduction of cloned fragments into a wide range of Gram-negative bacteria. *Gene* **24**, 299–308

24. Furste JP, Pansegrau W, Frank R, Blocker H, Scholz, P., Bagdasarian M, Lanka E (1987). Molecular cloning of the plasmid RP4 primase region in a multi-host range *tacP* expression vector. *Gene* **48**, 119–31

25. Gautier F, Bonewald R (1980) The use of plasmid R1162 and derivatives for gene cloning in the methanol-utilizing *Pseudomonas* AM1. *Mol Gen Genet* **178**, 375–80

26. Gay P, Le Coq D, Steinmetz M, Berkelman T, Kado CI (1985) Positive selection procedure for entrapment of insertion sequence elements in Gram-negative bacteria. *J Bacteriol* **164**, 918–21

27. Grinter NJ (1983) A broad-host-range cloning vector transposable to various replicons. *Gene* **21**, 133–43

28. Guerry P, van Embden J, Falkow S (1974) Molecular nature of two non-conjugative plasmids carrying drug resistance genes. *J Bacteriol* **117**, 619-30

29. Harkki A, Palva ET (1985) A *lam*B expression for extending the host range of phage lambda to other enterobacteria. *FEMS Microbiol Lett* **27**, 183–7

30. Harkki A, Hirst TR, Holmgren J, Palva ET (1986) Expression of the *Escherichia coli lam*B gene in *Vibrio cholerae*. *Microbial Pathogenesis* **1**, 283–8

31. Hawley DK, McClure WR (1983) Compilation and analysis of *Escherichia coli* promoter DNA sequences. *Nucl Acids Res* **11**, 2237–55

32. Hennam JF, Cunningham AE, Sharpe GS, Atherton KT (1982) Expression of eukaryotic coding sequences in *Methylophilus methylotrophus*. *Nature* **297**, 80–2

33. Hutter R, Nuesch J, Leisinger T (eds) (1981) Microbial Degradation of Xenobiotics and Recalcitrant Compounds, Academic Press, London.

34. Inouye S, Nakazawa A, Nakazawa T (1984) Nucleotide sequence of the promoter region of the *xylDEGF* operon on TOL plasmid of *Pseudomonas putida*. *Gene* **29**, 323–30

35. Ireland CR (1983) Detailed restriction enzyme map of crown gall-suppressive IncW plasmid pSA, showing ends of deletion causing chloramphenicol sensitivity. *J Bacteriol* **155**, 722–7

36. Ish-Horowicz, D. and Burke, J.F. (1981). Rapid and efficient cosmid cloning. *Nucl Acids Res* **9**, 2989–98

37. Knauf VC and Nester EW (1982) Wide host range cloning vectors: A cosmid clone bank of a *Agrobacterium* Ti plasmid. *Plasmid* **8**, 45–54

38. Kushner SR (1978) An improved method for transformation of *Escherichia coli* with ColE1 derived plasmids. *In* Genetic Engineering, (Boyer HW, Nicosia S, eds) Elsevier/North Holland, Amsterdam, p. 17–23

39. Leemans J, Langenakens J, DeGreve H, Deblaere R, Van Montagu M, Schnell J (1982) Broad-host-range cloning vectors derived from the W-plasmid pSa. *Gene* **19**, 361–4

40. Lundgren DG, Silver M (1980) Ore leaching by bacteria. *Ann Rev Microbiol* **34**, 263–83

41. Maniatis T, Fritsch EF, Sambrook J (1982) Molecular Cloning. A Laboratory Manual, Cold Spring Harbor Laboratory Press, New York.

42. Mermod N, Lehrbach PR, Reineke W, Timmis KN (1984) Transcription of the TOL plasmid toluate caboli plasmid toluate catabolic pathhway operon of *Pseudomonas putida* is determined by a pair of coordinately and postively regulated overlapping promoters. *EMBO J* **3**, 2461–6

43. McKenny K, Shimatke H, Court D, Schmeissner U, Brady C, Rosenberg M (1981) A system to study promoter and termination signals recognized by *Escherichia coli* RNA polymerase. *In* Gene amplification and analysis. Vol. II, (Chirikjian JC, Papas TS eds) Elsevier/North Holland, Amsterdam, pp 383–415

44. Osbourn AE, Barber CE, Daniels MJ (1987) Identifcation of plant-induced genes of the bacterial pathogen *Xanthomonas campestris* pathover *campestris* using a promoter-probe plasmid. *EMBO J* **6**, 23–8

45. Rogowsky P, Chimera JA, Close TJ, Shaw JJ, Kado CI (1987) Regulation of the *vir* genes of the *Agrobacterium tumefaciens* plasmid pTiC58. *J Bacteriol* **169**, 5101–12

46. Salmond GPC, Hinton JCD, Gill DR, Perombelon MCM (1986). Transposon mutagenesis of *Erwinia* using phage lambda vectors. *Mol Gen Genet* **203**, 524–9

47. Scherzinger E, Bagdasarian MM, Sholz P, Lurz R, Ruckert B, Bagdasarian M (1984) Replication of the broad-host-range plasmid RSF1010: Requirement for three plasmid-encoded proteins. *Proc Natl Acad Sci USA* **81**, 654–8

48. Serdar CM, Gibson DT (1985) Enzymatic hydrolysis of organophosphates : cloning and expression of a parathion hydrolase gene from *Pseudomonas diminuta*. *BioTechnology* **3**, 567–70

49. Sharpe GS (1984) Broad-host-range cloning vectors for Gram-negative bacteria. *Gene* **29**, 93–102

50. Shaw JJ, Kado CI (1986) Development of a *Vibrio* bioluminescence gene-set to monitor phytopathogenic bacteria during the ongoing disease process in a nondisruptive manner. *BioTechnology* **4**, 560–4

51. Shingler V, Franklin FCH, Tsuda M, Bagdasarian M (1988) Phenol degradation by *Pseudomonas* CF600 proceeds via a plasmid mediated meta-cleavage pathway. *J Bacteriol* (In press)

52. Smith HR, Humphreys GO, Anderson ES (1974) Genetic and molecular characterization of some nontransferring plasmids. *Mol Gen Genet* **129**, 229–42

53. Spooner RA, Lindsay K, Franklin FCH (1986) Genetic, functional and sequence analysis of the *xylR* and *xylS* regulatory genes of the TOL plasmid pWWO. *J Gen Microbiol* **132**, 1347–58

54. Spooner RA, Bagdasarian M, Franklin FCH (1987) Activation of the *xylDEGF* promoter of the TOL toluene-xylene degradation pathway by over-production of the *xylS* regulatory gene product. *J Bacteriol* **169**, 3581–6

55. Tait RC, Close TJ, Lundquist RC, Hagiya M, Rodriguez L, Kado KI (1983) Construction and characterization of a versitile broad host range DNA cloning system for Gram-negative bacteria. *BioTechnology* **1**, 269–75

56. Thomas CM (1981) Molecular genetics of broad-host-range plasmid RK2. *Plasmid* **5**, 10–9

57. Uhlin BE, Molin S, Gustafsson P, Nordstrom K (1979) Plasmids with temperature-dependent copy number for amplification of cloned genes and their products. *Gene* **6**, 91–106

58. Watanabe T, Furuse C, Sakaizumi S (1968) Transduction of various R-factors by phage P1 in *Escherichia coli* and by phage P22 in *Salmonella typhimurium*. *J Bacteriol* **96**, 1791–9

59. Weisshaar MP, Franklin FCH, Reineke W (1987) Molecular cloning and expression of the 3-chlorobenzoate-degrading genes from *Pseudomonas* sp. strain B13. *J Bacteriol* **169**, 394–402

60. Werneke JM, Sligar SG, Schuler MA (1985) Development of broad-host range vectors for expression of cloned genes in *Pseudomonas*. *Gene* **38**, 73–84

61. Windass JD, Worsey MJ, Pioli EM, Pioli D, Barth PT, Atherton KT, Dart EC, Byrom D, Powell K, Senior PJ (1980) Improved conversion of methanol to single cell protein by *Methylophilus methylotrophus*. *Nature* **287**, 396–401

INDEX

Note: figures in *italics*
tables in **bold**.

cos 152
CPG2 **237**
cro 90

D

2,4-D 57, 67
Dichlorophenoxyacetic acid 57, 67
Dihydrofolate reductase *29, 66*, 134,
 139, 140, 152
DNA 2, 5, 8, 14, 36, 104, 137
 DnaA 5, 6, *69*, 70, 106, 120
 DnaB (helicase) 8, 51, 106,
  117–18, *119*, 120, 121
 DnaC 106, 111, 120
 DnaG (primase) 8, 30, 42, **45**, 52,
  106, 111–12, 120
 DNA gyrase 8, 106, 120
 DNA helicase 8, 51, 106, 117–18,
  120, 121
 DnaK 232
 Dna PolI 111
 Dna PolIII 9, 106
 Dna primase 30, 52, 106, 111–12,
  120
 DnaT 106, 120
 DnaZ 106
 DNA polymerase 106, 112, 121
 rDNA 247
Drosophila 141
Dtr 28, 35, 51, 52

E

eex 170, 172
Enterobacter spp. 126, **146**, 165, **191**
Erwinia spp. **81, 146, 191**, 214, 216,
 234, **250**, 256
Escherichia coli 3, 6–10, 14–16, 28,
 30, 51, 57, 63, 70, **81, 87**, 104,
 106, 111, 112, 118, 120, 125–7,
 138, 141, 143, **146**, 147, 152, 153,
 165, 168–70, 173–4, 185–9, **191**,
 192, 194–6, 207–8, 211, 213–14,
 216–18, 220, 229–36, **235**, 239–40,
 247, 250, **250**, 255–6, 259–60,
 262–3

F

Fi 28, 186–7
fip 166, *167, 170*, 172
fiwA 64, 65

G

Galactokinase *82*, **84, 87**, 257
Galactose epimerase 143–5
β-galactosidase 257
Gentamycin (Gm) resistance **58**, 126
gln 233
Gluconobacter spp. **81, 250**

H

Helicase 8, 117
htpR 232
Hypomicrobium spp. **81**

I

inc 2, *5*, 10, 14, *64*
Insertion sequences 151–2
 IS*2* 187
 IS*21 58*, 63, 67, 186, 188–90, *189*,
  195–7
 IS*3* 187
 IS*8 29*, 186
 IS*46* 167
 IS*50* 219
 IS*70* 67
 IS*401* 190
 IS-trapping vector 151

K

Kanamycin (Km) resistance 32, 61,
 63, *64*, 125–6, 128–30, *129*, 143,
 186, 189, 217, **248**, 254
kcr 2, 9, *11*, 16
kfr3 9, *11*
*kik*1 *167*, 170, 171
kil 2, 7, 9, *11*, 16–17, 18, *64*, 69, 70,
 71, *167*, 169–71
Klebsiella 126, **146**, 170, 171, **235**
 K. aerogenes 165, **250**
 K. pneumoniae **81, 146, 167**, 168,
  169, 171, **191**, 232